Collins

Collins Revision

GCSE Higher Science

Revision Guide

FOR AQA A + B

Contents

B2a Discover Martian living!

B2b Discover DNA!

C2a Discover Buckminsterfullerene!

C2b Discover electrolysis!

P2a Discover forces!

P2b Discover nuclear fusion!

Grades

Coordination

The nervous system

D–C

- The nervous system coordinates the body's responses to internal and external changes. Information as electrical impulses travels:
  - from **receptors** to the **central nervous system** along **sensory neurones**
  - from the central nervous system to **effectors** along **motor neurones**.
- To touch the nose with the finger: eyes sense nose and finger positions; arm muscles contract.
  - This involves rapid, complex brain/muscle messages.

B–A*

brain
spinal cord
central nervous system
nerve

The human nervous system.

Different receptors detect light, sound, chemicals, touch, pressure, position change, pain and temperature.

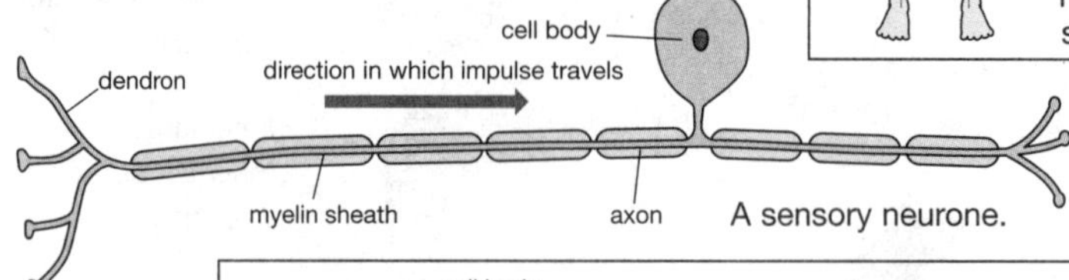

A sensory neurone.

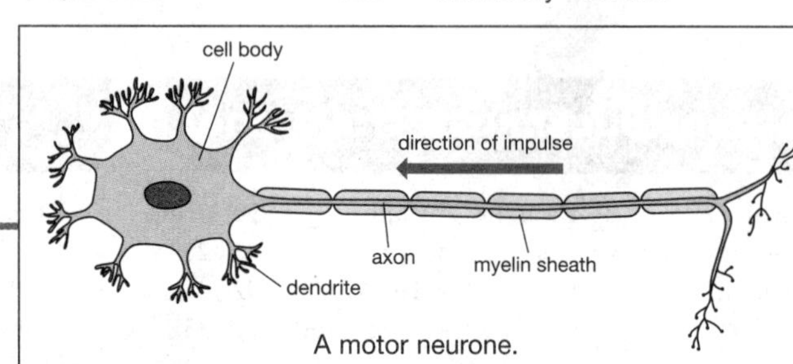

A motor neurone.

Receptors

Receptors, effectors and hormones

D–C

- Stimuli affect **receptors** in sense organs such as eyes and ears.
- Receptors **transform** energy (see page 50). They change one form of energy (a stimulus which is not an electrical impulse) into another form (an electrical impulse). Electrical impulses enable information to travel rapidly in the nervous system.
- Light energy reaching the cells of the eye's retina is transformed into electrical energy that travels as an electrical impulse to the brain.

B–A*

chemical energy in rod cell
light energy from Sun
electrical energy as impulses in optic nerve

In the eye, light energy is transformed into electrical energy.

- The organs that respond to stimuli are **effectors** – the **muscles** and **glands**:
  - a muscle responds rapidly to information the brain sends as electrical signals
  - glands **secrete** substances (e.g. nose detects food smells, glands secrete saliva).
- Some glands secrete **hormones**, chemicals that travel in the bloodstream to **target** cells or organs.

D–C

- The pituitary gland and the ovaries secrete hormones that regulate events in the menstrual cycle (see page 6).
- **Adrenaline** from the adrenal gland mobilises the body for action. It affects the heart, breathing muscles, eyes and the digestive system.
- Nerve impulses are fast and short lived. The **nervous system** can trigger an immediate short-term response (e.g. blinking).
- Hormones allow for a longer-term response. For example, adrenaline prompts a sustained fight-or-flight response to a threat (e.g. from fire), lasting long enough for escape. Anxiety fades as the liver breaks down adrenaline to products that are transferred to the blood and excreted in urine.

B–A*

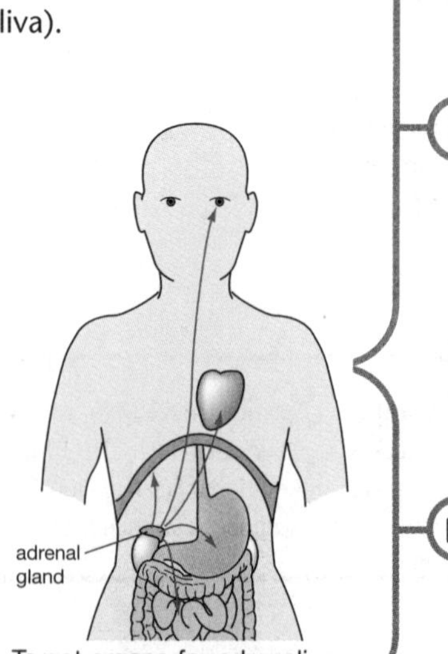

Target organs for adrenaline.

Questions

Grades D-C

1 Name the kind of neurone that carries impulses away from the central nervous system.

Grades B-A*

2 Receptors change energy from one form to another. Describe **one** example of this.

Grades D-C

3 Give an example of a hormone and its target organs.

Grades B-A*

4 Explain how the effect of adrenaline on the heart could help you in a 'fight-or-flight' situation.

Grades

Reflex actions

Reflex actions and synapses

Top Tip!
Do not say that neurones carry 'messages'. They carry electrical impulses.

D–C

- A **reflex action** is a rapid, automatic response to a stimulus, often to avoid harm.
- A reflex action involves a **reflex arc**. An electrical impulse travels from the receptor along a **sensory neurone**, to a **relay neurone** in the spinal cord. It leaves the spinal cord in a **motor neurone** and reaches an **effector** (e.g. a muscle or gland) which responds to the stimulus.

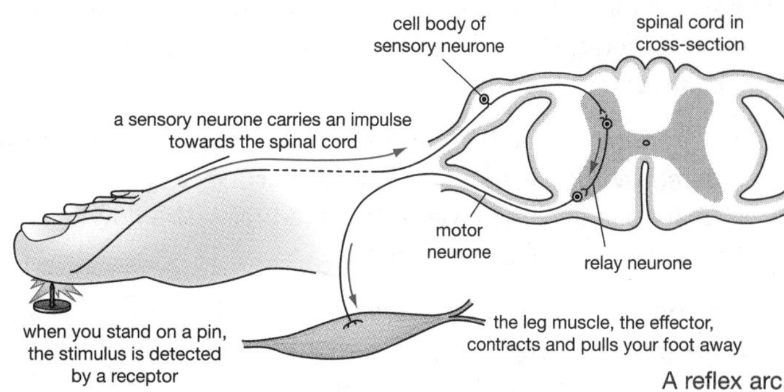

cell body of sensory neurone
spinal cord in cross-section
a sensory neurone carries an impulse towards the spinal cord
motor neurone
relay neurone
when you stand on a pin, the stimulus is detected by a receptor
the leg muscle, the effector, contracts and pulls your foot away

A reflex arc.

- Two neurones join at a **synapse**. The electrical impulse arriving at the end of one neurone causes a chemical to be secreted into the gap. In the next neurone, the chemical initiates an electrical impulse that continues onwards.

B–A*

- Synapses slow down nerve impulses, but a neurone can form synapses with several neurones, and this enables us to respond to a stimulus in more than one way.

impulse
neurone
neurone
transmitter chemical in vesicle
vesicle releases chemical

What happens at a synapse.

- A relay neurone in the spinal cord has synapses with a sensory and a motor neurone. It also has synapses with neurones that carry nerve impulses from the brain, and information from the brain can override a reflex action.
- As an example, you may stumble and reach for a rail which is hot. A reflex action would stop you grasping the rail. Instead, your brain sends an impulse to the relay neurone to block impulses to the motor neurone that would make you withdraw your hand. You grasp the hot rail rather than fall.

In control

How Science Works
You should be able to: evaluate the claims of manufacturers about sports drinks containing water, ions and glucose.

Controlling water and ions

D–C

- The body requires a steady water content, ion content, temperature and blood sugar level.
- The lungs lose water, and the kidneys excrete excess water and ions in urine.
- Sweat contains water, ions (including sodium and chloride) and urea. The evaporation of water in sweat takes heat from the skin and lowers body temperature.
- The blood transports **glucose** (an energy source) from digested food to all the body's cells.
- When there is excess blood glucose, the **pancreas** secretes **insulin** into the blood. Insulin makes the **liver** remove glucose from the blood and stores it until it is needed.

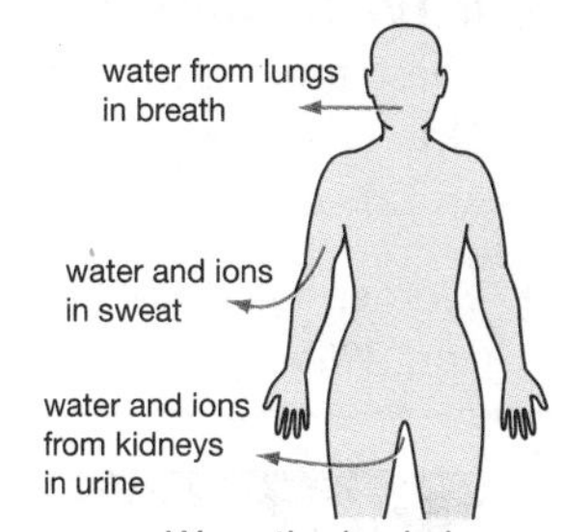

water from lungs in breath
water and ions in sweat
water and ions from kidneys in urine

Ways the body loses water and ions.

B–A*

- For anyone stranded in a desert, an SAS survival manual gives this advice to conserve water: avoid exertion; keep cool and stay in the shade; do not lie on the hot ground; do not eat, because digestion uses up fluids; talk as little as possible; breathe through your nose rather than your mouth.

Questions

Grades D-C

1 Describe what happens when a nerve impulse reaches a synapse.

Grades B-A*

2 Describe **one** advantage and **one** disadvantage of having synapses between neurones.

Grades D-C

3 Explain how sweating cools you down.

Grades B-A*

4 Explain why you need to drink more liquids on a hot day than a cold one.

Grades

Reproductive hormones

The menstrual cycle and hormones

- In the **menstrual cycle**, an **egg** is released from an ovary every 28 days. If fertilised, the egg embeds in the uterus lining. If unfertilised, the **uterus** lining breaks down and **menstruation** begins.
- The menstrual cycle is controlled by the hormones **oestrogen**, secreted by the ovaries, and **FSH** and **LH**, secreted by the pituitary gland at the base of the fore-brain. The flow-chart shows the roles of these hormones.

In the pituitary gland	In the ovary
FSH is secreted →	FSH causes an egg to mature
	FSH causes the ovary to secrete oestrogen
oestrogen stops the pituitary secreting FSH	
oestrogen causes the pituitary to secrete LH →	LH causes the mature egg to be released from the ovary (ovulation)

How hormones control the release of an egg.

D–C

- The concentrations of oestrogen, FSH and LH in the blood change during the menstrual cycle (see diagram). The uterus lining thickens as oestrogen concentration rises. As oestrogen concentration falls, the uterus lining breaks down.

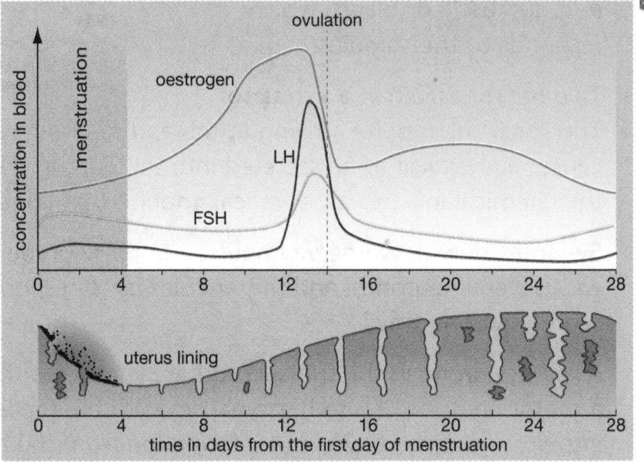

Levels of oestrogen, FSH and LH and changes in the uterus lining.

B–A*

Controlling fertility

How Science Works

You should be able to: evaluate the benefits of, and the problems that may arise from, the use of hormones to control fertility and IVF.

Fertility treatment

- A woman may not produce enough FSH to stimulate eggs to mature in her ovaries. Some **fertility drugs** contain FSH to bring one or several eggs to maturity during each menstrual cycle.
- In **IVF**, eggs are removed from the ovary and fertilised outside a woman's body.
- In **IVF treatment**, a woman first receives fertility drugs to make her ovaries produce several eggs. These are removed under anaesthetic and fertilised *in vitro* (in a dish). They develop into **embryos**. One or two embryos are transferred to the woman's uterus. Each becomes attached to the lining and develops into a baby.

D–C

- Women taking fertility treatment with FSH can have multiple births. About 20% of births are twins and 3% are quads or more.

B–A*

Oral contraceptives, benefits and drawbacks

- To avoid pregnancy, women can take oral contraceptive pills. They contain hormones including **oestrogen** which stop FSH production, so eggs do not mature.
- Oral contraceptives reduce unwanted pregnancies and abortions. But some cultures and religions prohibit their use. They may encourage women to have more sexual partners, increasing the danger of catching sexually transmitted infections.

Questions

Grades D-C

1 Describe **two** effects of FSH.

Grades B-A*

2 Describe how concentrations of oestrogen in the blood change during the menstrual cycle, and the effects that these changes have.

Grades D-C

3 State **one** advantage of oral contraceptives and **one** possible disadvantage.

Grades B-A*

4 Suggest why fertility treatment using FSH often results in multiple births.

Diet and energy

A balanced diet, energy and respiration

Top Tip!

The only nutrients that supply energy are carbohydrates, fats and proteins.

D–C

- A healthy, **balanced diet** contains:
  - **carbohydrates** for energy; **fats** for energy and making cell membranes; **proteins** for growth, repair, making enzymes, and energy; vitamins and minerals to keep healthy; roughage (fibre) for digestive system function; water.
- A person with an unbalanced diet is **malnourished**. They may be too fat, too thin, or have deficiency diseases.
- The rate of chemical reactions in cells – **metabolic rate** – is generally faster in people with a high muscle-to-fat ratio, and in those who exercise or do physical work. It can also be affected by inherited factors. Metabolic rate stays high for some time after activity.

B–A*

- **Basal metabolic rate** (**BMR**) is metabolic rate when resting, measured in megajoules (MJ) per day. Average values between ages 10 and 17:
  - female BMR = (body mass in kg × 0.056) + 2.898 MJ/day
  - male BMR = (body mass in kg × 0.074) + 2.754 MJ/day
- Cells release energy from food by **respiration**:

  glucose + oxygen → carbon dioxide + water + ENERGY
- The energy in 1 gram of fat is twice the energy in 1 gram of carbohydrate or 1 gram of protein. (Proteins are not main energy sources, but are used more for growth, repair and making enzymes.)

Obesity

Obesity and related illnesses

How Science Works

You should be able to: evaluate information about the effect of food on health.

D–C

- Very overweight people are described as **obese**. They have an increased risk of having arthritis, high blood pressure, heart disease and diabetes.
- The **joints** of someone with **arthritis** are worn and painful.
- Overweight people and those who eat too much salt often have **high blood pressure**. This can lead to **heart disease** by straining the heart and damaging blood vessels.
- Obese people are prone to **Type 2 diabetes**: insulin fails to regulate **blood glucose** level.

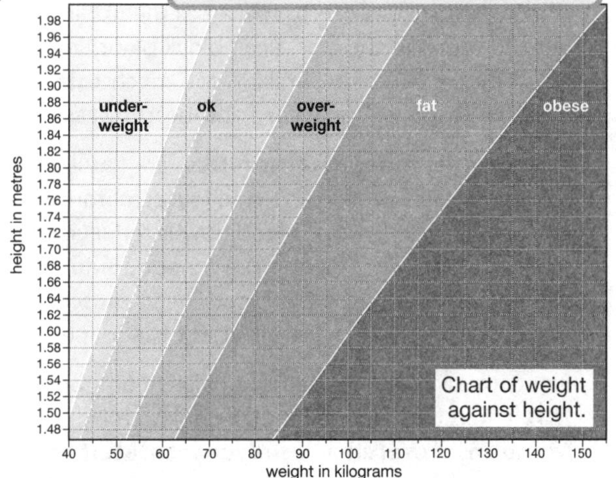

Chart of weight against height.

B–A*

- Those who get Type 2 diabetes are often overweight, or are fit but inherit faulty genes. Type 2 diabetes has become more common in children and young adults. Bad diet, obesity and smoking increase the risk.
- **Type 1 diabetes** is not related to lifestyle. It often starts in young people under 20 when the cells in the pancreas that make **insulin** are destroyed, possibly by a person's own immune system.
- Neither type of diabetes has a cure, but transplanting healthy cells into the pancreas may cure Type 1 diabetes in the future.

Questions

Grades D-C

1 How does the proportion of muscle to fat in the body affect metabolic rate?

Grades B-A*

2 Explain why eating fatty foods is more likely to make a person put on weight than eating carbohydrate-rich foods.

Grades D-C

3 Explain how arthritis makes movement difficult.

Grades B-A*

4 Can lifestyle changes prevent diabetes? Explain your answer.

Not enough food

Eating too little or too much

Grades

D–C

- In some developing countries, a poor diet and insufficient food lowers people's resistance to disease. Women's periods may be irregular or stop altogether.
- **Malnutrition** affects small children most. Their bodies are growing fast and cannot combat infections well.
- Children between 1 and 3 years fed mainly on carbohydrates (e.g. maize or rice) may suffer from **kwashiorkor**. Without protein, muscles cannot develop, and the abdominal tissues swell with accumulated fluid.
- In rich countries, overweight people often try **slimming programmes**, even though less food and more exercise would be more effective.

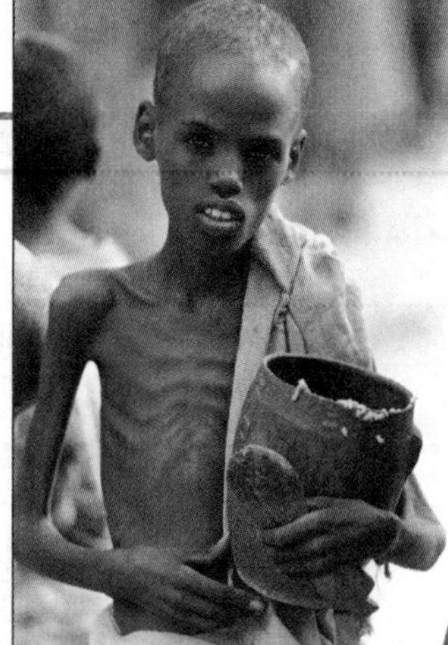

A child suffering from malnutrition.

B–A*

- The rule for losing weight is: use up more energy than you take in. Eat less overall but retain a balanced diet, and take extra exercise.
- Commercial slimming programmes work for some people, or work for only a short time. A high protein and fat diet with very little carbohydrate may lead to weight loss, but it can also increase the risk of heart disease.

How Science Works

You should be able to: evaluate claims made by slimming programmes.

Cholesterol and salt

Cholesterol, salt and heart disease

D–C

- **Cholesterol** is needed to make cell membranes. But too much in the blood can block blood vessels and cause **heart disease**. Similarly, too much **salt** can cause **high blood pressure** and heart disease.
- **Low density lipoproteins** (**LDLs**) are tiny balls of cholesterol and protein. When cholesterol forms **plaques** in artery walls, blood flow is reduced or can stop altogether (see diagram). Without a sufficient supply of oxygen from the blood, heart muscle stops beating properly, and a heart attack can follow.
- Blood should contain the right balance of 'good' **HDL** cholesterol and 'bad' **LDL** cholesterol. **Monounsaturated** and **polyunsaturated fats** in plant oils *lower* cholesterol and improve the balance. **Saturated fats** from animal products *raise* blood cholesterol level.

a healthy artery has a stretchy wall and a space in the middle for blood to pass through

plaque

sometimes, a substance called plaque builds up in the wall. This is more likely to happen if you have a lot of LDL cholesterol in your blood

blood clot

the plaque slows the blood down, and a clot may form. Or a part of the plaque may break away

A plaque in an artery.

B–A*

- If cholesterol intake in the diet is too low, the **liver** starts to make cholesterol.
- Genes may cause some people to make too much cholesterol, and even a good diet fails to reduce their blood cholesterol. Instead, they can take **statins**, drugs that inhibit the enzymes involved in making cholesterol.
- Statin tablets help anyone liable to a heart attack from high blood pressure, and people can now buy them in chemists.

How Science Works

You should be able to: evaluate the effect of statins on cardiovascular disease.

Questions

Grades D-C

1 What causes kwashiorkor?

Grades B-A*

2 Write a sentence giving a simple rule for losing weight.

Grades D-C

3 Which kind of cholesterol should you have more of in your blood – LDLs or HDLs?

Grades B-A*

4 Explain why the only way for some people to lower their blood cholesterol level is by taking statins.

Drugs

Drugs and addiction

Top Tip!

Just because a drug is legal does not mean it is harmless. Legal drugs kill more people than illegal drugs.

D–C

- Throughout history, people have used **drugs** from natural materials. Today's medical drugs combat disease and relieve pain. People benefiting from legal drugs far outnumber those damaged by illegal drugs.
- Many **recreational** drugs are illegal. Their harmful effects vary, but often include damage to the **liver**, which destroys harmful chemicals in the body. Illegal drugs change chemical processes in the brain, affecting how users feel and behave.
- People using drugs can become **addicted** (**dependent**), and may suffer **withdrawal symptoms** without them, particularly if they are taking cocaine or heroin.

B–A*

- Each year in Britain, illegal drugs kill thousands of people, many between the ages of 20 and 39. Some die of poisoning, others because the drug affected their brain and they took a fatal course of action.

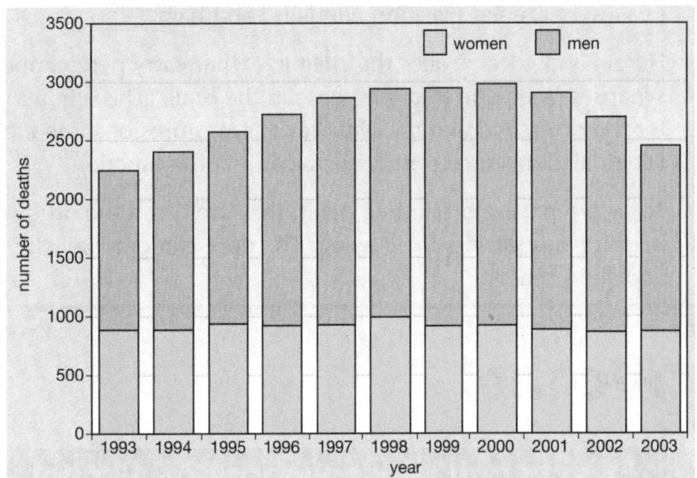

Deaths in Britain from using illegal drugs between 1993 and 2003.

How Science Works

You should be able to: evaluate the different types of drugs and why some people use illegal drugs for recreation.

Trialling drugs

Testing and trialling a drug

D–C

- Scientists **test** the safety of a new drug on human tissues or animals in the laboratory to check that it is not **toxic** (poisonous). Then it is **trialled** on human volunteers to check for side effects. The process can take years.
- Scientists failed to test **thalidomide**, a sleeping pill, on pregnant women who took the drug to relieve morning sickness. Users then had babies with deformed limbs. Thalidomide was banned. Now it is used again, to treat leprosy.

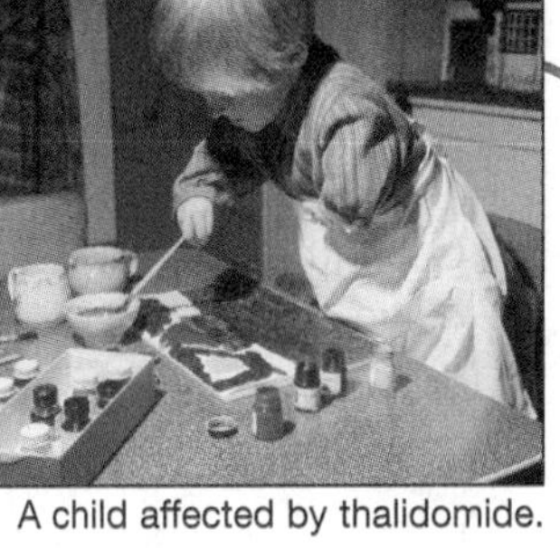
A child affected by thalidomide.

B–A*

- **Zanamivir** is a new, approved drug designed to lessen **flu** virus symptoms. It is available in Europe and the US.
- The National Institute for Clinical Excellence, NICE, tells National Health doctors about government policy on prescribing drugs.
- NICE ruled that zanamivir should not be available free on prescription except to those with flu who could develop life-threatening conditions. Eligible people include the elderly, or those with long-term kidney, lung or heart disease, or diabetes or a faulty immune system.
- The decision to restrict the free prescription of zanamivir is because National Health funds are limited.

Questions

Grades D-C

1 Describe what happens when a person addicted to a drug stops taking it.

Grades B-A*

2 Name **two** illegal, recreational drugs.

Grades D-C

3 Explain how the harmful effects of thalidomide were missed, even though it was extensively trialled.

Grades B-A*

4 Suggest why people now consider that it is safe to use thalidomide as a medical drug.

Grades

Illegal drugs

Cannabis, heroin and cocaine

How Science Works

You should be able to: evaluate claims made about the effect of cannabis on health and the link between cannabis and addiction to hard drugs.

D–C

- **Cannabis** can cause bronchitis and lung cancer. It may increase the risk of schizophrenia, or lead a person to take dangerously addictive hard drugs such as **heroin** and **cocaine**.
- Injecting **heroin** or **cocaine** directly into a vein hastens its effect on the brain. Sharing syringes spreads infections such as HIV/AIDS and hepatitis.
- Finding the money for drugs dominates a person's life and leads to crime.
- For someone trying to break their habit, **withdrawal symptoms** can include insomnia, sweating, vomiting and pain.

B–A*

- Heroin and cocaine alter the chemicals that carry nerve impulses across the synapses (see page 5) of neurones in the brain. This alteration causes the craving for the drug, and also the withdrawal symptoms of someone attempting to quit. Eventually the correct brain chemicals resume function.
- To help a person break their habit, they are given the drug methadone. It is less harmful and satisfies the craving. The dose can gradually be reduced until addiction ceases.

Alcohol

Effects and consequences of alcohol

D–C

- **Alcohol** affects the nervous system by slowing down reactions. It helps people relax. Too much may lead people to lose their self-control. They may become **unconscious**, go into a **coma**, or even die.
- In the long term, alcohol causes brain cells to shrink and kills liver cells that normally break down poisons.
- Its long history makes alcohol acceptable, but it is a dangerous drug.
- In moderation alcohol causes little harm, but every year excess drinking kills about 4000 women and 6500 men in the UK, and damages the organs of many more.
- Some people are **dependent** on alcohol. Like hard drugs, alcohol ruins lives and relationships.

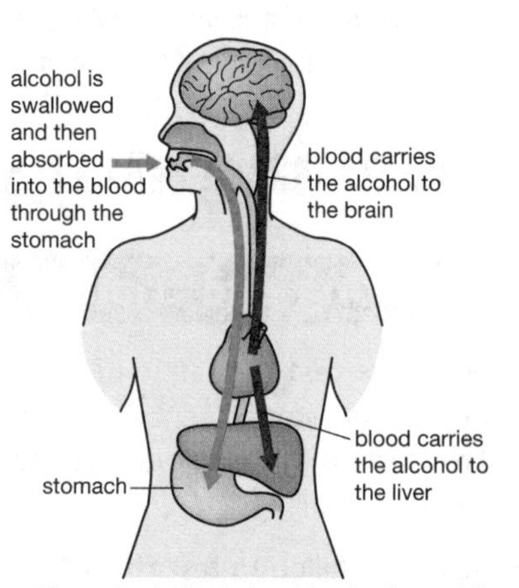
alcohol is swallowed and then absorbed into the blood through the stomach
blood carries the alcohol to the brain
blood carries the alcohol to the liver
stomach

How alcohol reaches the brain and liver.

B–A*

- Alcohol is a **depressant** since it slows down the brain's activity. It hinders the ability of:
  - the **cortex** to think clearly and make decisions
  - the **cerebellum** to coordinate body movements
  - the **medulla** to control breathing.

  Alcohol's interference with breathing can result in a coma and death.

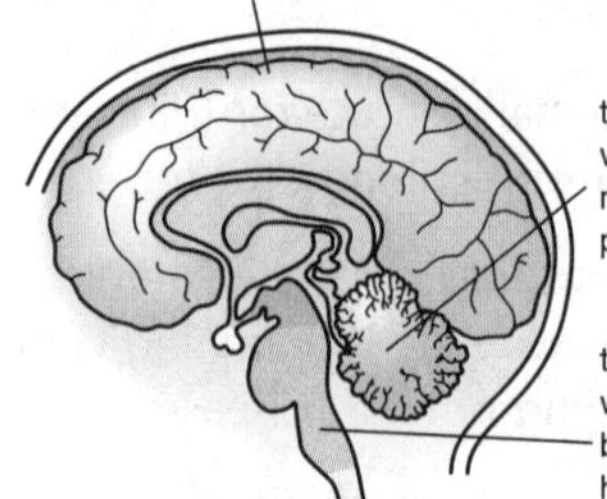
the **cortex** (the wrinkled surface layer of the brain), which is responsible for conscious thought and actions
the **cerebellum**, which controls movement and posture
the **medulla**, which controls breathing and heart rate

Parts of the brain that alcohol affects.

Questions

Grades D-C

1 Why do some drug users have a high risk of developing AIDS?

Grades B-A*

2 Explain why heroin and cocaine are addictive.

Grades D-C

3 Which organ is likely to be damaged by drinking alcohol excessively?

Grades B-A*

4 Alcohol is a depressant. What does this mean?

Grades

Tobacco

Tobacco poisons and disease

D–C

- **Tar** in tobacco smoke contains **carcinogens**. They cause **cancer** of the lungs. A quarter of all smokers die of smoke-related diseases including cancer. Cigarette smoke contains the addictive substance **nicotine**.
- **Carbon monoxide** gas in tobacco smoke stops red blood cells from carrying oxygen. Cells become starved of oxygen. Babies of smoking mothers often have a low birth weight.
- The evidence that Richard Doll gathered over 10 years showed the **correlation** (link) between smoking tobacco and lung cancer.

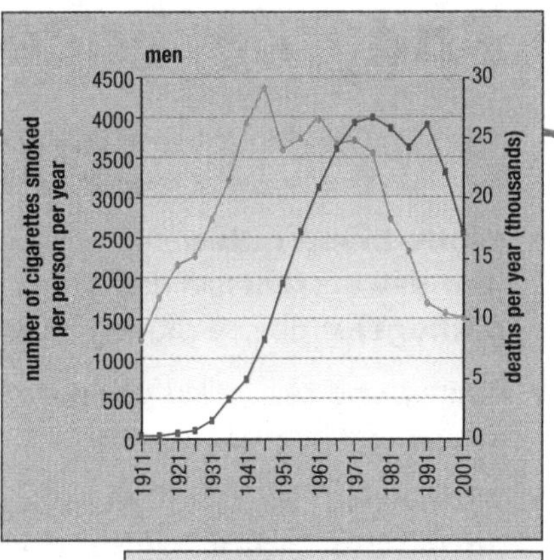

Key
cigarettes smoked per person per year
deaths

The link between smoking and deaths in men.

How Science Works

You should be able to:

- explain how the link between smoking tobacco and lung cancer gradually became accepted
- evaluate the different ways of trying to stop smoking.

Tar and cancer

B–A*

- In growth from a fertilised egg to an adult, cells **divide** many times. The timing and number of divisions is determined by genes in the nucleus of cells. Some genes instruct 'divide' and others instruct 'do not divide': the balance is finely controlled.
- **Carcinogens** are chemicals that cause **mutations** (changes) in the controlling genes. When genes which instruct 'do not divide' mutate and fail to stop division, a cell becomes cancerous. It divides uncontrollably and forms a tumour. Tobacco tar contains carcinogens.

Pathogens

Top Tip!

Do not use the word 'germs'. The correct scientific term is 'pathogens'.

Microorganisms and disease

D–C

- Disease-causing **microorganisms** called **pathogens** include **bacteria** and **viruses**. They reproduce rapidly inside the body.
- Bacteria produce **toxins** that enter the blood and make you ill.
- Viruses invade cells and reproduce in them. The new viruses destroy the cells as they burst out, then invade other cells.
- **Ignaz Semmelweis** was an Austrian doctor who thought that doctors' hands spread disease between patients in a hospital. In his time, no-one knew about disease-causing microorganisms. He made doctors wash their hands, and the death rate plummeted.

How Science Works

You should be able to: relate the contribution of Semmelweis in controlling infection to solving modern problems of the spread of infection in hospitals.

B–A*

- A stomach ulcer is a sore raw patch in the stomach wall. Its cause was assumed to be stress or excess stomach acid.
- In 1982, Australian researchers Marshall and Warren discovered a new bacterium in the stomachs of patients with ulcers. They named it *Helicobacter pylori*, 'spiral bacterium of the stomach'.
- Doctors did not believe their evidence until Warren swallowed some bacteria and developed an ulcer. Now, antibiotics (see page 12) are routinely used to treat ulcers.

Helicobacter (brown) on a stomach wall (green).

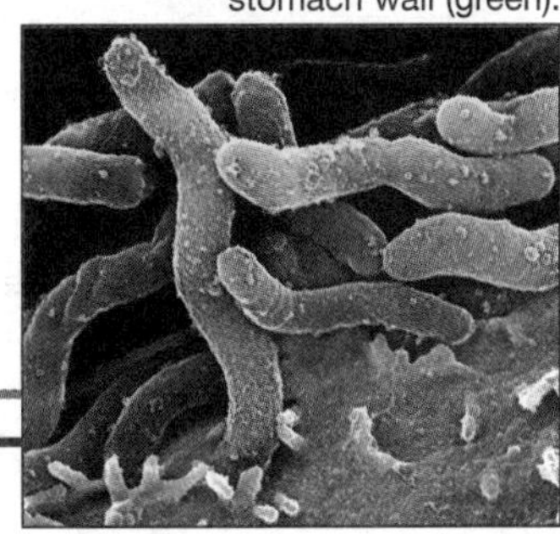

Questions

Grades D-C

1 There is a correlation between smoking and getting lung cancer. What does 'correlation' mean?

Grades B-A*

2 Explain how a carcinogen can cause cancer.

Grades D-C

3 Explain why there was less disease in the hospital where Semmelweis worked, after he made doctors wash their hands.

Grades B-A*

4 Suggest why doctors were initially unwilling to believe that a bacterium caused stomach ulcers.

Body defences

How white blood cells combat pathogens

- **White blood cells** are part of the body's **immune system** and defend it against **pathogens** (disease-causing organisms).
- White blood cells called **phagocytes ingest** bacteria in a wound by **phagocytosis**.
- White blood cells called **lymphocytes** produce **antibodies**. Each antibody is **specific** for a particular antigen on the surface of a pathogen. The antibody attaches to the antigen and destroys the pathogen.

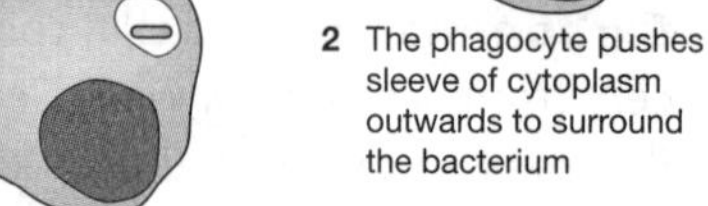

1 A phagocyte moves towards a bacterium
2 The phagocyte pushes a sleeve of cytoplasm outwards to surround the bacterium
3 The bacterium is now enclosed in a vacuole inside the cell. It is then killed and digested by enzymes

Phagocytosis.

D–C

- Lymphocytes also make **antitoxins** to destroy bacterial toxins (see page 11).
- In an **epidemic**, many people become infected because they are not **immune** to a new pathogen. A **pandemic** is a worldwide epidemic.

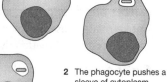

SCID

- The white blood cells involved in the body's immune system contain special **enzymes** required for their roles.
- Very rarely, a baby is born without one of the genes needed to make these enzymes. The child suffers from **SCID – Severe Combined Immunodeficiency Disease** – and has no defences against pathogens. Any infection would lead to death.
- Children with SCID are kept alive in a sterile plastic box or bubble that prevents any direct contact with objects carrying pathogens.

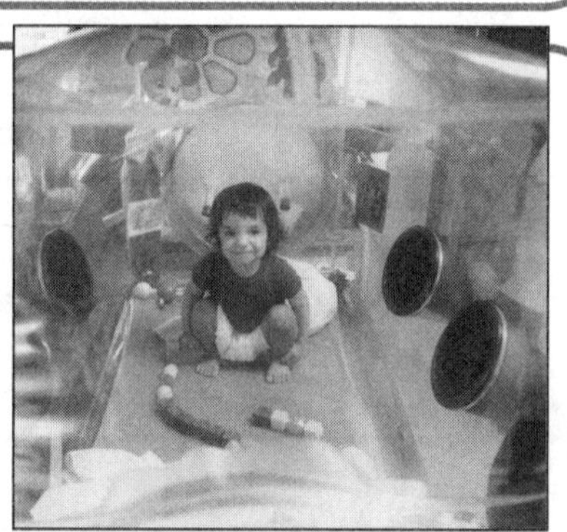
A baby with SCID.

B–A*

- **Gene therapy** has helped some of these children. They have been given the gene involved in making the missing enzymes and can live a normal life for as long as the gene is effective.

Drugs against disease

Antibiotics

- **Painkillers** reduce pain symptoms but have no effect on the cause.
- **Antibiotics** (e.g. penicillin) kill bacterial cells but not body cells, and not viruses.
- Since viruses invade body cells, **antiviral** drugs are liable to kill the body's cells as well as the virus.
- AZT is an antiviral that slows down the development of HIV/AIDS, but it often causes severe side effects.

D–C

- Over-use of antibiotics has made bacteria resistant to some antibiotics (see also page 13). Doctors have been urged to prescribe antibiotics for essential cases only.
- The table shows the response of doctors in Scotland. The figures are the number of prescriptions for each 100 members of the population in six different years. Column 2 shows all the prescriptions for any antibiotic, and column 3 shows just penicillin.

year	all antibiotics	penicillin
1992	95.6	51.4
1993	105.6	59.8
1994	86.1	47.5
1995	82.0	44.9
1996	81.9	44.7
1997	79.2	41.7

B–A*

Questions

Grades D-C

1 Which kind of blood cells produce antibodies?

Grades B-A*

2 Explain how a better understanding of immunity has led to successful treatment for SCID in some children.

Grades D-C

3 Explain why it is difficult to develop antiviral drugs.

Grades B-A*

4 By how much did antibiotic prescriptions decrease in Scotland between 1992 and 1997? Explain the reasons for this decrease.

Arms race

New viral diseases and resistance to antibiotics

D–C

- In 1993, the new, highly infectious **viral** disease **SARS** first appeared in China. A **pandemic** was feared. In an international effort, a **vaccine** was developed. By 2003, SARS was under control.
- **Bird flu** arose in 2004, killing people in Asia. It passes from poultry to humans, but human-to-human transfer is feared if the virus mutates. World health agencies continue to monitor this disease closely.
- Over-use of antibiotics has increased the chance of bacteria developing **resistance** to antibiotics by the process of **natural selection**. **MRSA** is a life-threatening mutant form of the bacterium ***S**taphylococcus **a**ureus* which is **r**esistant to the antibiotic **m**ethicillin.

B–A*

- The bacterium *Staphylococcus aureus* has always killed vulnerable people. Antibiotics have saved the lives of many more.
- Then *S. aureus* became resistant to penicillin. Stronger, more expensive antibiotics, including methicillin were used. Again, a resistant strain evolved, with less effective treatment available for it.
- Through mutation, lethal strains of bacteria and viruses may emerge that are resistant to all known drugs.

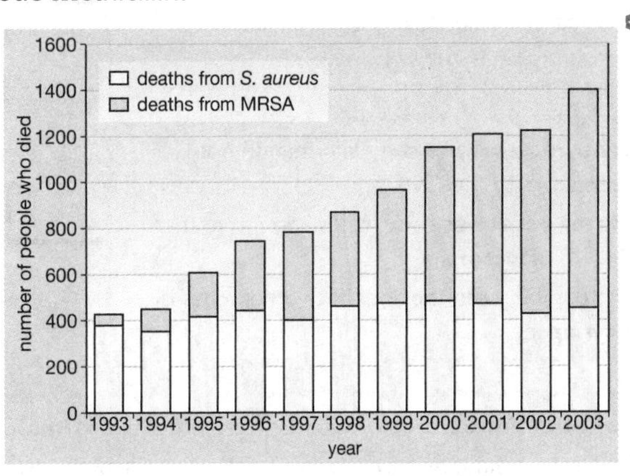

Deaths from *S. aureus*, 1993–2003.

Vaccination

Immunity through vaccination

D–C

- **Vaccination** gives people **immunity** to dangerous diseases. A **vaccine** of the dead or inactivated pathogen is injected into the bloodstream. White blood cells form **antibodies** specific to the pathogen that destroys it. If the live pathogen then infects the person, the same antibodies are rapidly produced and combat the infection.

How Science Works

You should be able to:

- explain how the treatment of disease has changed as a result of increased understanding of the action of antibiotics and immunity
- evaluate the consequences of mutations of bacteria and viruses in relation to epidemics and pandemics, e.g. bird flu
- evaluate the advantages and disadvantages of being vaccinated against a particular disease.

The MMR controversy

D–C

- **MMR** is a triple **antiviral** vaccine that protects children against measles, mumps and rubella. A research team suggested in 1998 that the vaccine caused autism, a personality disorder. Consequently, some parents failed to have their children vaccinated. Later, mumps and measles cases rose sharply.

B–A*

- In 1996, there were just 94 mumps cases. The MMR controversy broke 2 years later. In 2004, there were 8104 mumps cases, and in the first 4 months of 2005, the number was 28 470.
- The MMR jab was introduced in 1989, so students starting university in 2005 were born too early to have been vaccinated as small children. To avoid campus epidemics, universities asked students to ensure that they were vaccinated before arrival.

Questions

Grades D-C

1 Explain why scientists feared that SARS might become a pandemic.

Grades B-A*

2 Calculate the percentage increase in total deaths from *S. aureus* and MRSA between 1993 and 2003.

Grades D-C

3 Explain why the misinformation about the MMR jab had dangerous consequences.

Grades B-A*

4 Explain why the number of mumps cases rose sharply in 2004–2005.

B1a summary

Nerves and hormones

Nerves and **hormones** coordinate body processes and help control the body's:
- water content
- ion content
- blood sugar concentration
- temperature.

The **brain** and **spinal cord** make up the **central nervous system** (**CNS**).
Nerves made up of **neurones** connect the body's organs to the CNS.

Sensory neurones carry information from **receptors** to the CNS.
Motor neurones carry information from the CNS to **effectors**.
The gap between the ends of two neurones is a **synapse**.

Reflex actions are rapid, automatic responses to a **stimulus**.
An electrical impulse travels along a **reflex arc**: sensory neurone → spinal cord → **relay neurone** → motor neurone → **effector**

Different types of **receptors** detect light, sound, chemicals, touch, pressure, position change, pain and temperature.
Receptors transform energy from stimuli into electrical impulses, enabling information to travel rapidly along the nerves:
- to the brain
- to effectors (muscles and glands), which respond.

Glands secrete substances, e.g. saliva from the salivary gland.
Endocrine glands secrete **hormones** into the bloodstream to convey information to **target cells** or **organs**.

The **pituitary gland** and **ovaries** secrete hormones which control the **menstrual cycle**.
The concentrations of **FSH**, **LH** and **oestrogen** change during the cycle.
Glucose balance is maintained by **insulin** produced by the **pancreas**.

Diet

A **balanced diet** contains the right amounts of:
- **carbohydrates** – **fats** – **proteins**
- **vitamins**, **minerals**, **fibre** and **water**.

Obesity increases the chance of getting: arthritis; diabetes; high blood pressure; heart disease.

Too much **cholesterol** or **salt** increases the chance of heart disease. LDL cholesterol is 'bad' and HDL cholesterol is 'good'. LDL cholesterol levels are:
- *raised* by **saturated fats**
- *lowered* by **monounsaturated** and **polyunsaturated fats**.

The **liver** makes cholesterol if intake is too low.

Drugs

Medicinal drugs combat disease and relieve pain.
Recreational drugs can be:
- **legal** (e.g. alcohol, tobacco)
- **illegal** (e.g. cannabis, heroin, cocaine).

Hard drugs such as heroin are very **addictive** and cause severe health problems.

Alcohol slows down reactions, shrinks brain cells and kills liver cells.
It is a **depressant**, slowing down brain activity, and hindering the function of the **cortex**, **cerebellum** and **medulla**.

Tobacco smoke contains:
- **nicotine** (addictive)
- **carcinogens** (cause cancer)
- **carbon monoxide** (prevents blood from carrying oxygen)
- **irritants** (cause bronchitis, emphysema and diseases of the heart and blood vessels).

Disease

Pathogens are microorganisms that cause infectious diseases.
Bacteria and **viruses** produce **toxins** and damage cells, which makes people ill.

Body defences:
white blood cells
- ingest bacteria
- produce **antibodies**
- produce **antitoxins**.

Drugs against disease: **painkillers** relieve symptoms but do not cure diseases; **antibiotics** can kill bacteria, but bacteria can develop resistance to them; **immunisations** and **vaccinations** offer protection from infectious diseases.
Over-use of antibiotics increases the chance of bacteria developing **resistance** to antibiotics by **natural selection**.

MMR is an **antiviral vaccine** that protects against **measles**, **mumps** and **rubella**.
Failure to vaccinate most of a population against a disease can cause a sharp rise in its incidence.

Grades

Hot and cold

Animals and plants are adapted to different temperatures

- In **hot** places, including deserts, some **animals** lose heat from blood circulating through bare skin and through large thin ears. **Plants** reduce water loss with swollen stems, a thick surface, spines not leaves and a spreading, shallow root system to catch limited rainfall.

D–C

- The camel is adapted to desert conditions: its stomach can store 20 litres of water; it produces very little urine; fat is stored in the hump, keeping fat stores away from the rest of the body so that heat can be lost more easily; and its long legs hold it well above the hot ground.

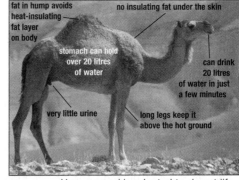

fat in hump avoids heat-insulating fat layer on body
no insulating fat under the skin
stomach can hold over 20 litres of water
can drink 20 litres of water in just a few minutes
very little urine
long legs keep it above the hot ground

How a camel is adapted to desert life.

- In **cold** places, thick fur **insulates** mammals from heat loss. Small ears and compact bodies reduce **surface area**. Plants grow low to absorb heat from the ground warmed by sunshine and to avoid cold winds.

B–A*

- The fennec fox, a desert mammal, and the Arctic fox have different adaptations to their extreme **habitats**.

A fennec fox and an Arctic fox.

How Science Works

You should be able to:
- suggest how organisms are adapted to the conditions in which they live
- suggest the factors for which organisms are competing in a given habitat
- suggest reasons for the distribution of animals or plants in a particular habitat.

Adapt or die

Competition and adaptations against predators

D–C

- To **survive**, organisms take materials from their surroundings and from other living organisms.
- Animals **compete** for food, territory and a mate. Only those well **adapted** to their **environment** are likely to survive.
- Plants compete with each other for light. Leaves are arranged to maximise the sunlight reaching them. Their root systems are large to maximise water and nutrient intake.
- Sharp **thorns** protect plants from grazing animals.

The kingsnake (top) mimics the coral snake.

B–A*

- Plant and animal **poisons** and warning colours deter **predators**.
- To avoid being eaten, animals (including insects) are camouflaged or **mimic** more aggressive creatures.
- The harmless kingsnake survives because predators mistake it for the venomous coral snake.

Questions

Grades D-C

1 Name the habitat of a camel, and state **two** ways in which it is adapted to live there.

Grades B-A*

2 Compare the adaptations of a fennec fox and an Arctic fox to their environments.

Grades D-C

3 Describe how a plant could be adapted to cope with competition for light.

Grades B-A*

4 How does the colour of a kingsnake help it to survive?

Two ways to reproduce

Grades

Top Tip!

There can be sexual reproduction with only one parent – for example in some plants where a single flower produces both male and female gametes.

Sexual and asexual reproduction

D–C

- **Genes** carry information about characteristics from parents to offspring.
- In **sexual reproduction**, each parent produces **sex cells** or **gametes**: **eggs** (female) and **sperms** (male). An egg and a sperm fuse at **fertilisation**.
- An individual **inherits** half its genes from each parent. Different mixtures of genes make each individual different from all others. This is **variation**.
- **Asexual reproduction** involves only one parent. There are no gametes, no fertilisation and no variation in offspring. Parent and offspring are all genetically identical: they are **clones**.

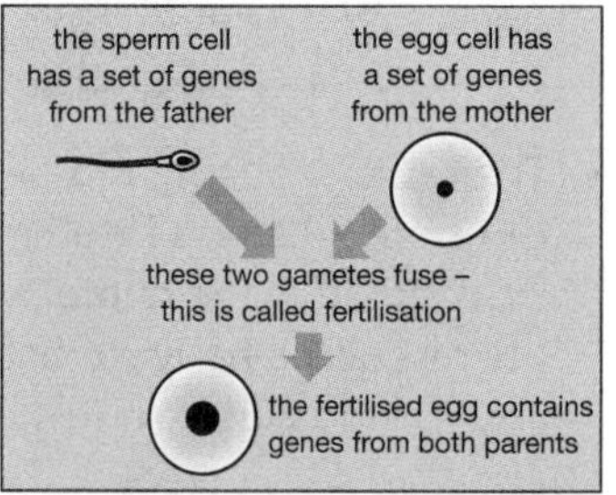

How an egg is fertilised in sexual reproduction.

B–A*

- Dolly the sheep was the first **cloned mammal**. The nucleus of the **egg cell** that developed into Dolly was replaced with the **nucleus** of an adult body cell (see also page 17).
- Normal sheep live for 10 to 16 years, but at six, Dolly was put down after developing arthritis and lung disease, conditions of ageing.
- Chromosomes are thought to age a little each time a cell divides. Since Dolly's first nucleus came from an adult, at birth her chromosomes were already old.
- Similar experiments could, in theory, be done on human eggs. Producing clones such as Dolly poses ethical questions about cloning in humans and other mammals.

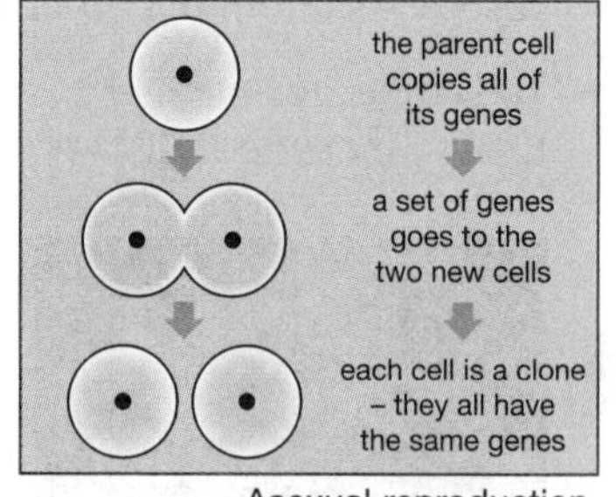

Asexual reproduction.

Genes and what they do

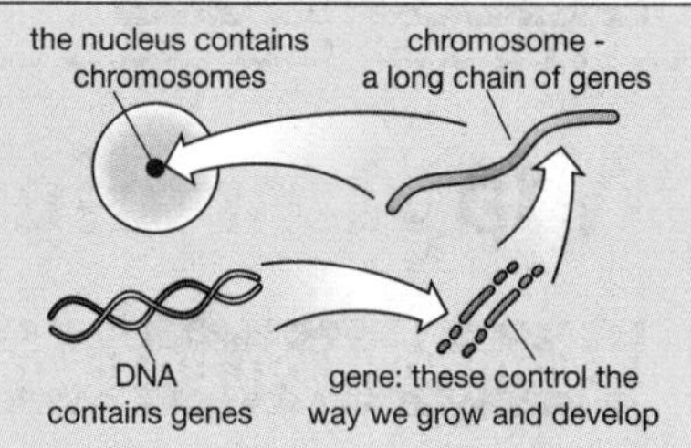

Looking at the detail – from genes to nucleus.

Genes and chromosomes

D–C

- A **gene** is a stretch of **DNA** on a **chromosome** which contains many genes. The **nucleus** of each cell contains a set of chromosomes.
- The fertilised egg contains genes from both parents. One or more genes determine each characteristic that an organism inherits. Non-inherited characteristics are acquired during an organism's life.
- The monk Gregor Mendel studied the inheritance of characteristics in generations of pea plants. From his observations, he correctly proposed how genes were passed from parents to offspring.

B–A*

- Scientists working on the **Human Genome Project** unravelled the entire genetic code of a human being. Similar techniques now routinely show whether individuals carry genes for inherited disorders.
- It would be possible for life insurance companies and employers to ask for such information when setting insurance costs or offering jobs. Whether this should be allowed is a social, ethical and economic question.

Questions

Grades D-C

1 Describe **one** difference between sexual reproduction and asexual reproduction.

Grades B-A*

2 Explain how the production of Dolly the sheep opened up new ethical questions for debate.

Grades D-C

3 Explain the difference between inherited characteristics and non-inherited characteristics, and give an example of each.

Grades B-A*

4 Explain your point of view in respect of employers being able to obtain genetic information about their employees.

Grades

Cuttings

Growing plants from cuttings

D–C

- Plants can be produced asexually from **cuttings** of the **stem**, **leaf** or **root** of a young, actively growing plant. The procedure is quick and cheap, and all plants should grow at about the same rate.
- The parent and new plants are **genetically identical** (clones).

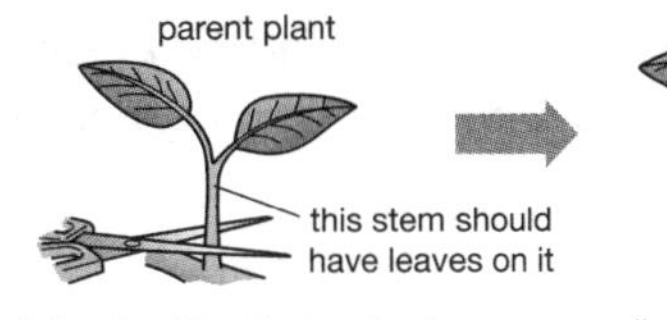
parent plant
this stem should have leaves on it
take a healthy plant and cut off a small length of stem

dip the end of the cut stem into hormone rooting powder

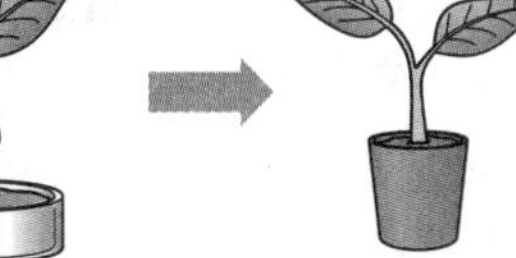
put the stem into a flowerpot full of damp compost

cover the pot with a plastic bag to keep it moist

this will grow into a new plant

Taking cuttings.

B–A*

- **Tea** is a valuable export of countries such as India and Sri Lanka.
- Hundreds of leaf cuttings can be taken from the best tea bush on a plantation to produce new plants that replace old and dead bushes.
- Plants from cuttings are better than plants from seed because they all have the characteristics of the parent plant, all develop at the same rate and can be transplanted together.

Clones

Tissue culture, cloning and embryo transplants

D–C

- In **plant tissue culture**, many small groups of **cells** from a plant grow on a jelly containing nutrients (**growth medium**) into new plants that are clones.
- A sheet of human skin can be **cultured** in a growth medium from a few healthy skin cells, and used in **skin grafts** (e.g. for people with burns).
- In **adult cell cloning**, the nucleus of an **egg** cell is replaced with the nucleus of an adult **body cell** (see diagram). The resulting individual is a clone of the body cell donor.
- In **embryo transplantation**, the cells of a developing animal embryo are **split** apart before they specialise. Each cell develops into an embryo that is transplanted into a host mother that gives birth to the young animal.

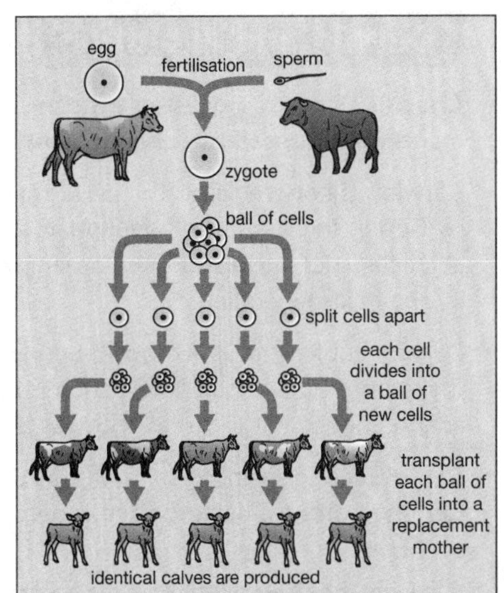
egg
fertilisation
sperm
zygote
ball of cells
split cells apart
each cell divides into a ball of new cells
transplant each ball of cells into a replacement mother
identical calves are produced

Embryo splitting and transplantation – many identical offspring from one zygote.

B–A*

- As the survival of many species is threatened, so it becomes more necessary to preserve the populations of these species in zoos.
- Scientists at the Zoological Society of London transplanted the embryo of a rare species of zebra into a horse. This **surrogate** mother successfully produced a zebra foal.
- This technique of 'cross-species' embryo transplantation may help to preserve many other endangered species. When their safety can be assured in the wild, they can be reintroduced to their natural habitats.

Endangered mammals can be successfully bred by cross-species transplantation.

Questions

Grades D-C

1 List **two** advantages of growing new plants from cuttings.

Grades B-A*

2 Explain why taking leaf cuttings is the best way of producing new tea bushes.

Grades D-C

3 Describe how adult cell cloning is done.

Grades B-A*

4 Explain how embryo splitting and transplantation can help to conserve rare species of mammals.

Grades

Genetic engineering

GM crops and the manufacture of insulin

- **Genetic engineering** is the technique of transferring a gene from the chromosome of one organism to another at an early stage of its development. The organism is then **genetically modified** (**GM**).
- Soya is a rich protein source. To enable a **herbicide** spray to kill weeds and not the soya plants, scientists inserted a gene for **resistance** to the herbicide into a soya chromosome. These modified soya plants are now a **GM crop**.
- Many people take insulin to control their **diabetes**. The gene for producing insulin was cut from human DNA by a special enzyme and inserted into a bacterium. Cultures of the bacterium now provide insulin on an industrial scale.

How Science Works

You should be able to:

- interpret information about cloning techniques and genetic engineering techniques
- make informed judgements about the economic, social and ethical issues concerning cloning and genetic engineering, including GM crops.

D–C

- Genetic modification is widespread. Added genes have made: plants resistant to microbe diseases; salmon grow faster; food plants less toxic; fruit and vegetables ripen faster; and cattle grow faster after being injected with GM hormones.
- Concerns about genetic engineering include: Are GM plants safe? Will the modified genes jump to other species? Are there long-term consequences of eating GM foods?

B–A*

Theories of evolution

How Science Works

You should be able to:

- identify the differences between Darwin's theory of evolution and conflicting theories
- suggest reasons for the different theories
- suggest reasons why Darwin's theory of natural selection was only gradually accepted.

Darwin and evolution

- **Charles Darwin** suggested that the process of **natural selection** made **species** gradually change (**evolve**) from one form to another.
- Darwin's theory replaced the belief that all organisms were created by God in the form they have today, and Lamarck's idea that characteristics **acquired** in an organism's lifetime can be inherited by the next generation.
- Darwin studied the varied beaks of finches from several islands in the Galapagos. He suggested that their common ancestor was a finch from the mainland, and that the different beaks had evolved as adaptations to eat the food that was available on each island.

D–C

- The 15 races of giant tortoise on the Galapagos islands form two groups, based on the vegetation available to them:
  - **dome-backed** tortoises have an arch at the front of the shell that allows them to lift their heads up; they are able to reach plant food well above ground level
  - **saddle-backed** tortoises have no arch, restricting how far they can raise their heads; they are restricted to eating plants on or near the ground.

  Natural selection has caused these two different groups to evolve.

A saddle-backed giant Galapagos tortoise.

B–A*

Questions

Grades D-C

1 What kind of GM organism is used to produce insulin?

Grades B-A*

2 Suggest why some people do not want to eat foods containing GM soya.

Grades D-C

3 Why do the Galapagos finches have different kinds of beaks?

Grades B-A*

4 Explain the differences between Lamarck's and Darwin's theories of evolution.

Natural selection

Genes, environment and survival

D–C

- In **natural selection**, a **mutation** (change) to a gene within an organism, or a change in the environment outside an organism, can alter the organism's chances of survival and successful breeding.
- A **mutation** that increases an organism's survival chances is more likely to be inherited. Mutations can therefore cause a **species** to change, sometimes rapidly.
- Before the industrial revolution, **peppered moths** showed **variation** in colour. A *pale*, speckled colour **camouflaged** the moths well against pale tree bark. Predators easily spotted *dark* moths, and they were much rarer.
- The industrial revolution blackened buildings and trees with soot. Dark moths were then better camouflaged and became the commoner type. Now, with cleaner air, pale peppered moths are common again.

Top Tip!

Remember that organisms like the moths do not change on purpose – it happens through natural selection.

Can you see the light-coloured moth in this photograph?

B–A*

- A tornado that struck Canada in 1890 killed large numbers of sparrows.
- An ornithologist measured the leg lengths of dead and surviving sparrows. She noticed that the dead birds had either very long or very short legs, and that the survivors had medium-length legs. Short legs cannot hold on to twigs very well and long legs break easily. Medium-length legs are able to hold on to twigs well and do not snap easily. This is an example of natural selection.

Fossils and evolution

Fossils and the origins of life

D–C

- **Fossils** provide evidence of changes to organisms over time, and of how fast they occurred. Similarities and differences between fossils and living organisms give clues as to how species are related.
- Fossils and the **theory of evolution** suggest that all present-day species evolved from simple **life-forms** which appeared about 3.5 billion years ago.

B–A*

- There is disagreement as to where life on Earth originated.
- Some say that the combination of early atmospheric gases, sunlight and shallow seas enabled life to begin. Others say that a meteorite or other object first brought life to Earth. Yet others suggest that life started in the deep oceans in the absence of light, where lava and hot gases well up through the Earth's crust.
- Space researchers continue to seek evidence of life, now or in the past, in the Solar System beyond Earth, particularly on Mars. No evidence yet convinces the majority of scientists that life exists beyond Earth.

How Science Works

You should be able to: suggest reasons why scientists cannot be certain about how life began on Earth.

Questions

Grades D-C

1 Explain why the black variety of the peppered moth became more common in the industrial revolution.

Grades B-A*

2 Suggest how the effects of the tornado in Canada could have affected the characteristics of sparrows in future generations.

Grades D-C

3 Explain how fossils provide evidence for evolution.

Grades B-A*

4 List **three** possible places in which life on Earth may have begun.

Grades

Extinction

The dodo, a flightless bird, became extinct 300 years ago when people introduced rats which were predators of the dodo.

Causes of extinction

D–C

- A species may become **extinct** if it cannot **adapt** to a change to its **physical** (non-living) **environment**, or to the arrival in its **habitat** of other organisms.
- An **environmental** change may be a change in climate or a change of land use by humans. Only organisms that can adapt or move away will survive.
- A **predator** arriving in a new habitat may kill and eat all individuals of a species not adapted to escape it. Predators include human hunters.
- Successful **competitors** for food may survive, while less successful competitors become extinct.
- A species may not be adapted to withstand a new **disease**.

B–A*

- Once living in Tasmania to the south of Australia, the **Tasmanian wolf** had stripes like a tiger and the shape of a wolf. But, like kangaroos and wallabies, it had a pouch for its young.
- This unique animal was hunted almost to extinction. The remaining few individuals died from disease. The last died in captivity in 1936.

Hunting drove the Tasmanian wolf to extinction.

More people, more problems

Human populations and population growth

D–C

- Worldwide industrial development has improved the living standards of many people. People live longer with improved food, housing, health care and safe water. By 2075, the world's **population** may reach 10 billion.
- Whereas populations in developed countries are now stable, there is a **population explosion** in many less developed countries.
- Rising populations require more land for buildings, roads and food production, so habitats are being destroyed.
- Already, **non-renewable energy resources** such as **crude oil** and **coal**, and raw materials such as **metal ores** and **rocks**, are running out. **Waste** and **pollution** will increase if they are not regulated by laws.

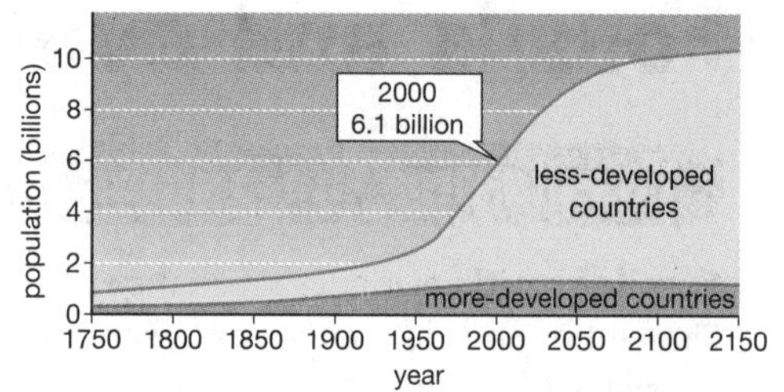

Human population growth – will it rise as predicted?

Superquarries

B–A*

- New roads are required in areas of expanding population in the UK, particularly in the Midlands and South-east. For road building, vast quantities of crushed rock – aggregate – are mined from superquarries on the west coast of Scotland where few people live.
- Iron ore is also mined, for the construction of new bridges and buildings.

A quarry such as this leaves a huge crater in which nothing lives.

Questions

Grades D-C

1 Explain how a new competitor could cause a species to become extinct.

Grades B-A*

2 Suggest why natural selection did not help the Tasmanian wolf to adapt to resist invasion of its habitat by humans.

Grades D-C

3 Look at the graph of world population growth. What is happening now to the *rate* of population growth?

Grades B-A*

4 Explain how population growth in one part of the world can cause habitat destruction in another, even if this part has only a small human population.

Grades

Land use

> **Top Tip!** Properly treated sewage is absolutely safe – it is untreated sewage that causes problems.

Damage to the natural environment

D–C

- Humans reduce the natural habitats of other animals and plants by: **building** homes, workplaces and roads; **quarrying** for building materials; expanding the land used for **farming**; and using landfill sites in which to dump **waste**.
- **Incinerating** waste can be a health hazard if not carried out carefully.
- **Intensive farming** methods increase food production but also reduce the variety of species in a habitat (**biodiversity**):
  - herbicides and pesticides are toxic chemicals that kill unwanted plants and pests
  - fertilisers increase yield but can pollute waterways if not used carefully
  - removing hedgerows enlarges fields but reduces habitats.

Incineration – a safe waste disposal method?

B–A*

- When plastics are burned, the gases produced may contain **dioxins**, toxic chemicals that damage the liver, nervous system, endocrine system, reproductive system and the immune system. Dioxins are also suspected of causing cancer.
- Careful design and maintenance of incineration plants is required, to avoid the release of dioxins to the environment.

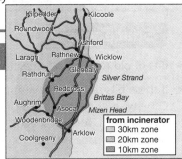

Plans to build an incinerator at Redcross on the east coast of Ireland were met with protests.

Pollution

How water is polluted

D–C

- Because raw **sewage** is a health hazard, protective clothing should be worn by all who work in sewers, toilets, water treatment plants and all who transport sewage sludge.
- Unless **waste** is properly treated, it may cause **pollution**:
  - **sewage** contains microorganisms that cause human diseases, and toxic chemicals from industry that can kill aquatic life
  - if excess **fertiliser** reaches waterways, water plants grow rapidly, cover the surface and then die; organisms below the surface die as light fails to reach them, and oxygen is used up by bacteria living on dead plants
  - **toxic chemicals** illegally dumped in waterways can poison living organisms.

How DDT is taken up the food chain.

Death in the Caspian Sea

B–A*

- The Caspian Sea is a huge landlocked body of water in Eastern Europe; 130 rivers flow into it but no water flows out. Only evaporation regulates its level.
- The rivers carry run off from agricultural land into the Caspian Sea. Fertilisers have already become highly concentrated in the north-east (coloured green). Here, the cycle of rapid surface plant growth and death, and then bacteria living off the dead plants, has killed the fish and other aquatic organisms in that area.

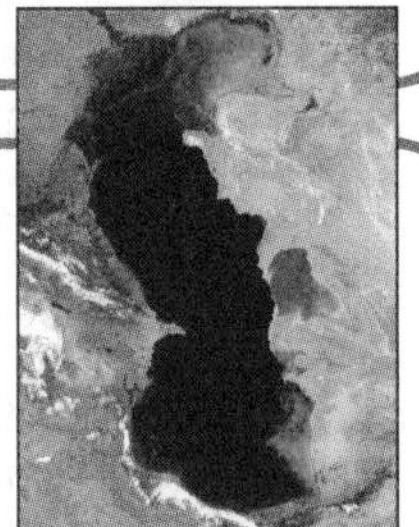

The Caspian Sea, photographed from space in 2004.

Questions

Grades D-C

1 What is meant by 'intensive farming'?

Grades B-A*

2 Give **one** advantage and **one** potential disadvantage of disposing of plastics by incineration rather than landfill.

Grades D-C

3 Explain why DDT killed many predatory birds.

Grades B-A*

4 Explain how pollution is affecting the Caspian Sea.

Air pollution

Pollution by smoke and sulfur dioxide

- Power stations, factories and traffic produce **smoke** particles. Fossil fuels produce **sulfur dioxide**, an acidic gas which helps form **acid rain**.
- **Sulfur dioxide** and **smoke** cause respiratory illnesses. Sulfur dioxide damages leaves and erodes limestone on buildings. Smoke settles as black **soot**.
- To reduce air pollution, factories fit filters to clean the smoke they emit (page 120), power stations use gas scrubbers (page 35), and cars have catalytic converters (see below).

D–C

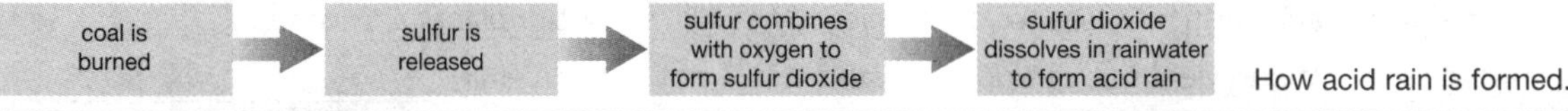

coal is burned
sulfur is released
sulfur combines with oxygen to form sulfur dioxide
sulfur dioxide dissolves in rainwater to form acid rain

How acid rain is formed.

Space waste

- Bits of old rockets and satellites litter space high above the Earth.
- If these bits re-enter the atmosphere, they are likely to burn up. Otherwise, they stay in orbit, moving at between 30 000 and 45 000 km/h. At such speeds, the smallest bit of debris could seriously damage a spacecraft or satellite.

Space waste – a threat to satellites and spacecraft?

B–A*

What causes acid rain?

Top Tip!

Do not confuse acid rain with the greenhouse effect! One is caused by sulfur dioxide, and the other by carbon dioxide.

Acidic gases and the effects of acid rain

- Power stations that burn **fossil fuels**, especially coal, produce the gas **sulfur dioxide**. It reacts with oxygen and water in the air to produce **sulfuric acid**.
- Oxides of nitrogen in traffic exhaust form **nitric acid**.
- Rainwater is naturally weakly acidic: **carbon dioxide + water → carbonic acid**
- These acids dissolve in rainwater and form **acid rain**, which:
  - kills **trees**. It strips the waxy surface from leaves. Their cells shrivel and cannot photosynthesise. Acid rain also washes nutrients and minerals from the soil, and releases aluminium which is toxic
  - makes **lakes** so acidic that fish die
  - reacts with the calcium carbonate of buildings made of **limestone** and **sandstone**, breaking them down.

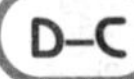

Car exhaust emissions

- **Catalytic converters**, fitted to all new cars, convert most polluting gases to non-polluting ones, but not 100%.
- The table shows exhaust products from cars. Oxygen and nitrogen are also produced. (Most petrol is now unleaded.)

exhaust content	its effects
water	no harmful effect
carbon dioxide	greenhouse gas
carbon monoxide	stops blood carrying oxygen
nitrogen oxides	cause acid rain
sulfur dioxide	causes acid rain
benzene	can cause leukaemia
lead	affects nervous system

B–A*

Questions

Grades D–C

1 How can sulfur dioxide affect people's health?

Grades B–A*

2 Suggest why it may be very difficult to get countries to reduce or clean up pollution by 'space waste'.

Grades D–C

3 How does acid rain harm limestone buildings?

Grades D–C

4 Catalytic converters convert nitrogen oxides to nitrogen, and carbon monoxide to carbon dioxide. Suggest why nitrogen oxides and carbon monoxide are still released into the air from motor vehicles.

Grades

Pollution indicators

How Science Works

You should be able to: analyse and interpret scientific data concerning environmental issues.

Organisms that indicate pollution levels

D–C

- **Lichens** can be used as indicators of sulfur dioxide **air pollution**. **Invertebrate animals** are indicators of **water pollution** by, for example, **sewage**.

distance downstream from sewage entry point (m)	what the water is like	invertebrates found	oxygen levels
sewage enters here 0–10	dark and cloudy; very smelly	*Chironomus* larva; rat-tailed maggot	falling quickly
10–100	cloudy; bad smell	tubifex worm (sludge worm); mosquito larva	very low
100–200	starts to clear; slight smell	flatworm; caddis fly larva	gradually rising
200+	clear	freshwater shrimp; mayfly larva	back to normal

The effects of untreated sewage in a river.

Pollution in the Thames

B–A*

- From 1975, the water quality of the Thames improved. Now over 120 fish species inhabit the river.
- In August 2004, raw sewage was accidentally discharged into the Thames. The authorities tried to **oxygenate** that part of the river by churning up the water. But some 100 000 fish died as oxygen levels fell to almost zero in a slow-flowing part.

How Science Works

You should be able to:
- evaluate methods used to collect environmental data and consider their validity and reliability as evidence for environmental change
- weigh evidence and form balanced judgements about some of the major environmental issues facing society, including the importance of sustainable development (see page 24).

Deforestation

Deforestation and the carbon cycle

D–C

- In the **deforestation** of **tropical rainforests**, trees are cut for **timber** and burned to make way for **agriculture**.
- In the **carbon cycle**, all living things **respire** and give out carbon dioxide. In **photosynthesis**, carbon dioxide is 'locked up' in plant material. But deforestation 'unlocks' it, increasing the rate at which carbon dioxide is returned to the air.

deforestation
blocks this
CO_2 added
industrialisation
carbon dioxide in the air
respiration
green plants are eaten by animals
green plants use carbon dioxide to make their food
dead remains of plants and animals
decay by fungi and bacteria

The carbon cycle.

- Deforestation also reduces **biodiversity**: some species become **extinct**, the survivors show reduced **variation**, and **habitats** are destroyed. Some lost organisms may have been beneficial to humans.

B–A*

- About 25% of all medicines originate from the world's rainforests.
- Among plants with **anti-cancer** properties, the Madagascar periwinkle contains chemicals used in **chemotherapy** for some forms of child leukaemia.
- It is possible that cures for currently incurable diseases such as **HIV/AIDS** may be found in rainforests.

Chemicals in the Madagascar periwinkle are used in cancer treatment.

Questions

Grades D-C

1 If a river contains rat-tailed maggots, is it polluted or unpolluted?

Grades B-A*

2 Explain why oxygen levels dropped in the Thames following the pollution incident in 2004.

Grades D-C

3 How can rotting trees affect carbon dioxide levels in the atmosphere?

Grades B-A*

4 Suggest why rainforests are good places to hunt for new cures for diseases.

The greenhouse effect – good or bad?

The greenhouse effect

- The **atmosphere** traps (absorbs) some of the Sun's heat energy in the **greenhouse effect**.
- The greenhouse effect is thought to be increasing, with less heat escaping from the atmosphere back into space. One suggested cause is a rise in the level of the heat-absorbing gases: **carbon dioxide** from burning fossil fuels, and **methane** from cattle and rice fields.
- As climates change throughout the world, sea levels will rise as the Earth's great ice caps melt.

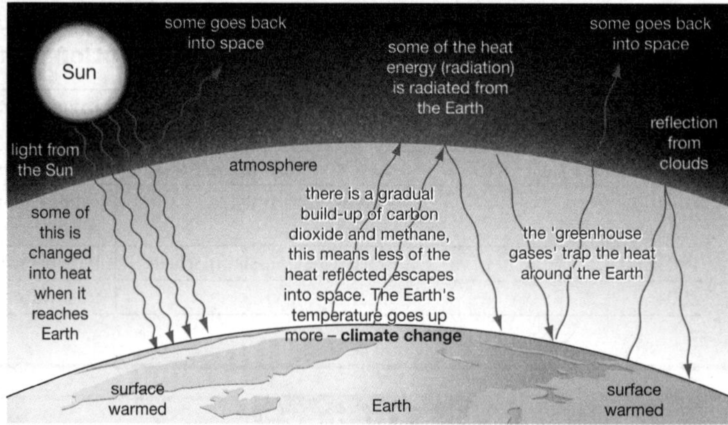

Sun
some goes back into space
some of the heat energy (radiation) is radiated from the Earth
some goes back into space
reflection from clouds
light from the Sun
atmosphere
some of this is changed into heat when it reaches Earth
there is a gradual build-up of carbon dioxide and methane, this means less of the heat reflected escapes into space. The Earth's temperature goes up more – **climate change**
the 'greenhouse gases' trap the heat around the Earth
surface warmed
Earth
surface warmed

How an increase in the greenhouse effect leads to climate change.

Grades D–C

Methane from rice paddies

- A given quantity of methane is over 20 times more effective as a greenhouse gas than the same quantity of carbon dioxide. In waterlogged rice paddy fields, anaerobic bacteria (that survive without oxygen) produce up to 100 million tonnes of methane a year.
- Experiments show that when the fields are drained for part of the rice-growing season, more rice is produced and less methane is released.

Top Tip!

The greenhouse effect is good because without it the Earth would be too cold for life. But if it increases, that is *not* good.

For good productivity, rice does not need to be under water all season.

Grades B–A*

Sustainability – the way forward?

Means of sustainable development

- For the long-term future of humans, steps must be taken at local, regional and global levels to find ways of **sustainable development**, in which **resources** are not used up faster than they can be replaced.
- Rubbish must be treated as a **resource** and **recycled**. People must reduce travel by powered vehicles and use **public transport** where possible. **Buildings** must be better insulated.

Grades D–C

- U-values tell builders and architects how good are energy insulating methods. The smaller the U-value, the better the method.
- The formula for the heat energy that a building loses in watts (1 W = 1 J/s) is:
heat energy loss (watts) = U-value × area (m^2) × temperature difference

structure	U-value
tiled roof, no insulation	2.0
tiled roof with insulation	0.3
cavity wall, no insulation	1.4
cavity wall, foam insulation	0.4
single-glazed window	5.7
double-glazed window	2.9

Grades B–A*

Questions

Grades D-C

1 Name **two** gases that contribute to the greenhouse effect.

Grades B-A*

2 Explain why, although less methane is released into the atmosphere than carbon dioxide, methane is contributing greatly to climate change.

Grades D-C

3 State **one** way that *you* could help to reduce carbon dioxide emissions.

Grades B-A*

4 A building with an area of 500 m^2 has an average U-value of 3.5. The temperature outside is 12 °C, and the temperature inside is 19 °C. How much heat energy is lost each second?

B1b summary

Animal and plant adaptations

Animals and plants are adapted:
- to **survive** in different **habitats**
- to **compete** for resources
- for **protection** against predators.

Animals compete for food, territory and a mate.

Plants compete for light, water and nutrients from the soil.

The fennec fox and Arctic fox have adapted to extreme environments.

Reproduction and genes

Reproduction can be:
- **sexual** (using **gametes**, variation in offspring)
- or **asexual** (one parent, identical offspring).

Genetically identical **clones** can be produced:
- from **cuttings** and **tissue culture** in plants
- by **embryo splitting** and **transplantation** in animals.

Cross-species embryo transplantation may help to preserve endangered wild species in captivity so that they can be later reintroduced into their natural habitat.

DNA contains **genes** on **chromosomes** that control our **inherited characteristics**.

In **genetic engineering**, a gene from one organism is inserted into another.

Genetically modified (**GM**) **organisms** include: plants resistant to microbe diseases; bacteria that make human insulin; plants and animals that produce our food faster.
Concerns about GM organisms include: their **safety** for use; genes jumping to other species; the long-term consequences of eating them.

Evolution

Charles Darwin proposed the theory of **evolution** of **species** by **natural selection**.

Comparing **fossils** with present-day plants and animals can show us how species have evolved.
Life on Earth may have **originated**:
- as a result of early air/light/water conditions
- from a meteorite
- in the deep oceans.

Mutation of a gene that causes an advantageous change to a characteristic can increase an organism's chance of **survival**.

Factors that contribute to **extinction** include:
- environmental change
- new predators or diseases
- successful competitors
- adverse human impact on habitats.

Human impact on the environment

Increase in the human **population** has affected the environment by:
- using up resources more rapidly
- increasing waste and pollution.

Land is used up for:
- buildings and quarrying materials to build them
- producing food
- waste disposal.

Intensive farming methods and **deforestation** destroy habitats, reducing **biodiversity**.

Increases in the heat-absorbing gases responsible for the **greenhouse effect** can contribute to **climate change**.

Water pollution is caused by toxic chemicals, raw sewage and excess fertiliser.

Air pollution caused by burning fossil fuels results in smoke, smog and acid rain.

Sulfur dioxide and **oxides of nitrogen** dissolve in rainwater to give **acid rain** which damages plants, pollutes lakes and erodes limestone and sandstone structures.

Lichens, invertebrates and fish can be used as **pollution indicators**.

Sustainable development includes using renewable energy sources, recycling and conserving energy and fuels.

Elements and the periodic table

The elements and patterns in properties

D–C

- An **element** contains only one sort of atom. Each element has a **chemical symbol**. O = oxygen; Na = sodium.
- Elements are arranged in the **periodic table**. Each vertical column is a **group**, containing elements with similar properties.
- The **mass number** of an element = number of protons + number of neutrons.

The periodic table of the elements.

Key: mass number, atomic number — $^{1}_{1}$H hydrogen

Group I	II											III	IV	V	VI	VII	0
																	$^{4}_{2}$He helium
$^{7}_{3}$Li lithium	$^{9}_{4}$Be beryllium											$^{11}_{5}$B boron	$^{12}_{6}$C carbon	$^{14}_{7}$N nitrogen	$^{16}_{8}$O oxygen	$^{19}_{9}$F flourine	$^{20}_{10}$Ne neon
$^{23}_{11}$Na sodium	$^{24}_{12}$Mg magnesium											$^{27}_{13}$Al aluminium	$^{28}_{14}$Si silicon	$^{31}_{15}$P phosphorus	$^{32}_{16}$S sulfur	$^{35}_{17}$Cl chlorine	$^{40}_{18}$Ar argon
$^{39}_{19}$K potassium	$^{40}_{20}$Ca calcium	$^{45}_{21}$Sc scandium	$^{48}_{22}$Ti titanium	$^{51}_{23}$V vanadium	$^{52}_{24}$Cr chromium	$^{55}_{25}$Mn manganese	$^{56}_{26}$Fe iron	$^{59}_{27}$Co cobalt	$^{59}_{28}$Ni nickel	$^{64}_{29}$Cu copper	$^{65}_{30}$Zn zinc	$^{70}_{31}$Ga gallium	$^{73}_{32}$Ge germanium	$^{75}_{33}$As arsenic	$^{79}_{34}$Se selenium	$^{80}_{35}$Br bromine	$^{84}_{36}$Kr krypton
$^{85}_{37}$Rb rubidium	$^{88}_{38}$Sr strontium	$^{89}_{39}$Y yttrium	$^{91}_{40}$Zr zirconium	$^{93}_{41}$Nb niobium	$^{96}_{42}$Mo molybdenum	$^{99}_{43}$Tc technetium	$^{101}_{44}$Ru ruthenium	$^{103}_{45}$Rh rhodium	$^{106}_{46}$Pd palladium	$^{108}_{47}$Ag silver	$^{112}_{48}$Cd cadmium	$^{115}_{49}$In indium	$^{119}_{50}$Sn tin	$^{122}_{51}$Sb antimony	$^{128}_{52}$Te tellurium	$^{127}_{53}$I iodine	$^{131}_{54}$Xe xenon
$^{133}_{55}$Cs caesium	$^{137}_{56}$Ba barium	$^{139}_{57}$La lanthanum	$^{178}_{72}$Hf hafnium	$^{181}_{73}$Ta tantalum	$^{184}_{74}$W tungsten	$^{186}_{75}$Re rhenium	$^{190}_{76}$Os osmium	$^{192}_{77}$Ir iridium	$^{195}_{78}$Pt platinum	$^{197}_{79}$Au gold	$^{201}_{80}$Hg mercury	$^{204}_{81}$Tl thallium	$^{207}_{82}$Pb lead	$^{209}_{83}$Bi bismuth	$^{210}_{84}$Po polonium	$^{210}_{85}$At astatine	$^{222}_{86}$Rn radon
$^{223}_{87}$Fr francium	$^{226}_{88}$Ra radium	$^{227}_{89}$Ac actinium															

- The **atomic number** = number of protons = number of electrons. An atom has no net charge.

B–A*

- In 1869, Mendeleev put the elements known then in a table, mostly in order of **atomic mass** (found by experiment). But, as a priority, he considered their **properties**. He placed tellurium (atomic mass 128) *before* iodine (atomic mass 127), so that iodine was with the other halogens, fluorine, chlorine and bromine.
- Because Mendeleev deliberately grouped elements according to their properties, scientists found his table very useful, and based the modern periodic table on it.

Atomic structure 1

Particles in an atom

D–C

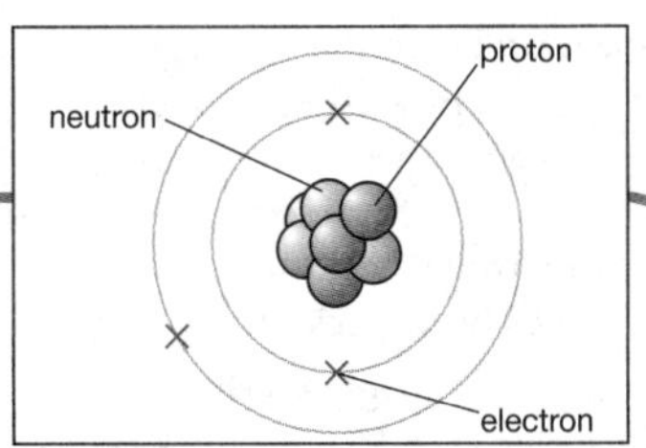

The structure of an atom.

- An atom has a central **nucleus** of **protons** and **neutrons**, surrounded by orbiting **electrons**. (Hydrogen's nucleus is just 1 proton.)
- The masses of protons, neutrons and electrons are incredibly small. **Atomic mass units** are used to express their masses relative to one another – **relative masses**.

particle	relative mass	charge
proton	1	+
neutron	1	neutral
electron	almost 0	–

Electronic configuration and chemical properties

B–A*

- The **electronic configuration** of an atom gives the number of its **electrons** and their arrangement in **shells**.
- The diagrams show the electronic configurations of lithium, sodium and potassium.
- Because they have the same number of outer-shell electrons (one), all three elements have similar chemical properties, and all appear in the same group in the periodic table.

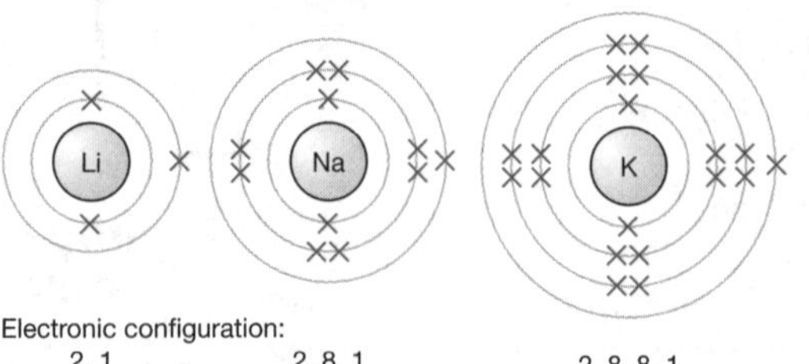

Electronic configuration:
2, 1 — 2, 8, 1 — 2, 8, 8, 1

Questions

(Grades D-C)

1 Use the periodic table to find the chemical symbol for zinc.

(Grades B-A*)

2 Explain why Mendeleev placed iodine *after* tellurium in his table of the elements, even though the atomic mass of iodine is less than tellurium.

(Grades D-C)

3 Use the periodic table to find the atomic number of magnesium. What does this tell you about the structure of a magnesium atom?

(Grades B-A*)

4 Name **two** elements that will have similar properties to calcium, and explain why.

Grades

Bonding

Bonds in elements and compounds

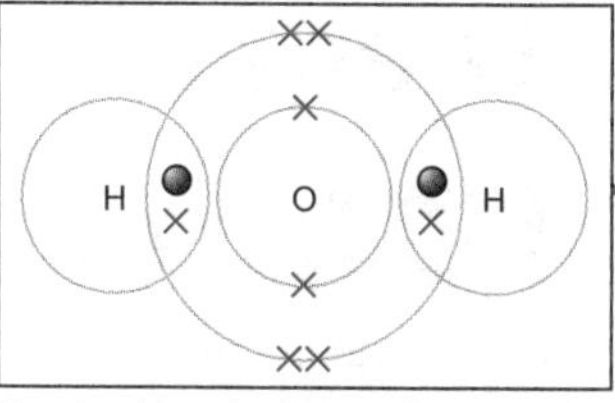

By sharing electrons all atoms in a water molecule have full outer electron shells: the hydrogen atoms have **two** electrons in shell 1 and oxygen has **eight** electrons in shell 2.

- Atoms with a full outer shell of electrons (see page 26) are **stable**. An atom will react chemically with another atom that either gives, takes or shares electrons so that both atoms have a full outer shell.
- We use atoms and symbols to represent and explain what happens to substances in chemical reactions.

D–C

- When atoms of two (or more) *different* elements react chemically, they form a **compound**.
  - In water, H_2O, each hydrogen atom *shares* a pair of electrons with the oxygen atom. Each shared pair of electrons is a **covalent bond**.
  - A magnesium atom joins an oxygen atom to form magnesium oxide, MgO. Magnesium *gives* two electrons, and oxygen *takes* them. Both then have a charge – they are **ions**. Their opposite charges attract, forming an **ionic bond**.
- Two atoms of the *same* element can join to form **molecules**: in chlorine, Cl_2, the atoms share electrons in a **covalent bond**.

Representing covalent bonds

B–A*

- **Bonds** are represented by lines between element symbols. For example:
  Cl–Cl represents a chlorine molecule – a single bond joins the atoms (one shared pair of electrons).
  O=O represents an oxygen molecule – a double bond joins the atoms (two shared pairs of electrons).

Extraction of limestone

How Science Works

You should be able to: consider and evaluate the environmental, social and economic effects of exploiting limestone and producing building materials from it.

Extraction and uses of limestone

D–C

- **Limestone** is **calcium carbonate**, $CaCO_3$. It is the main compound in chalk and marble which, like limestone, are mined from quarries.
- Limestone is a building material. It is also used to make cement, concrete and glass (page 28) and to extract iron from its ore (page 29).
- A **limestone quarry** is likely to supply jobs for local people, have a good road system and local income to pay for health care, recreational and other facilities.
- It may also scar the landscape and eliminate natural habitats, cause dust pollution and heavy traffic that damages buildings and roads, and produce noise from quarry blasting.

Establishing new quarries

B–A*

- Limestone rocks were formed at the bottom of warm seas, millions of years ago.
- The Peak District National Park, an area of outstanding natural beauty, has extensive high quality limestone deposits which have been mined since Roman times. Today, all new quarries require the approval of the local planning authority.

The Peak District, a place for people to enjoy or an area to be quarried?

Questions

Grades D-C

1 An atom of oxygen has six electrons in its outer shell. How many does it need to make this shell full?

Grades B-A*

2 In nitrogen gas, pairs of nitrogen atoms join together by sharing three pairs of electrons – a triple bond. How could we represent this?

Grades D-C

3 List **two** advantages and **two** disadvantages of having a limestone quarry near a small town.

Grades B-A*

4 Is limestone a renewable or non-renewable resource? Explain your answer.

Thermal decomposition of limestone

Grades

Reactions of limestone and its properties

- Limestone is mainly **calcium carbonate**. Its formula $\mathbf{CaCO_3}$ tells you the type and ratio of atoms joined together. The ratio is: 1 calcium atom : 1 carbon atom : 3 oxygen atoms.
- When strongly heated, limestone undergoes a **thermal decomposition** reaction, decomposing into simpler compounds. The word and **symbol** equations are:

  calcium carbonate → calcium oxide + carbon dioxide

  $\mathbf{CaCO_3 \rightarrow CaO + CO_2}$

  The equation is **balanced**, with the same number of each type of atom on either side. In this and all reactions, no atoms are lost or made, and: mass of product(s) = mass of reactant(s)
- Calcium oxide is **quicklime**. It reacts with water to produce calcium hydroxide, $Ca(OH)_2$, **slaked lime**:

  calcium oxide + water → calcium hydroxide
- Heat also decomposes other metal carbonates: **metal carbonate → metal oxide + carbon dioxide** (D–C)
- Adding water to quicklime forms solid slaked lime, calcium hydroxide, $Ca(OH)_2$. Adding more water dissolves it, giving calcium hydroxide solution – **limewater**.
- Limewater is used to test for **carbon dioxide**: when the gas is bubbled through limewater, a **suspension** of calcium carbonate is formed: (B–A*)

  $\mathbf{CO_2 + Ca(OH)_2 \rightarrow CaCO_3 + H_2O}$
- **Slaked lime** mixed with sand and water makes slow-setting **lime mortar**, once widely used to join bricks.

Uses of limestone

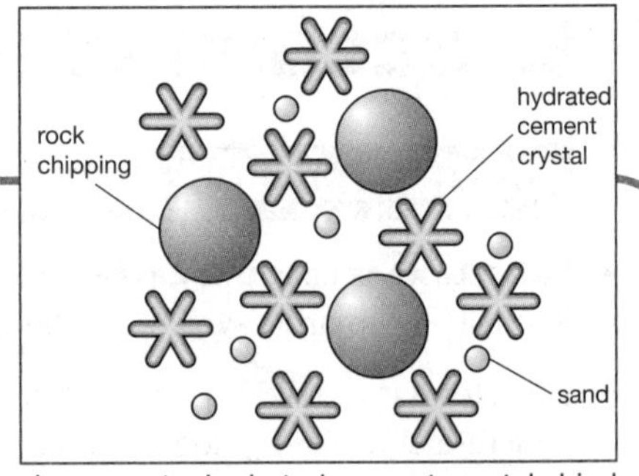

In concrete, hydrated cement crystals bind together the rock chippings and sand.

Limestone in building materials

- **Cement** is formed when crushed **limestone** and **clay** are strongly heated. Cement reacts with water: it is 'hydrated'.
- **Cement mortar** is a mixture of cement, sand (silicon dioxide) and water used to join bricks. Cement mortar has largely replaced lime mortar: it sets faster and is stronger, so lasts longer.
- **Concrete** is cement with sand and rock chippings added. The mixture reacts with water and sets to a much stronger material than cement mortar. (D–C)
- To make **glass**, **sand** and **sodium carbonate** are heated with limestone.
- A very high temperature melts the sand (silicon dioxide) and breaks down the sodium carbonate and limestone (calcium carbonate) to sodium oxide and calcium oxide.
- When cooled slowly, pure molten silicon dioxide gives glass with a regular crystalline **lattice** structure (see page 29). It shatters easily when hit. (B–A*)
- When cooled rapidly, the glass mixture fails to form a lattice structure. As ions, the sodium and calcium make the structure more irregular, increasing the toughness of the glass.

How Science Works

You should be able to:
- evaluate the developments in using limestone, cement, concrete and glass as building materials
- evaluate their advantages and disadvantages over other materials.

Questions

Grades D-C

1 Write the balanced symbol equation for the thermal decomposition of limestone.

Grades B-A*

2 Explain how the limewater test for carbon dioxide works.

Grades D-C

3 What is the difference between cement and concrete?

Grades B-A*

4 Why does adding sodium and calcium ions when making glass produce glass that does not easily break?

Grades

CIA PRODUCTS FROM ROCKS

The blast furnace

Extraction of iron from iron ore

D–C

- The raw materials for making iron are abundant and economical to extract and use. **Haematite**, iron ore, is mainly **iron(III) oxide**, Fe_2O_3. **Coke** is purified coal, C (carbon). **Limestone** is calcium carbonate, $CaCO_3$. Air contains **oxygen**, O_2. Iron ore is **reduced** to iron in a **blast furnace**.
- The diagram shows the raw materials and reactions that form iron.
- Carbon reacts with oxygen, forming carbon dioxide:
  $C + O_2 \rightarrow CO_2$ (+ heat) (1)
- More carbon converts the carbon dioxide to carbon monoxide:
  $C + CO_2 \rightarrow 2CO$ (2)

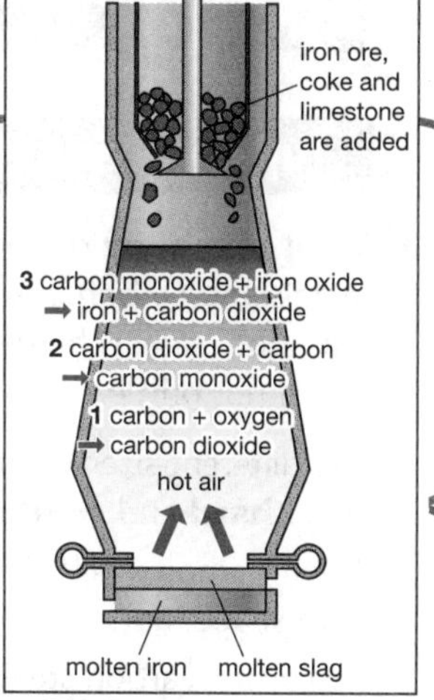

The chemical reactions in a blast furnace.

B–A*

- Metals that are *less reactive* than carbon can be extracted from their oxides by **reduction** with carbon. This applies to iron(II) oxide: carbon in carbon monoxide reduces it to iron (carbon monoxide is oxidised to carbon dioxide): **$3CO + Fe_2O_3 \rightarrow 3CO_2 + 2Fe$ (3)**
- Limestone helps remove impurities (e.g. sand) that form slag.
- **Molten iron** runs off at the bottom.
- **Oxidation** (addition of oxygen) and **reduction** (removal of oxygen) always occur together in a reaction.
- (**1**) In $C + O_2 \rightarrow CO_2$, carbon is oxidised and oxygen is reduced.
- (**2**) In $C + CO_2 \rightarrow 2CO$, carbon is oxidised and carbon dioxide is reduced.
- (**3**) In $3CO + Fe_2O_3 \rightarrow 3CO_2 + 2Fe$, carbon monoxide is oxidised and iron oxide is reduced.

Using iron

Different types of iron

D–C

- Iron from the blast furnace is 96% iron. It is remelted and moulded to form **cast iron** objects (e.g. drain covers) which are **hard** and **strong**.
- **Impurities** that make cast iron **brittle** are removed to give **wrought iron** (pure iron) that is softer and easily bent and shaped.
- **Carbon** and other **metals** are added in different quantities to pure iron, to make different types of **steel**.

Structure of pure iron and of steel

B–A*

- The atoms in pure (wrought) iron are bonded together in a 3-D regular, closely packed **lattice**. Layers of atoms can slide over each other, so wrought iron is relatively soft and easily bent and shaped.
- In **steels**, the different-sized atoms of added carbon and other metals distort the lattice. Without regular layers of atoms to slide over one another, steels are much harder and more rigid than pure iron.

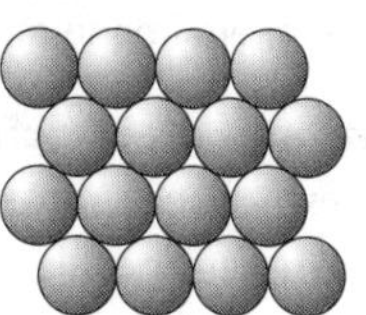

The regular lattice arrangement of atoms in wrought iron.

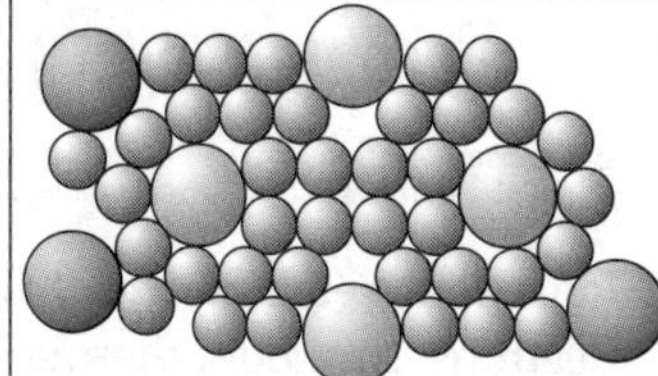

Different-sized atoms give steels hardness and strength.

Top Tip!

Pure metals are soft and easily shaped because the atoms form a regular arrangement. The layers of atoms can then pass easily over each other.

Questions

Grades D-C

1 Iron ore is *reduced* in the blast furnace. What does this mean, and why does it happen?

Grades B-A*

2 Explain why iron can be obtained from its ore by reduction with carbon.

Grades D-C

3 Explain why most iron is converted into steel.

Grades B-A*

4 Explain why steel is harder than pure iron.

Using steel

Grades

Types of steel

The hammer is strong (not brittle). The scissors have a hard cutting edge and are rust-proof.

D–C

- **Steel** is an **alloy** of iron. Types of steel contain other metals and carbon in varying amounts. Compared with pure iron, steel is harder, stronger, resists corrosion and is not brittle.

B–A*

- The different-sized atoms in steels (see page 29) make alloys **hard** and **rigid**:
  - *low* carbon steels (0.4% carbon or less) are soft and easy to shape, and are used in car bodies
  - *medium* carbon steels (about 0.8% carbon) are harder, more rigid and stronger, suitable for hand tools
  - *high* carbon steels (1.0–1.5% carbon) are very hard, ideal for knives, but may snap if bent
  - *stainless* steel is 70% iron, 20% chromium, 10% nickel. It is very hard and resistant to corrosion (rust). Stainless steel is generally more costly to produce than other steels.

Low carbon steels are easy to shape but are also soft.

Transition metals

The structure and properties of transition metals

D–C

- In the periodic table (see page 26), the **transition metals** are all the elements in the nine columns headed scandium (Sc) to copper (Cu).
- They are very good **structural** materials, and are drawn into wires and hammered into shapes.
- **Delocalised** electrons in the outer shells of transition elements are *free* to move through the metal lattice (see the diagram). The oppositely charged particles attract each other, holding the structure together strongly.

Copper can be drawn into wires.

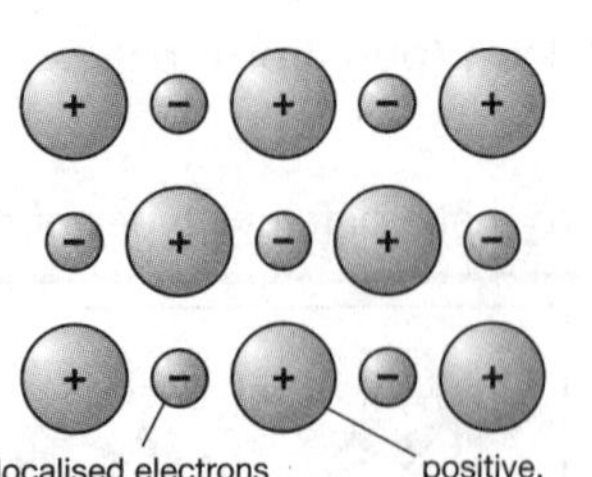

Metals have delocalised electrons that are free to move.

B–A*

- Transition metals **conduct heat** and **electricity** well because the delocalised electrons carry both thermal energy (heat) and charge (electrical current) rapidly through the metal.
- Metals are cold to the touch because the delocalised electrons rapidly carry away (conduct) heat from the surface, whereas non-metals have no delocalised electrons, and the warmed surface stays warm.

Top Tip!

Remember that metals are on the left-hand side of the periodic table, non-metals are on the right.

Questions

Grades D-C

1 Explain why steel containing a lot of carbon is used for making knives.

Grades B-A*

2 What kind of steel would you use for making a spanner? Explain your answer.

Grades D-C

3 Explain why copper is a very good conductor of electricity.

Grades B-A*

4 Explain why a plastic object feels warmer to touch than a metal object.

Aluminium

Extraction, properties and uses of aluminium

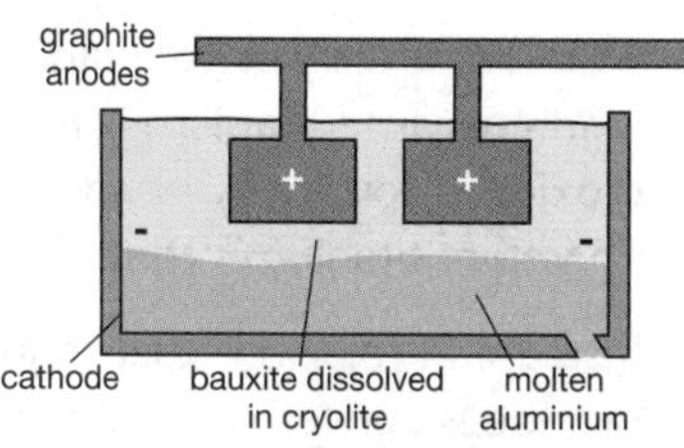

Electrolysis of bauxite. Aluminium forms at the cathode, oxygen at the anode.

- Aluminium ore is **bauxite**, aluminium oxide. Carbon will not reduce the oxide because aluminium is *more reactive* than carbon. Instead, extraction is by **electrolysis**, which has high energy costs.
- Melting bauxite allows its ions to move. Cryolite is added because it lowers the temperature needed for the process. Aluminium oxide dissolves in molten cryolite.

D–C

- Cathode: **aluminium ions + electrons → aluminium atoms**
  Anode: **oxide ions – electrons → oxygen molecules**
- **Alloys** of aluminium with other metals are hard, strong, light (low density), easy to shape, good thermal and electrical conductors and resist corrosion. They are used in drinks cans, bicycles, cars and cooking utensils.
- A thin film of aluminium oxide quickly forms on the surface, and stops oxygen reaching deeper. In this way, aluminium resists corrosion.

B–A*

- Bauxite, aluminium oxide, is Al_2O_3. In electrolysis, it forms Al^{3+} and O^{2-} ions.
- To become metal atoms, each aluminium ion must gain three electrons. It is **reduced**: $\mathbf{Al^{3+} + 3e^- \rightarrow Al}$
- To become oxygen gas, each oxygen ion must lose two electrons. It is **oxidised**: $\mathbf{2O^{2-} \rightarrow 4e^- + O_2}$
- Since reduction and oxidation reactions always happen together, they are called **redox** reactions.

Top Tip!

Remember '**OIL RIG**':
Oxidation **I**s **L**oss
Reduction **I**s **G**ain
(of electrons).

Aluminium recycling

Impact of bauxite mining and advantages of recycling

D–C

- Bauxite is mined from open-cast (surface) **pits** covering huge areas. The vegetation is cleared and burned, producing carbon dioxide that contributes to **climate change**.
- Mining displaces local people, who lose their land and livelihood. Mine workers pass on **diseases** to which local people have no immunity. Mining brings **pollution** such as oil and dust that cause health problems.
- Aluminium products can be **recycled** more than once. Recycling saves the high **energy** costs of electrolysis, and lessens the demand for **bauxite** ore and for waste disposal in **landfill** sites.

How Science Works

You should be able to: consider and evaluate the social, economic and environmental impacts of exploiting metal ores, of using metals and of recycling metals.

B–A*

- Most bauxite mining is in tropical and sub-tropical countries. It often requires large areas of rainforest to be cleared for the pits and access roads.
- Thousands of species of animals and plants in rainforests are not yet known about. They represent **biodiversity**, and some could be the source of useful medicines.
- By destroying habitats, mining activities render animals and plants extinct before they have even been discovered.

Questions

Grades D-C

1 Why can aluminium *not* be extracted from its ore by heating with carbon, as iron can?

Grades B-A*

2 Explain how aluminium ions are reduced at the cathode during the electrolysis of bauxite.

Grades D-C

3 List **two** ways in which bauxite mining can harm the environment.

Grades B-A*

4 Explain how recycling aluminium could help to conserve biodiversity.

Grades

Titanium

Extraction, properties and uses of titanium

- Titanium ore is **rutile**, titanium dioxide. Carbon will not reduce the oxide because titanium is *more reactive* than carbon. Titanium dioxide is covalently bonded, so does not conduct electricity. Therefore titanium cannot be extracted by electrolysis.
- Instead, in reactions that require a high energy input, titanium dioxide is converted to titanium chloride and then to the metal.
- Titanium is light, strong, withstands high temperatures and is easily shaped. Titanium is very reactive but does not corrode because a thin layer of titanium dioxide forms on its surface, protecting the metal beneath.
- Because of high extraction costs, titanium is used for high-value items including replacement joints, aircraft and missiles.

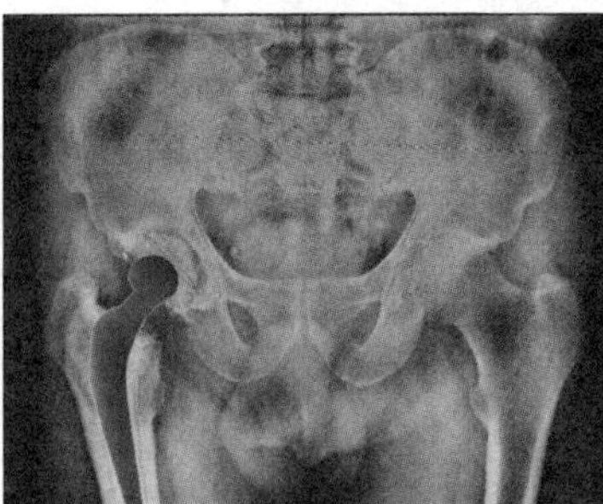

Titanium's low density, high strength and resistance to corrosion make it suitable for hip replacements.

D–C

- In the extraction process, titanium dioxide is first converted to titanium chloride (see below). Then, the more reactive metal magnesium **displaces** titanium from titanium chloride to form metallic titanium (and magnesium chloride).

$$TiO_2 \xrightarrow{\text{ore heated with carbon and chlorine}} TiCl_4 \xrightarrow{\text{titanium chloride reacts with molten magnesium}} Ti$$

titanium dioxide → titanium chloride → titanium

Hot titanium quickly reacts with oxygen, so the second reaction takes place in a vacuum or in an inert atmosphere.

B–A*

Copper

Extraction, properties and uses of copper

- Copper-rich ores are scarce. Ores include **chalcocite** (Cu_2S) and **chalcopyrite** ($CuFeS_2$).
- The ores, extracted from open-cast pits, are low in copper, so large quantities of ore produce little copper but cause extensive environmental damage.
- In a traditional extraction process, the ore is crushed, washed and heated in air. But the copper it produces is impure, and the by-product sulfur dioxide pollutes the atmosphere.
- In a method that increases the yield of copper from low-grade ore and causes less environmental damage, the copper is first **leached** out of the ore – chemically treated to form a solution containing copper ions. Then, electrolysis produces pure copper at the cathode.
- Copper is a dense, **unreactive** metal with a high melting point. It is a good conductor of heat and a very good **conductor of electricity**. Copper is used for electrical wiring and for plumbing pipes.

D–C

Bronze and brass

- Pure copper is too soft for many uses, but is useful in copper **alloys**.
- **Bronze** is copper with tin added. It is strong, hard and resists corrosion and is commonly used to make bells and statues.
- **Brass** is copper with zinc added. It is also strong, and is a good electrical conductor. It is used to make electrical connections (e.g. the pins of plugs) and musical instruments.

B–A*

Questions

Grades D-C

1 Explain why titanium cannot be extracted from its ore (titanium dioxide) by electrolysis.

Grades B-A*

2 Explain why the extraction of titanium from its ore is a very expensive process.

Grades D-C

3 Explain why people are researching new ways of extracting copper from low-grade ores.

Grades B-A*

4 Use what you know about the structure of metals and alloys to explain why bronze is harder than pure copper.

Smart alloys

Summary: useful metals and alloys

- **Metals** are reactive, so they exist naturally as **compounds** (ores). (Gold is unreactive, found naturally as an **element**.) Most metals in everyday use are **transition** elements.

metal (cost)	ore	extraction	properties and uses
iron (cheap to extract)	haematite, iron(III) oxide, Fe_2O_3	Reduction of ore with carbon monoxide: carbon monoxide + iron oxide → carbon dioxide + iron	Wrought iron: easily shaped. Steels are rigid, very strong and hard, so used for construction, bodywork, tools.
aluminium (costly to extract)	bauxite, aluminium oxide, Al_2O_3	By electrolysis of molten ore. Aluminium is more reactive than carbon, so cannot be extracted by heating with carbon (coal).	Light, does not corrode, so used for aircraft, spacecraft, drinks cans, bike frames.
titanium (very costly)	rutile, titanium dioxide, TiO_2	Many-stage process. Titanium cannot be extracted by heating ore with carbon (coal) or by electrolysis (since covalently bonded).	Very strong and light, does not corrode, so used for replacement joints, rockets, missiles and aircraft.
copper (moderately costly)	chalcocite, copper sulfide, Cu_2S	Electrolysis of solutions containing copper compounds.	Ductile (can be drawn into wires), very good conductor of electricity. Used for electrical wiring and water piping.

D–C

- Pure metals are usually too soft for use. **Alloys** with other metals (and carbon in steels) have improved properties (e.g. hardness).
- **Smart alloys** have a **shape memory**. They can be **deformed**, but warming returns them to their original shape. Expensive spectacle frames use nickel-titanium alloys.

B–A*

- To help badly broken bones mend, titanium or stainless steel plates are normally inserted.
- Plates of **smart alloys** (e.g. titanium alloys) can be **stretched** (deformed), and retain their stretched state at low temperatures. A deformed smart alloy plate is attached across broken bones. As it warms up in the body it contracts, pulling the bones tightly together. This speeds up the rate of healing.

How Science Works

You should be able to:
- evaluate the benefits, drawbacks and risks of using metals as structural materials and as smart materials
- explain how the properties of alloys (other than smart alloys) are related to models of their structures
- evaluate developments in the production and uses of better fuels, for example ethanol and hydrogen.

Fuels of the future

Fuel use: impact and remedies

D–C

- **Petrol** and **diesel** fuels (derived from oil) burn to give **carbon dioxide** and **water vapour**.
- With incomplete combustion, vehicle exhaust contains polluting gases and particles such as **unburnt hydrocarbons**, **carbon monoxide**, **sulfur dioxide** and **nitrogen oxides**.
- **Ethanol** (**alcohol**), a **renewable** resource, produces less carbon monoxide than petrol. In Brazil, ethanol made from sugar cane is added to petrol.
- Burning **hydrogen** as a vehicle fuel causes no pollution, producing water vapour only. But hydrogen is produced by electrolysis (see page 31), a costly process.
- A **catalytic converter** reduces a vehicle's harmful emissions by converting: carbon monoxide to carbon dioxide; unburnt hydrocarbons to carbon dioxide and water vapour; nitrogen oxides to nitrogen.

B–A*

- **Electric vehicles** are designed to reduce urban pollution. However, the batteries are heavy, bulky, must be recharged, and provide slow, short journeys only.
- To be sustainable, the electricity needs to have been generated from renewable sources.

Questions

Grades D-C

1 Explain what is meant by a 'smart alloy' and give an example.

Grades B-A*

2 Apart from its property as a smart alloy, suggest a property of a titanium alloy that would be important when it is used inside the body to help bones to heal.

Grades D-C

3 Why would using hydrogen as a fuel cause less air pollution?

Grades B-A*

4 Why have electric vehicles not become popular?

Crude oil

Crude oil and fractional distillation

Grades D–C

- **Crude oil** is a **mixture** of hundreds of **hydrocarbon** compounds whose molecules contain only **hydrogen** and **carbon** atoms. Molecule size affects the **properties**.
- Crude oil is heated to vapour point to separate all its compounds. The vapour rises up the **fractional distillation column**. **Fractions** contain molecules with *similar* numbers of carbon atoms and similar boiling points. As the vapour cools, fractions **condense** at different temperatures and leave the column.

fraction	no. of carbon atoms	uses
petroleum gas	1–4	heating and cooking
naphtha	5–9	making other chemicals
petrol	5–10	motor fuel
kerosene	10–16	jet fuel
diesel	14–20	diesel fuel and heating oil
oil	20–50	motor oil
bitumen	over 50	bitumen (tar)

A fractional distillation column.

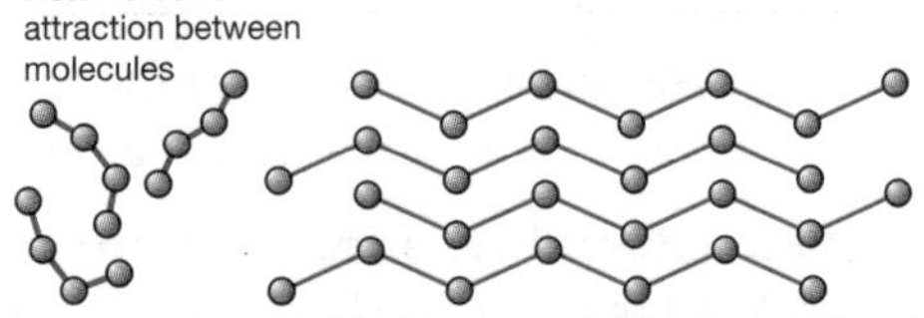

Intermolecular forces differ in small and large molecules. (Carbon atoms only are shown.)

Grades B–A*

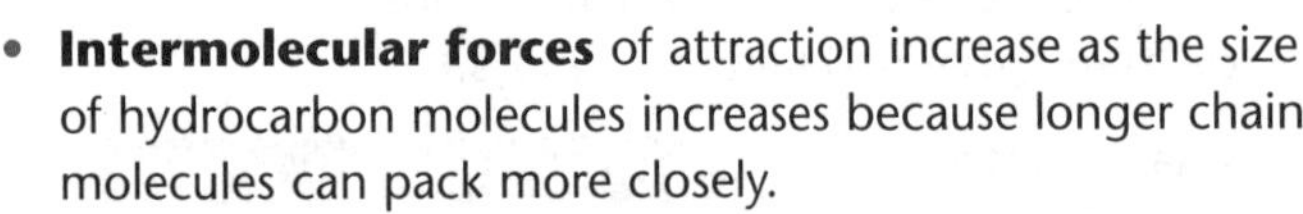

- **Intermolecular forces** of attraction increase as the size of hydrocarbon molecules increases because longer chain molecules can pack more closely.
- Thermal energy (heat) weakens intermolecular forces. Large molecules take more heat to break these forces than small molecules. So the larger the hydrocarbon, the higher its boiling point.

Alkanes

The alkane family of hydrocarbons

Grades D–C

- The **general formula** for the **alkane** family is: C_nH_{2n+2} (*n* = number of carbon atoms). All alkanes have similar chemical properties.
- **Covalent** bonds (see page 27) join the atoms in all hydrocarbons. In alkanes they are all **single** bonds (see table): alkanes are **saturated** compounds.

Grades B–A*

- Atoms in methane CH_4, ethane CH_3–CH_3 and propane CH_3–CH_2–CH_3, can only be arranged one way.
- From butane onwards, branched forms are possible. Butane can be CH_3–CH_2–CH_2–CH_3, or CH_3–CH(CH_3)–CH_3.
- Both are **isomers** of butane with the same formula, C_4H_{10}. But molecules of the branched form pack more loosely. Its melting and boiling points are lower and its chemical properties differ slightly from the straight-chain form.

name	formula	structure
methane	CH_4	H–C–H with one H above and one H below the C
ethane	C_2H_6	H–C–C–H with H above and below each C
propane	C_3H_8	H–C–C–C–H with H above and below each C
butane	C_4H_{10}	H–C–C–C–C–H with H above and below each C
pentane	C_5H_{12}	H–C–C–C–C–C–H with H above and below each C

Questions

Grades D-C

1 Which hydrocarbon fractions come out at the top of the fractionating column – those with big molecules, or those with small molecules?

Grades B-A*

2 Explain why long-chain hydrocarbons have higher boiling points than short-chain ones.

Grades D-C

3 Write down the chemical formula for an alkane with **five** carbon atoms.

Grades B-A*

4 Explain what is meant by the term 'isomer'.

Pollution problems

The consequences of burning fossil fuels

- **Coal** (carbon), **oil** and **natural gas** (hydrocarbons) are non-renewable **fossil fuels**. When burnt in air:

  carbon + oxygen → carbon dioxide
  hydrocarbon + oxygen → carbon dioxide + water vapour

  – Carbon dioxide is a **greenhouse gas** (see page 24). In excess, it contributes to **climate change**.

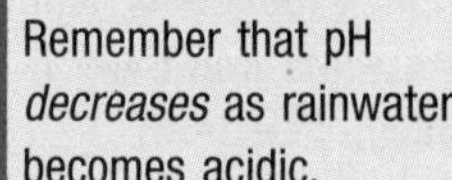

Remember that pH *decreases* as rainwater becomes acidic.

D–C

- Normal rainwater is slightly **acidic** (pH 5.5):

  carbon dioxide + water → carbonic acid

  All fossil fuels contain some sulfur. When burnt: **sulfur + oxygen → sulfur dioxide**

  Sulfur dioxide makes rainwater more acidic (pH 2.5):

  sulfur dioxide + water → sulfuric(IV) acid (= sulfurous acid)

- **Acid rain** (see page 22) damages vegetation and pollutes waterways. Aquatic organisms die.
- Tiny particles from burnt fuels may cause **global dimming** by reducing the amount of sunlight reaching the Earth's surface, and may possibly lower its temperature.

Acid rain affects fish

B–A*

- Fishermen found they were catching fewer but larger fish in lakes affected by acid rain.
- Increasingly, young fish were found to have deformities and to die before adulthood. Older fish therefore had less competition and grew larger, but were unable to reproduce. It was likely that such fish populations would decline irretrievably.

Reducing sulfur problems

Capturing the sulfur in fossil fuels

D–C

- Sulfur occurs in fossil fuels as **hydrogen sulfide** gas and as sulfur. **Sulfur** is a valuable chemical **resource** that is worth extracting:
  – Before natural gas and oil fuels are burned, hydrogen sulfide can be removed by dissolving it in a solvent.
  – After burning fuels in power stations, the flue gases (including sulfur dioxide) are sprayed with a mixture of powdered limestone and water. About 95% of the sulfur dioxide reacts to form solid calcium sulfate that is collected and sold to farmers to improve their soil.

Pollution from shipping

B–A*

- Fuels burnt by international shipping generate twice the levels of air pollution that the world's airlines produce. Yet the shipping industry has done little to reduce their output of carbon dioxide and sulfur dioxide.
- Wales has lost much of its aquatic life. This may be because of acid rain caused by air pollution from the numerous ships in the North Atlantic.

How Science Works

You should be able to:
- evaluate the impact on the environment of burning hydrocarbon fuels
- consider and evaluate the social, economic and environmental impacts of the use of fuels.

Questions

Grades D-C

1 Why is normal rain, formed in unpolluted air, slightly acidic?

Grades B-A*

2 Why can acid rain cause there to be fewer, larger fish in lakes and rivers?

Grades D-C

3 Describe how sulfur is removed from waste gases at coal-burning power stations.

Grades B-A*

4 Suggest why there has been more concern about air pollution from aircraft than from ships, even though ships cause more air pollution.

C1a summary

Atoms and rocks

All substances are made from **atoms**, which have:
- **protons** and **neutrons** in a small **nucleus**
- **electrons** orbiting round the nucleus.

The **mass number** of an element = number of protons + number of neutrons.

The **atomic number** = number of protons = number of electrons.

An atom has no net charge.

An **element** consists of one type of atom. Elements are arranged in the **periodic table** in vertical **groups**, containing elements with similar properties.

Two atoms of the same element can join together to form a **molecule**.

A **compound** consists of atoms of two or more different elements joined together.

A **covalent bond** is formed when atoms **share** electrons.

An **ionic bond** is formed when atoms lose or gain electrons to become oppositely charged **ions** that attract each other.

The **electronic configuration** of an atom shows the number of its electrons and their arrangement in **shells**.

Atoms with a full outer **shell** of electrons are **stable**.

Limestone is **calcium carbonate**.

When heated, it undergoes **thermal decomposition**:

calcium carbonate → **calcium oxide (quicklime)** + **carbon dioxide**

Limestone can be used to make cement, mortar, concrete and glass.

Carbon dioxide bubbled through **limewater**, calcium hydroxide, gives a milky suspension of calcium carbonate:

$$CO_2 + Ca(OH)_2 \rightarrow CaCO_3 + H_2O$$

Metals

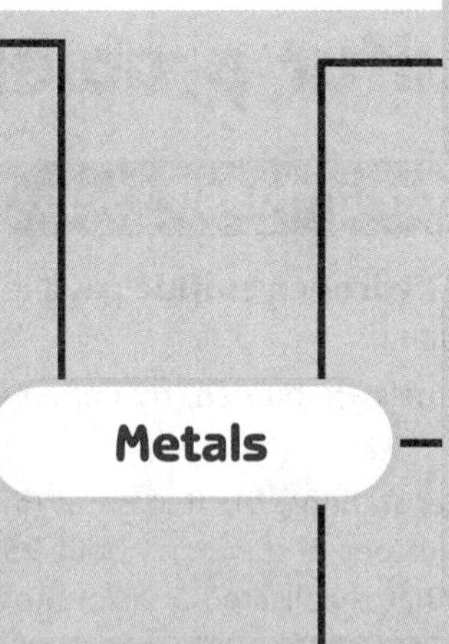

Metals are elements that are extracted from their **ores**, often oxides, by:
- chemical reactions (iron, titanium, copper)
- electrolysis (aluminium).

Iron is made in a **blast furnace** from **haematite**, iron ore, **iron(III) oxide**, Fe_2O_3:
- impurities make **cast iron** from the furnace hard but brittle
- removing impurities makes **wrought iron**, which has atoms arranged in a regular lattice and is soft and easy to shape
- adding metals and carbon to pure iron distorts the lattice and makes different types of **steel** (an **alloy** of iron), which is hard and rigid.

Aluminium is extracted by **electrolysis** from **bauxite**, aluminium oxide, Al_2O_3:
- it has a low density and is soft
- it forms strong, low density alloys with other metals.

Transition metals, such as **titanium** and **copper**:
- are found in the central block in the periodic table
- are hard and strong
- are good conductors of heat and electricity
- can be drawn into wires and hammered into sheets.

Smart alloys can be **deformed**, but return to their original shape when heated.

Crude oil

Crude oil is a **fossil fuel**. It is a mixture of **hydrocarbon** (i.e. hydrogen and carbon only) compounds that can be separated by **fractional distillation**.

Alkanes are an important type of hydrocarbon with the **general formula** C_nH_{2n+2}

All bonds in alkanes are **single**, **covalent** bonds.

Scientists are working to reduce pollution and develop alternative, cleaner fuels for the future, e.g. **ethanol**.

Burning fossil fuels releases useful **energy**, but harmful substances are also made:
- **carbon dioxide** contributes to **climate change** by increasing the **greenhouse effect**
- **sulfur dioxide** causes **acid rain**
- **smoke** particles contribute to **global dimming**.

Incomplete combustion produces unburnt hydrocarbons, carbon monoxide and nitrogen oxides. All cause problems for health and the environment.

Grades

Cracking

Top Tip!

Monkeys Eat Peanut Butter helps you count the number of carbons in molecules.

1 = meth-

2 = eth-

3 = prop-

4 = but-

Cracking alkanes to obtain smaller molecules

- **Alkanes** are **saturated** hydrocarbons: all carbon atoms are joined by **carbon-carbon single bonds**, C–C (see page 34). Since each carbon atom is joined to its maximum of four other atoms, alkanes are relatively unreactive.

D–C

- Short-chain alkanes are more useful as **fuels**. Alkanes in petrol are 5 to 12 carbon atoms long.
- The **cracking** process makes shorter-chain molecules from long-chain ones in a **thermal decomposition** reaction:
  - crude oil is heated to form a vapour that passes over a hot catalyst
  - the molecules 'crack', splitting into shorter molecules that include **alkenes**:
    long-chain alkanes → shorter-chain alkanes + alkenes

Cracking decane

B–A*

- The straight-chain alkane (general formula C_nH_{2n+2}) with 10 carbon atoms is **decane**. One way that decane can be **cracked** gives an alkane of 8 carbon atoms and an alkene of 2 carbon atoms:
- Decane can crack in other places, to give many different pairs consisting of an alkane plus an alkene.

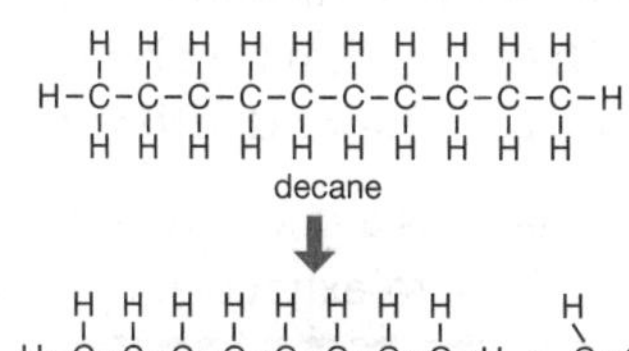

Decane cracks to form an alkane and an alkene.

Alkenes

Top Tip!

In all molecules, a carbon atom always forms four bonds. This is a good way to check if your structures are correct.

The family of alkenes

D–C

- **Alkenes** are a family of hydrocarbons of **general formula C_nH_{2n}** Alkenes are more **reactive** than alkanes because, being **unsaturated**, they contain a **carbon-carbon double bond**. One bond can break so that each carbon joins to an additional atom.
- Alkenes are produced by cracking alkanes (see above).
- Alkenes are the starting materials for making many useful compounds including **plastics** which are **polymers** formed when alkene molecules join in very long chains (see page 38).

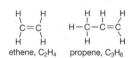

ethene, C_2H_4 propene, C_3H_6

These two alkenes are gases.

Ethene helps fruit to ripen

B–A*

- Ethene is a growth substance produced by plants that causes fruit to ripen and leaves and flower petals to fall.
- A ripe banana will speed up the ripening of unripe bananas in the same container.

Ripe fruit is best kept away from cut flowers.

Questions

Grades D-C

1 What is 'thermal decomposition'?

Grades B-A*

2 Write a balanced equation to show how decane is cracked to form an alkene with three carbon atoms and an alkane.

Grades D-C

3 Explain why alkenes are more reactive than alkanes.

Grades B-A*

4 Most growth substances act inside an organism, and are carried within it in solution. Why is ethene an unusual growth substance?

Grades

Making ethanol

Top Tip!

Ethanol is the chemical name for alcohol, formula C_2H_5OH. It is not a hydrocarbon as it contains oxygen as well as carbon and hydrogen.

Making ethanol from crude oil and plant crops

D–C

- Ethene is produced by cracking crude oil. When a mixture of ethene and steam is passed over a **catalyst**, the colourless liquid **ethanol**, C_2H_5OH, is formed:

ethanol: H–C(H)(H)–C(H)(H)–O–H

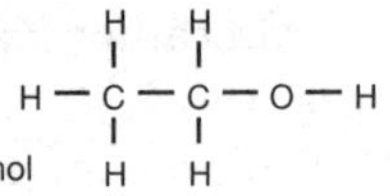

- Ethanol is in the **alcohol** family of compounds. Alcohols contain a reactive –OH group. Ethanol is highly **flammable**. It burns to give carbon dioxide and water.
- Ethanol ('alcohol') is also made from sugar dissolved in water and **fermented** with **yeast**.
- Sugar comes from **renewable** crops such as grapes, sugar cane and sugar beet. In some countries, including Brazil (see page 33), ethanol is a **biofuel** that replaces or is added to petrol.

B–A*

- In **fermentation**, yeast aids the conversion of sugar in solution to ethanol: **sugar** $\xrightarrow{\text{yeast}}$ **ethanol + carbon dioxide**
- The yeast is filtered off, leaving a mixture of water and ethanol. These liquids have different boiling points and so can be separated by **fractional distillation** (see page 34).
- Ethanol is a good solvent for many substances that do not dissolve in water. Research chemists use ethanol to extract compounds they make (e.g. **pharmaceuticals**), from reaction mixtures.

How Science Works

You should be able to: evaluate the advantages and disadvantages of making ethanol from renewable and non-renewable sources.

Plastics from alkenes

How polymers are formed

D–C

- In the process of **polymerisation**, small molecules called **monomers** join together in very long chains to form **polymers**. The **carbon-carbon double bond** breaks, allowing each double-bonded carbon atom to join to an additional atom.
- (**1**) To make the polymer **poly(ethene)** (**polythene**), the gas ethene is heated with a catalyst under very high pressure.
- (**2**) The alkene, **propene** can polymerise to form poly(propene).

(**1**) three ethene molecules → poly(ethene)

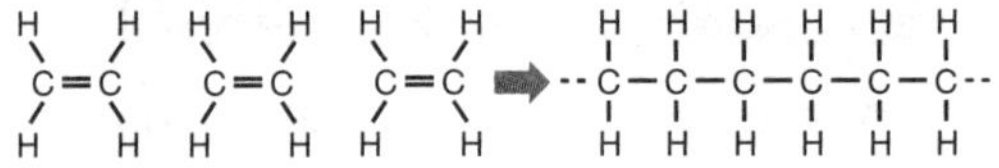

(**2**) 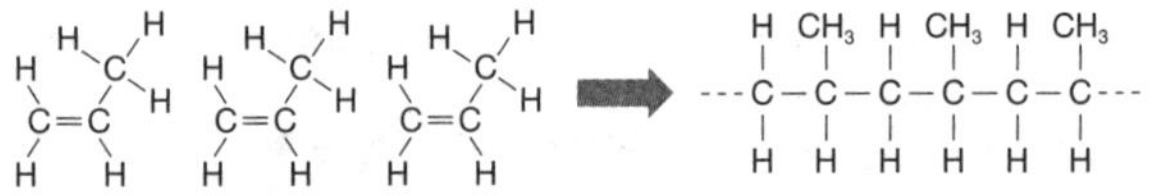 three propene molecules (monomers) → a section of a strand of poly(propene)

Polymers as smart materials

B–A*

- The properties of polymers include: low density; resistance to corrosion by acids and alkalis; flexibility; softening when heated (useful in recycling processes).
- New polymers with additional properties are called **smart materials**. Some conduct electricity, emit light or change shape when electricity passes through them, or become hard when near a magnet.
- New products are being made to exploit these properties. For example, thin flexible sheets are being developed as television and computer screens.

Questions

Grades D-C

1 Explain why ethanol is reactive.

Grades B-A*

2 Describe how you would use fractional distillation to get ethanol from a mixture of ethanol and water.

Grades D-C

3 Explain why alkenes can form polymers.

Grades B-A*

4 Suggest why, if a polymer that was a really good conductor was available, it might be preferable to using copper wires for conducting electricity.

Polymers are useful

Properties of polymers match their uses

- Particular monomers and reaction conditions give **polymers** (plastics) with particular **properties**.
- Packaging plastics are strong and flexible; tent fabrics repel water; dental crowns harden under ultraviolet light; wound dressings breathe; and hydrogels retain moisture for rapid healing.
- The polymer **Teflon®** is like poly(ethene) with a fluorine atom replacing every hydrogen atom.

D–C

- **Teflon®** is used for artificial hip sockets (see page 32) because an almost frictionless surface lets the ball swivel freely, and because it is non-toxic, inert and unaffected by temperature changes.
- **Slime** forms a ball which seems to melt when tipped, and breaks when stretched quickly. It is made by mixing aqueous solutions of the polymer **polyethenol** and of the salt borax. Borax forms **cross-links** between the polymer chains. The slime's **viscosity** depends on the solution concentrations and the polymer chain lengths.

Top Tip!

Viscosity is a measure of the thickness or gooeyness of a liquid. The thicker it is, the higher its viscosity.

Polymers with a memory

B–A*

- A **shape memory polymer** can be distorted, then resume its original shape when warmed.
- In surgery, it can be useful to distort an implant into a convenient shape for insertion into a patient's body. Then, in the warmth of the body, it can change to its original shape.
- Such shape memory polymers can be made **biodegradable** so that they gradually disappear from the patient's body.

Disposing of polymers

How Science Works

You should be able to:

- evaluate the social and economic advantages and disadvantages of using products from crude oil as fuels or as raw materials for plastics and other chemicals
- evaluate the social, economic and environmental impacts of the uses, disposal and recycling of polymers.

Problems and solutions

D–C

- Polymers are made from **non-renewable** crude oil. They can be burned to generate heat energy, but can produce particle **pollution** and toxic fumes that damage the **environment**. Instead, they can be **recycled** by shredding, melting and forming into pellets used to make new products such as fleeces.
- Most polymers are **non-biodegradable**. But microorganisms will break them down if they are mixed with cellulose or starch.

Plastic bones

B–A*

- A gel of **biodegradable** polymer has been developed to replace the permanent metal pins that have been used to repair broken bones.
- The polymer glues the bones together. Then, new bone grows into the spaces left as the polymer gradually biodegrades and disappears.

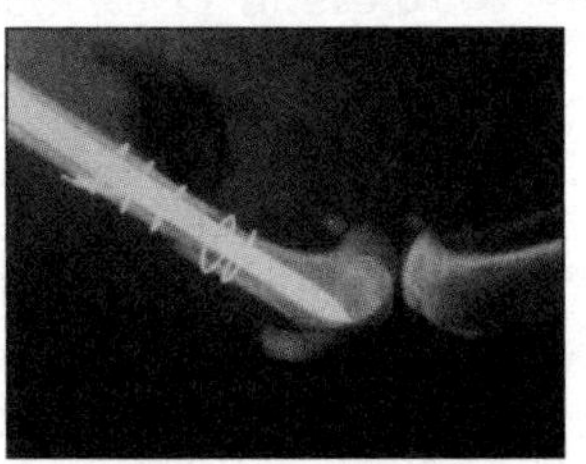

Polymer pins used to fix a broken bone.

Questions

Grades D-C

1 Slime is a polymer. State **one** property of it that varies according to the conditions under which it is made.

Grades B-A*

2 Give **one** example of using a shape memory polymer.

Grades D-C

3 Describe how polymers can be made biodegradable.

Grades B-A*

4 Explain why using a biodegradable polymer can be useful in mending broken bones.

Grades

Oil from plants

Vegetable oils, their value and extraction

Top Tip!

Distillation works because liquids have different boiling points.

D–C

- **Vegetable oils** are high energy sources used widely in foods. The oils from some nuts are excellent sources of **vitamins** and **trace elements**.
- Palm fruits, olives, avocados, coconuts and groundnuts are all rich in oils. The seeds of cotton, rape, mustard, sunflower and sesame also supply oils.
- The crop is crushed and any hard coatings discarded, leaving a mixture of oil, water and plant dust.
- Such a mixture can be **distilled** in the apparatus shown. The oil evaporates with the water. This mixture of pure oil and water condenses and is collected. The oil layer can then be separated.

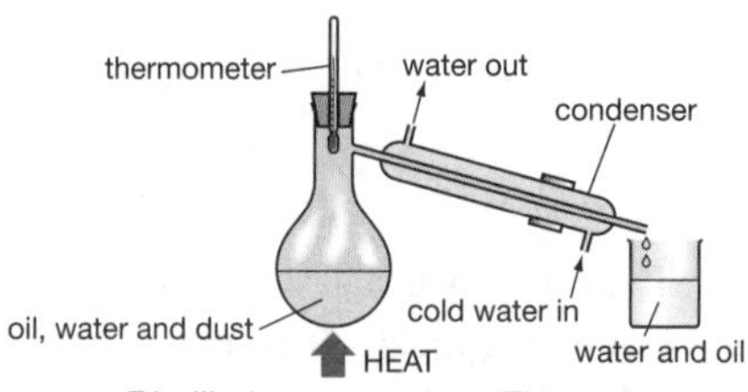

Distillation apparatus. The water and oil boil off.

Are walnuts good for the heart?

B–A*

- Research suggests that **omega-3 fatty acids** lower the level of fats in blood, reduce the risk of blood clots, lower blood pressure and generally prevent heart disease.
- English walnuts and oily fish are high in these compounds. The omega-3 fatty acids in walnuts differ from similar substances in fish oils, and research continues to compare the benefits of both types.

Green energy

How Science Works

You should be able to:

- evaluate the effects of using vegetable oils in foods and the impacts on diet and health
- evaluate the benefits, drawbacks and risks of using vegetable oils to produce fuels.

Biofuels

D–C

- Some vegetable oils such as rapeseed and sunflower oil can be used as **biofuels**, called **green fuels** because they are **renewable**. These fuels are either used alone or are thinned with petrol or diesel to use in traditional engines.
- Biofuels are 'greenhouse neutral' (carbon neutral) because the carbon dioxide *given out* when they burn is *taken in* by plants grown to replace them.
- As fossil fuels run low, biofuels will increasingly be used, and more vehicles will be developed that can use them.

Biodiesel – a fuel of the future

B–A*

- **Biodiesel** is a clean-burning renewable fuel produced from plant oils.
- The plant oil is reacted with methanol, an alcohol, to form a mixture of compounds called 'methyl esters'. This mixture is the biodiesel. The reaction also forms glycerol that separates out and is used to make soap, so all the reaction products are useful.
- Burning biodiesel produces less greenhouse gases than burning diesel, it is non-toxic and is biodegradable.

A tractor fuelled by rapeseed oil.

Questions

Grades D-C

1 Describe how a vegetable oil can be purified.

Grades B-A*

2 Why is a diet containing omega-3 fatty acids good for your health?

Grades D-C

3 Explain why we will need to use more biofuels in the future.

Grades B-A*

4 Suggest **one** disadvantage of using biofuels.

Grades

Emulsions

Making an emulsion

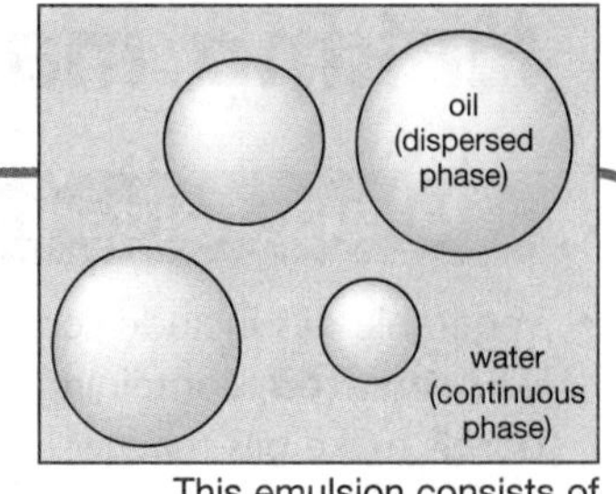

This emulsion consists of tiny oil droplets suspended in water.

D–C

- Oils do not dissolve in water. Water and oil form separate layers – they are **immiscible**.
- However, when shaken hard, oil forms an **emulsion** of tiny droplets of oil (dispersed phase) **suspended** in water (continuous phase).
- Milk is an emulsion of fat droplets in water. Butter is an emulsion of water droplets in fat.
- Emulsions are thicker than oil or water alone. Emulsions include ice creams, sauces and dressings (which pour slowly) and emulsion paints (which cover surfaces well and do not drip).
- Emulsions are opaque or milky because the tiny droplets scatter light.

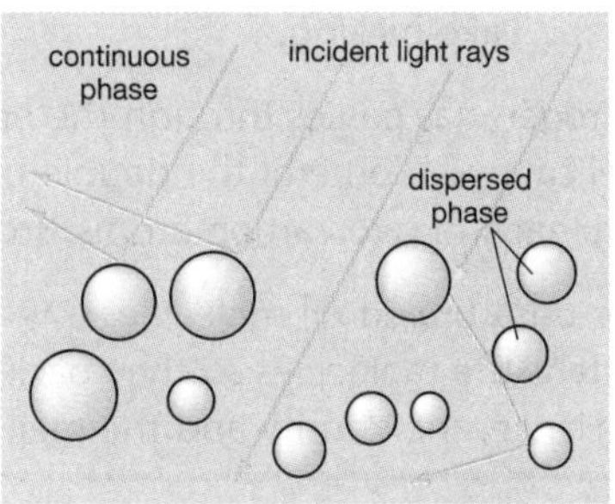

Light rays are scattered by the droplets in an emulsion, making it look opaque.

Top Tip!

Oils are liquids at room temperature, and fats are solids.

Mayonnaise

B–A*

- Mayonnaise is an emulsion of tiny oil droplets suspended in water-based ingredients that include vinegar or wine.
- Egg yolk, which contains the **emulsifier** lecithin, is added to prevent the mayonnaise from separating into oil and water-based layers.

Top Tip!

Liquids that do not mix together are called immiscible liquids. Those that do mix together are called miscible liquids.

Polyunsaturates

Top Tip!

Do not say that the bromine becomes clear. It is clear anyway, even when it is orange! Say that it changes its colour from orange to colourless.

Saturated and unsaturated oils and fats

D–C

- Fats and oils with single bonds only between carbon atoms are **saturated**.
- **Unsaturated** fats and oils have one or more double bonds between carbon atoms.
- Olive oil has one double bond, so it is a **monounsaturated** oil. Fats with more than one double bond are **polyunsaturates**.
- **Bromine water** (or iodine water) is used to test whether a fat or oil is saturated or not. If there are no double bonds, the orange colour of the bromine is unchanged. If double bonds are present, the bromine reacts with the carbons at the double bond and becomes colourless.

Unsaturated fats lower cholesterol

B–A*

- Cholesterol is a substance that can build up in arteries and block them, causing heart attacks.
- Blood cholesterol levels are lowered by eating polyunsaturates (in sunflower, soya and corn oils), and to a lesser extent by monounsaturates (in olive and rapeseed oils), so it is beneficial to include them in the diet, and to avoid saturated fats.

Questions

Grades D-C

1 Explain why emulsions are not transparent.

Grades B-A*

2 Foods that are emulsions usually have emulsifiers added to them. Why is this done?

Grades D-C

3 Describe how you test a liquid to find out if it contains unsaturated compounds.

Grades B-A*

4 Which of these types of fat or oil are best for you, and why: fats from animals or unsaturated oils from plants?

Making margarine

Making margarine

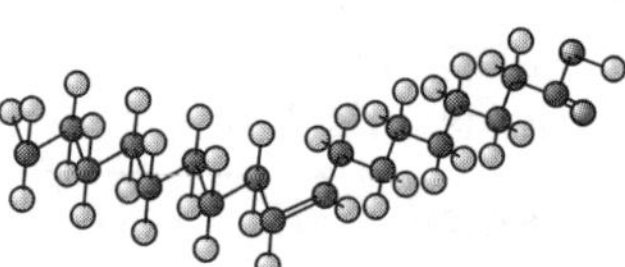

The vegetable oil compound oleic acid has one double bond. In hardening, two hydrogen atoms are added at this bond and the molecule straightens out.

D–C

- Vegetable oils include corn, sunflower and olive oils. Their molecules are **unsaturated**, containing **carbon-carbon double bonds**. These bonds make oils reactive.
- The oils can be chemically hardened by **hydrogenation** to form soft, solid **margarine**, with a higher melting point than oils, making it suitable for spreading and cake-making.
- In hydrogenation, hydrogen gas passes through the oil warmed to about 60 °C and with a nickel catalyst. Some of the double bonds break and an extra hydrogen atom joins to each of two carbon atoms, so the double bond becomes single bonds.

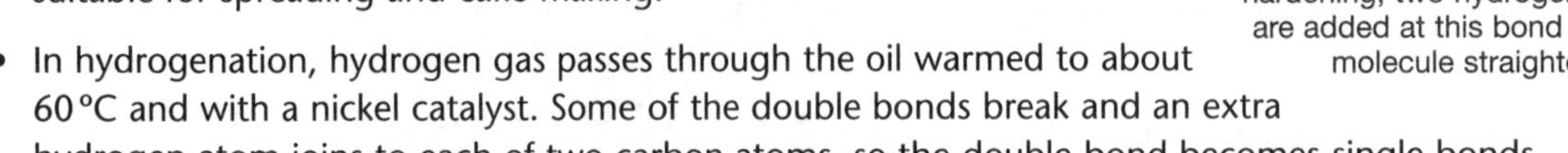

- The double bond gives unsaturated oil molecules a V-shape. Molecules cannot pack closely. Hydrogenation straightens the molecules so they pack closely in parallel and with stronger intermolecular forces. Hence, the density and the melting point of the solid fat are raised.

Is margarine or butter better for health?

B–A*

- Having 'good' HDL cholesterol in the blood is thought better than 'bad' LDL cholesterol which is linked to heart disease. (See page 8.)
- Hydrogenation of oils can form trans fats. Evidence suggests eating these fats raises LDL cholesterol levels and lowers HDL cholesterol levels.
- For this reason, some manufacturers are developing 'trans fat free' margarines. These are softer than margarines with high trans fat levels.

Food additives

How Science Works

You should be able to: evaluate the use, benefits, drawbacks and risks of ingredients and additives in food.

What are additives?

D–C

- Food **additives** keep food fresh longer or make it more appetising. Additives tested for safety and permitted by regulating organisations are not harmful, but no synthetic additive is essential for a healthy diet.
- Each additive has an **E-number** which identifies its chemistry and what it does. Additives must be listed on food labels.
- Permitted types include: antioxidants; colourings; emulsifiers, stabilisers, gelling agents and thickeners; flavourings; preservatives; and sweeteners.

Unhealthy additives

The synthetic dyes colouring these sweets are not necessary for health.

B–A*

- **Sudan dyes** are substances once widely used to colour food, until they were linked to cancers. Now they are banned.
- **Tartrazine** (E102) is a bright orange dye used to colour many foods including drinks, ice cream, crisps, custards, jams, marmalades and cereals. It is known to make some children hyperactive and is linked to allergies, asthma, migraines and types of cancer.

Questions

Grades D-C

1 Explain why hydrogenation makes liquid oils become solid margarine.

Grades B-A*

2 What are 'trans fat free' margarines, and why are they made?

Grades D-C

3 Explain why additives listed on food labels have an E-number.

Grades B-A*

4 Suggest why manufacturers are still allowed to add tartrazine to foods, even though there is evidence that it makes some children hyperactive.

Analysing chemicals

Chemicals in food and paper chromatography

D–C

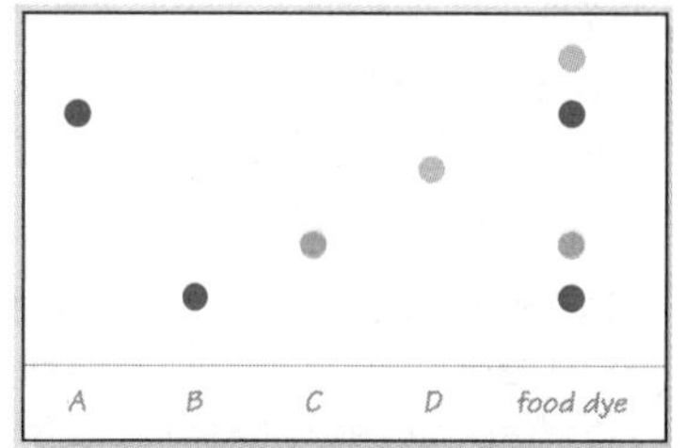

Chromatogram of known dyes and an unknown food dye.

- **Artificial colourings** and other food additives can be identified by techniques of **chemical analysis**. Techniques include **paper chromatography** which separates substances in a mixture.
- An unknown food dye is spotted on a pencil line alongside samples of known dyes. A solvent (e.g. water or ethanol) carries the dyes up the paper, separating them.
- The paper is removed from the solvent and dried. The food dye spots are matched to known dyes, and unidentified spots can be analysed further.

Retention factors are used to identify samples

B–A*

- The orange food dye spot on the chromatogram above does not match any of the known dyes *A* to *D*.
- Such a substance can be identified by making two measurements: the distance from the pencil line of the spot and of the solvent, and calculating the **retention factor** (unique for each substance) from:

$$\frac{\textbf{distance moved by the substance}}{\textbf{distance moved by the solvent}}$$

- Suppose a substance moved up 6 cm and the solvent 10 cm. The retention factor is: 6 cm ÷ 10 cm = **0.6**. This value is then looked up in a database of values, and the substance is identified.

The Earth

The Earth, its surface and interior

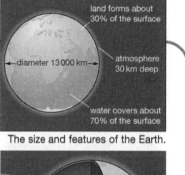

land forms about 30% of the surface
diameter 13 000 km
atmosphere 30 km deep
water covers about 70% of the surface

The size and features of the Earth.

D–C

- The Earth is a rocky planet (see page 67) aged about 4.6 billion years.
- The Earth has three main layers: the **crust**, the **mantle** and the **core**.
- The **crust** is a thin layer of solid rock.
- The **mantle** occupies half the diameter. Heat from radioactive decay drives convection currents in the 'plastic' rock of the mantle.
- The **core** occupies the centre and is about half the Earth's diameter:
  - the *outer core* is a fluid mixture of iron and nickel; the Earth's magnetic field is thought to be generated here
  - the *inner core* is solid iron and nickel.

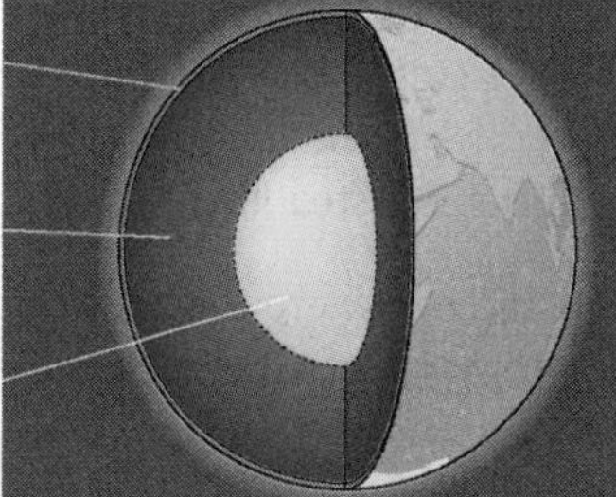

The layers of the Earth.

- The Earth's magnetic field protects life on Earth by repelling dangerous electrically charged particles that stream from the Sun as the 'solar wind'.
- Periodically in history, the magnetic field has faded, disappeared and re-emerged in the reverse direction. Today, the field is about 90% of its strength 150 years ago. Scientists estimate that it will disappear in 1500 to 2000 years, then re-emerge switched over, with 'north' at the south.
- During this sequence of events, it is possible that many species will become extinct.

Questions

Grades D-C

1 Explain how chromatography separates the colours in a mixture.

Grades B-A*

2 Calculate the retention factor for the green food colour in the chromatogram, assuming that the solvent has just reached the top of the paper.

Grades D-C

3 Name the **three** main layers that make up the Earth.

Grades B-A*

4 Explain why the Earth's magnetic field is important for life.

Earth's surface

Theories of the Earth's formation

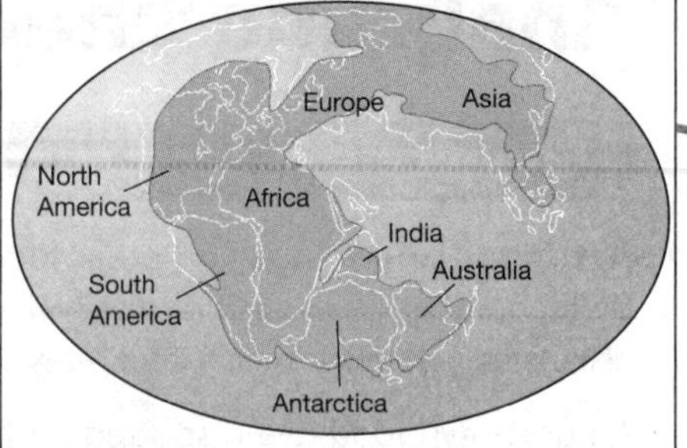

Pangaea, over 200 million years ago.

D–C

- Scientists once thought that the Earth's surface solidified, shrank and broke into today's pattern of **continents** and oceans; and that the heat inside the Earth was just heat left over from Earth's formation.
- In 1915, the German Alfred Wegener suggested that the first landmass was one **supercontinent** (**Pangaea**) which broke up into pieces (plates) that 'drifted' to their present positions.
- Today's scientists agree with Wegener, except for how he thought continental drift happens.
- The Earth's crust and the semi-solid upper mantle cracked into **tectonic plates** (see below). Heat from **radioactive** processes inside the Earth drives **convection currents** in the 'plastic' rock of the **mantle**. The tectonic plates ride on this circulating material, moving a few centimetres a year.

B–A*

- At **mid-ocean ridges**, adjacent plates are moving apart. Molten **magma** from the mantle fills the gap, solidifies underwater and moves apart with the plates, allowing more magma to well up.
- Because of this process at the Mid-Atlantic Ridge, the Atlantic Ocean, once a tiny stretch of water between Europe, Africa and the Americas, is now a vast ocean.

Earthquakes and volcanoes

Plate boundary processes

D–C

- The mantle circulating below tectonic plates exerts huge pressures on them. Eventually, plates are forced to move suddenly along a **plate boundary**, and an **earthquake** occurs.
- **Volcanoes** form at points of weakness. Molten magma is forced up through a **vent** in the crust. In some kinds of volcanoes, the magma blasts out as **lava** and gases. In others, the magma runs out slowly and quite steadily.

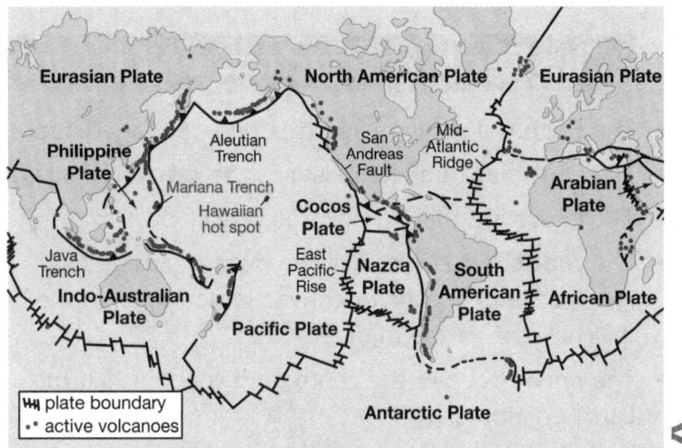

The main plate boundaries and active volcanoes.

B–A*

- A region where one plate moves under another is a **subduction zone**.
- Oceanic crust is heavier (denser) than continental crust. Where they collide, the oceanic crust moves under the continental crust. A volcano often forms in the continental crust pushed up by this process. Where the oceanic plate sinks, a very deep sea trench forms.

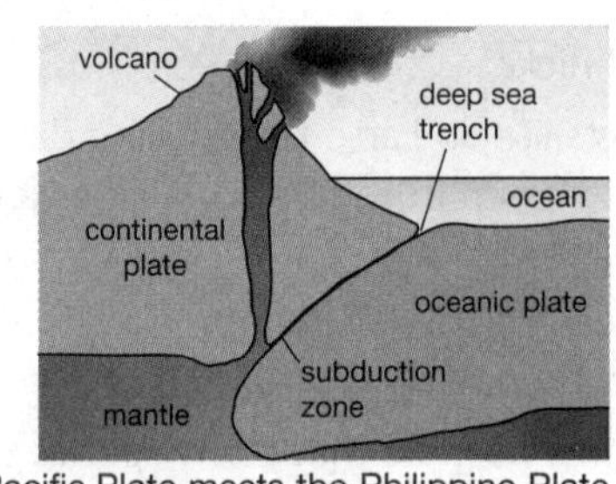

The Pacific Plate meets the Philippine Plate at a subduction zone where the Mariana Trench is 11 000 m deep.

How Science Works

You should be able to:
- explain why the theory of crustal movement (continental drift) was not generally accepted for many years after it was proposed
- explain why scientists cannot accurately predict when earthquakes and volcanic eruptions will occur.

Questions

Grades D-C

1 What makes the inside of the Earth hot?

Grades B-A*

2 Explain why the Atlantic Ocean is very slowly getting wider.

Grades D-C

3 Explain why earthquakes often happen at subduction zones.

Grades B-A*

4 Why do deep trenches occur at subduction zones?

Grades

The air

Composition of the atmosphere

D–C

- The Earth's atmosphere is a **mixture** of gases, about four-fifths (78%) **nitrogen**, one-fifth (20%) **oxygen** and varying amounts of **water vapour**. It has very small amounts of **carbon dioxide** and the **noble gases**.
- The noble gases, in Group 0 of the periodic table (page 26), have a full outer shell of electrons, so are unreactive. They are used in filament lamps and in electrical discharge tubes. Helium is used in air balloons because it is much less dense than air.

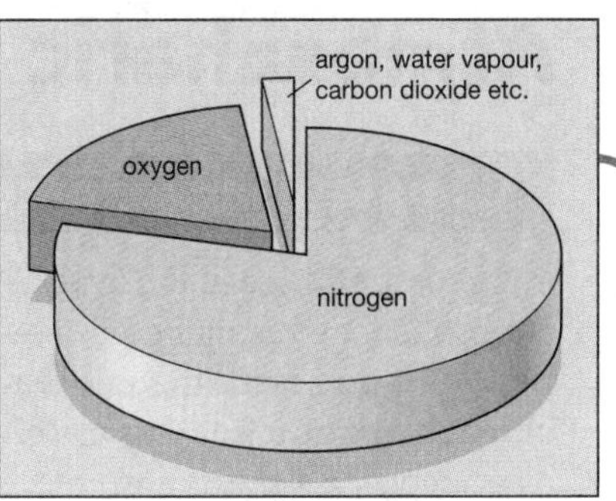

Gases in the atmosphere.

How airbags work

B–A*

- Nitrogen is a safe, relatively unreactive gas. It is the gas formed in airbags, which have reduced deaths in head-on road crashes by 30%.
- A collision triggers a small electrical charge which starts a very rapid chemical reaction in the airbag. In a fraction of a second, the solid sodium azide decomposes to form nitrogen gas and sodium metal:

$2NaN_3$ (sodium azide) → $3N_2$ (nitrogen gas) + 2Na (sodium metal)

Evolution of the air

The early atmosphere

D–C

- Intense volcanic activity during the Earth's first billion years produced an atmosphere of **carbon dioxide**, **methane** and **ammonia**, and **water vapour** which **condensed** to form the first seas.
- The very high level of carbon dioxide dropped as it dissolved in the seas.

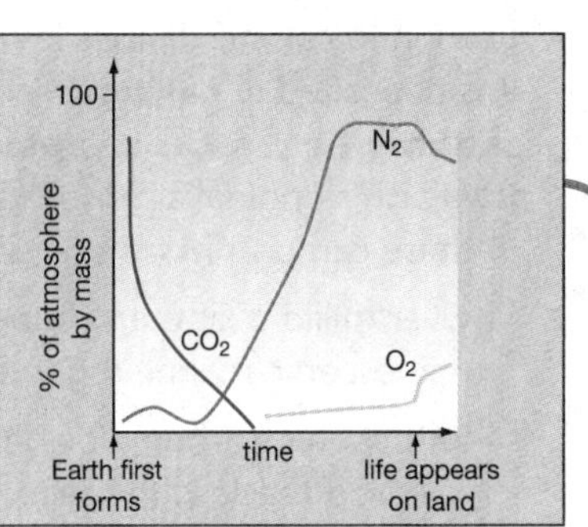

Changes in the levels of nitrogen (N_2), oxygen (O_2) and carbon dioxide (CO_2) since the Earth's formation.

- Early life forms evolved in water. Some of these photosynthesised: their waste product was **oxygen** gas.
- **Ammonia**, NH_3, reacted with oxygen and produced **nitrogen**, which gradually built up in the atmosphere. Bacteria also produced nitrogen from ammonia.
- Some oxygen was converted to **ozone**, O_3, forming a shield against ultraviolet rays. This enabled plants to evolve on land. Plants photosynthesised, removing carbon dioxide and adding more oxygen. Animals evolved, taking in oxygen and giving out carbon dioxide.
- The gases in the atmosphere eventually became balanced about 200 million years ago.
- The graph shows the change during Earth's history in levels of nitrogen, oxygen and carbon dioxide.

The greenhouse effect on Venus

B–A*

- Venus has a surface temperature of about 500 °C, hotter than Mercury which is nearer the Sun.
- The reason is Venus's extreme greenhouse effect: its atmosphere is 97% carbon dioxide and is nearly 100 times denser than Earth's. Also, it has a very thick cloud cover of sulfuric acid droplets.

Questions

Grades D-C

1 Explain why noble gases are unreactive.

Grades B-A*

2 Explain why the decomposition of sodium azide causes an airbag to inflate.

Grades D-C

3 Why is there oxygen in our atmosphere?

Grades B-A*

4 Suggest how Venus's atmosphere helps us to understand what it happening to our atmosphere today.

Atmospheric change

Grades

Climate change, greenhouse gases and carbon sinks

- In the past 100 years, the average atmospheric temperature has risen by 1 °C. Most scientists agree that **climate change** is taking place. As evidence, ice sheets are melting and the sea level is rising; the weather is becoming stormier and more extreme.

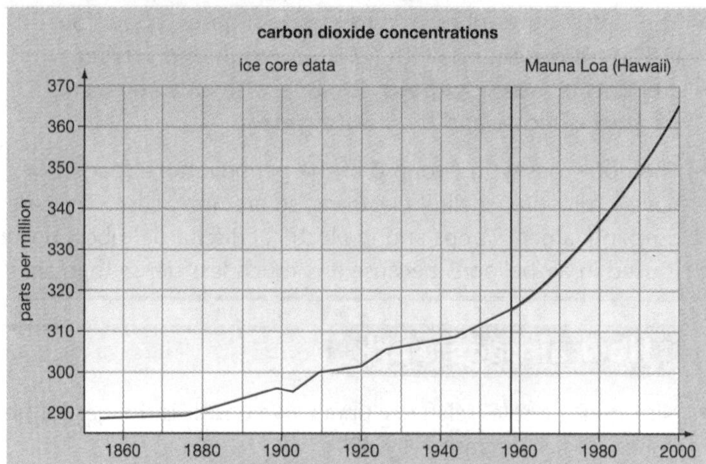

Atmospheric concentrations of carbon dioxide from 1860 to 2000.

- One reason proposed for climate change is a rise in the level of **greenhouse gases**, in particular of **carbon dioxide**. Its concentration in the atmosphere has increased by about 30% since 1860 (see graph). It is thought that this is due to the burning of **fossil fuels**.

D–C

Chalk cliffs like this are carbon sinks.

- Carbon in carbon dioxide leaves the atmosphere when it is deposited in **carbon sinks**. These include **fossil fuels**, **sedimentary rocks** and **plants**. Most of the carbon in the Earth's original atmosphere (see page 45) is locked up in these carbon sinks:
  - Over millions of years, buried plants form **coal**, and microscopic marine organisms form **oil**.
  - With water, carbon dioxide forms **carbonic acid**. The acid reacts with calcium-containing materials in seawater to form **calcium carbonate**. Being insoluble, this builds up as sediment on the sea floor. Over millions of years, sediments became **sedimentary rocks** including **limestone** and **chalk**.
  - During **photosynthesis** (see page 73), plants take in carbon dioxide and change it to carbon compounds which are 'locked up' in the **wood** of trees.

How Science Works

You should be able to:
- explain and evaluate theories of the changes that have occurred and are occurring in the Earth's atmosphere
- explain and evaluate the effects of human activities on the atmosphere.

Carbon dioxide and climate change

- Over the Earth's lifetime, global temperatures have fluctuated widely, ranging from long ice ages to shorter periods in which temperatures were much hotter than today's.
- Explanations for these hot/cold cycles include natural effects such as the Sun's output of heat, the Earth's tilt, its orbital path round the Sun and other factors that humans cannot control.
- Few doubt that climate change is happening. Many believe that it is the result of burning fossil fuels. The questions being asked are: What are the likely consequences of a further increase in carbon dioxide? and, Whether or not humans are the cause of climate change, what, if anything, can and should be done to halt this change?

B–A*

Questions

Grades D-C

1 How does carbon dioxide contribute to climate change?

Grades D-C

2 What is a 'carbon sink? Name **three** carbon sinks.

Grades B-A*

3 How does the Earth's past history increase uncertainty about whether burning fossil fuels really *is* the cause of climate change?

Grades B-A*

4 Explain why, whatever the cause of climate change, we should still try to reduce the quantity of fossil fuels that we burn.

C1b summary

New substances from crude oil

Crude oil is made up of **long-chain hydrocarbons** that can be **cracked** by **thermal decomposition** to form **shorter-chain alkanes** (useful as fuels) and **alkenes** (used to make polymers and plastics).

Alkenes are **unsaturated**. They contain carbon-carbon **double bonds**.
The **general formula** for alkenes is C_nH_{2n}

Polymers are made when small molecules called **monomers** are joined together to make very long **chains** in a process called **polymerisation**. The **carbon-carbon double bonds** in alkenes are broken, allowing each double-bonded carbon atom to join to an additional atom.

Polymers can be used to make many **useful substances**, e.g. Teflon®, packaging materials, dental compounds, waterproof materials, wound dressings and hydrogels.
A **shape memory polymer** can be distorted, then resume its original shape when warmed.

Many polymers are **not biodegradable** and are produced from **non-renewable** crude oil.
To dispose of polymers, they can be:
- **recycled** to make other products
- burnt to generate heat (but this can result in pollution)
- mixed with starch or cellulose to make them biodegradable.

Oil from plants

Many **plants** contain useful **natural oils** that can be **extracted**, e.g. from fruits, seeds and nuts. **Omega-3 fatty acids**, found in walnuts, lower levels of fats in blood, reduce the risk of blood clots, lower blood pressure and generally prevent heart disease.

Biofuels can be made from plant oils and other substances. These **'green fuels'**:
- include rapeseed oil, sunflower oil and wood
- are **renewable**
- are **'greenhouse neutral'**.

Water and oil form separate layers – they are **immiscible**.
An **emulsion** is formed when tiny droplets of oil (dispersed phase) are **suspended** in water (continuous phase).

Fats and **oils** can be **saturated**, containing only **single bonds**. **Unsaturated** fats and oils contain one (**monounsaturated**) or more (**polyunsaturated**) C–C **double bonds**.
Blood **cholesterol** levels are lowered by eating polyunsaturates and monounsaturates.

Hydrogenation (breaking some of the double bonds and joining extra hydrogen atoms to the carbon) is used to **harden** unsaturated liquid vegetable oils in the manufacture of **margarine**.

Food **additives** keep food fresh longer or make it appetising. **E-numbers** identify permitted chemicals, which include: antioxidants; colourings; emulsifiers, stabilisers, gelling agents and thickeners; flavourings; preservatives; and sweeteners.

Artificial colourings and other food additives can be identified by **chemical analysis** including **paper chromatography**, which separates the substances in a mixture.

The Earth

The **Earth** has three main layers: the **crust**, the **mantle** and the **core**.

The Earth's **lithosphere** is cracked into **tectonic plates**, which ride on the 'plastic' rock of the mantle. **Earthquakes** can occur at the **plate boundaries** where they collide with each other. **Volcanoes** form at weak points in the Earth's crust.

The **air** is a mixture of mainly **nitrogen**, **oxygen**, variable amounts of **water vapour** and very small amount of other gases such as **carbon dioxide** and the **noble gases**.

The Earth's **atmosphere** has changed over millions of years. Many of the gases that make up the atmosphere came from volcanoes, e.g. water vapour, ammonia, carbon dioxide and methane. **Ozone** (O_3) forms a shield against ultraviolet rays.

The levels of the **greenhouse gases** in the atmosphere are thought to be rising, resulting in **climate change**.

Sedimentary rocks, **plants** and **fossil fuels** are **carbon sinks**, locking up the carbon from carbon dioxide that was once in the atmosphere.

P1A ENERGY AND ELECTRICITY

Heat energy

Heat (thermal energy) and temperature

Grades

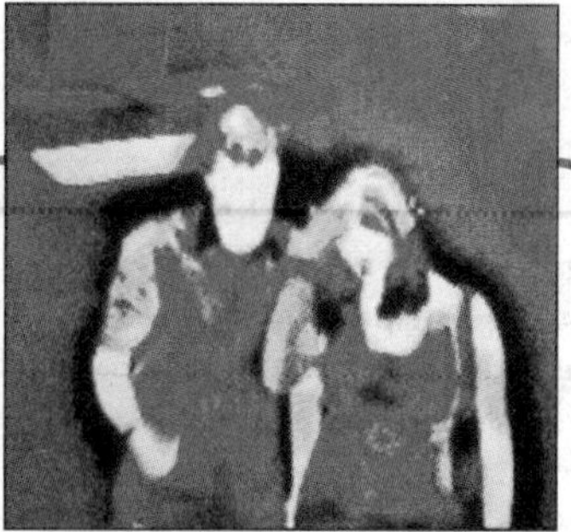
An infrared camera produces an image from thermal radiation.

D–C

- **Heat** is **thermal energy** that matter possesses and that can **transfer** from one material to another.
- One means of thermal energy transfer is by **thermal** (**infrared**) **radiation** which travels as **electromagnetic waves** (see page 59) in straight lines at the speed of light. It can travel through a vacuum.
- All materials **absorb** (take in) and **emit** (give out) thermal radiation.
- An **infrared camera** forms an image from people's thermal radiation.
- **Temperature** is a measure of the heat energy in a solid, liquid or gas.
- **Thermal energy** makes the particles in an object vibrate (see also page 49). The particles have differing energies, and the object's temperature measures the *average* energy of all the particles.

Top Tip!

Remember that 'cold' is not transferred! It is always heat energy that is transferred, from a hotter object to a colder one.

B–A*

- **Total thermal energy** depends on number of particles: 1 dm^3 of iron and 1 dm^3 of air may be at the *same temperature*, but with far more particles in iron, the total thermal energy in the iron is much greater than in the air.
- Similarly, at the same temperature, a *large mass* of a substance contains *more* thermal energy than a *small mass* of the same substance.

Thermal radiation

Transfer of thermal radiation

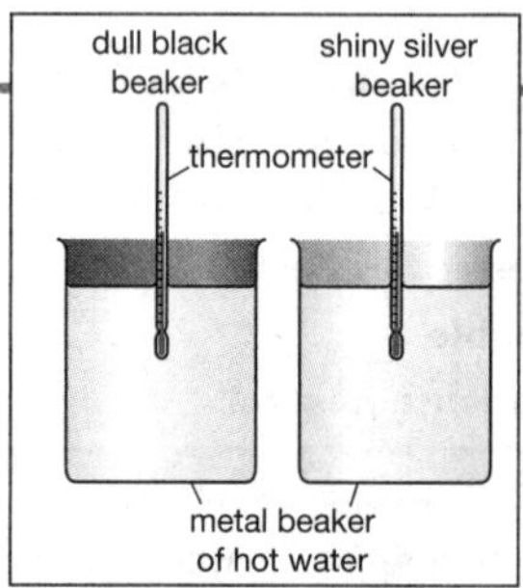

Comparing the emission and absorption of thermal radiation by different surfaces.

D–C

- Hot objects emit more thermal radiation than cold objects. The greater the difference between the temperatures of an object and its surroundings, the faster the *rate* of heat transfer.
- The amount of thermal radiation an object **emits** is affected by:
  - its shape and dimensions: a cube-shaped building emits less thermal radiation than an H-shaped building of the same volume and materials
  - the type of surface it has: the black grid behind a refrigerator emits thermal radiation fast, while silver 'space blankets' slow down emission.
- Experiments with the apparatus in the diagram show that:
  - the dull black beaker **emits** thermal radiation from hot water faster than the silver beaker
  - the dull black beaker **absorbs** thermal radiation (to heat cold water) *faster* than the silver beaker.
- Black surfaces emit more thermal radiation than white surfaces; dull (matt) surfaces emit more than shiny surfaces.

B–A*

- Polar bears have white fur but black skin. The fur looks white because it **reflects** all the Sun's light rays. But other rays from the Sun penetrate the fur and the black skin readily absorbs their energy.
- The fur is such a good thermal insulator that a thermal imaging camera picks up no image of the polar bear's shape: only its breath appears.

Questions

Grades D-C

1 Explain how an infrared camera can detect a person hiding among trees.

Grades B-A*

2 A large beaker of water and a small beaker of water are at the same temperature. Explain why the water in the large beaker has more thermal energy than the water in the small beaker.

Grades D-C

3 Explain why silver objects heat up *and* cool down more slowly than black objects.

Grades B-A*

4 Explain why a thermal imaging camera has to be very sensitive before it will show an image of a polar bear against snow.

Grades

Conduction and convection

Thermal energy transfer by conduction and convection

D–C

- Thermal energy flows rapidly through thermal **conductors** (e.g. metals), but flows only slowly through thermal **insulators** (e.g. plastic, air).
- In **convection**, a gas or liquid (a fluid) moves, carrying thermal energy with it. When heated, fluids expand, become less dense, and rise.

B–A*

- Water has a very high heat capacity – it absorbs a lot of heat without its temperature rising greatly. Also, being a liquid, it can circulate to carry thermal energy away from the equipment. Engines and other machines often use water as a coolant that circulates in pipes.

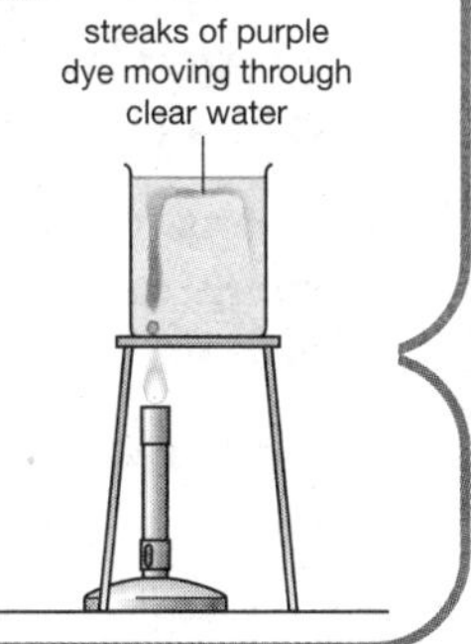
streaks of purple dye moving through clear water

Dye shows convection in heated water.

Heat transfer

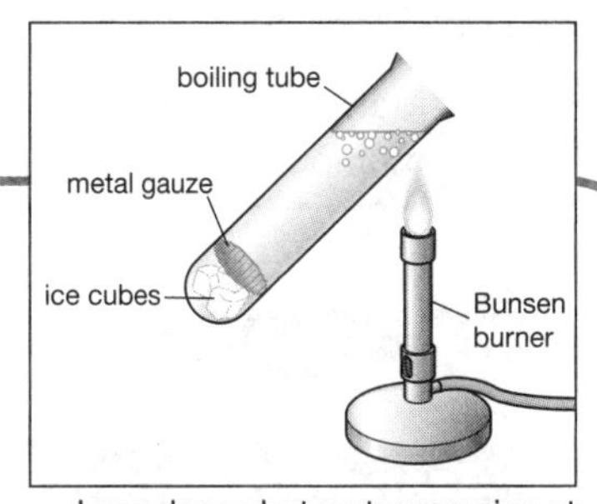
boiling tube
metal gauze
ice cubes
Bunsen burner

Less dense hot water remains at the top, and ice cubes take a long time to melt.

Energy transfer and the particle model

D–C

- Liquids are poor thermal **conductors**: their particles transfer heat slowly (see diagram).
- Gases are very poor thermal conductors. Air trapped in a cold-weather jacket greatly slows heat transfer from a person's body.
- Particles in **solids** are in fixed positions. In **conduction**, particles at the heated end of a rod gain thermal energy and vibrate more vigorously. They transmit vibrations to their neighbours, transferring thermal energy along the rod.
- Only particles on the *surface* of a solid (or liquid) can transfer energy to their surroundings.
- **Metals** are good conductors because their **free electrons** carrying thermal energy can move and transfer energy to the fixed particles.
- Particles in **liquids** are not held together firmly, so energy is not readily transferred between particles, and heat transfer by conduction is slow. Particles in **gases** are far apart. Heat energy is transferred only when particles collide, so heat transfer by conduction is very slow.
- In **convection**, the particles of liquids (and gases) move more vigorously when heated. They occupy more space and the liquid **expands**. Being less dense, it rises. Cooler, denser fluid falls under gravity, and **convection currents** are set up. (Convection cannot occur without gravity.)

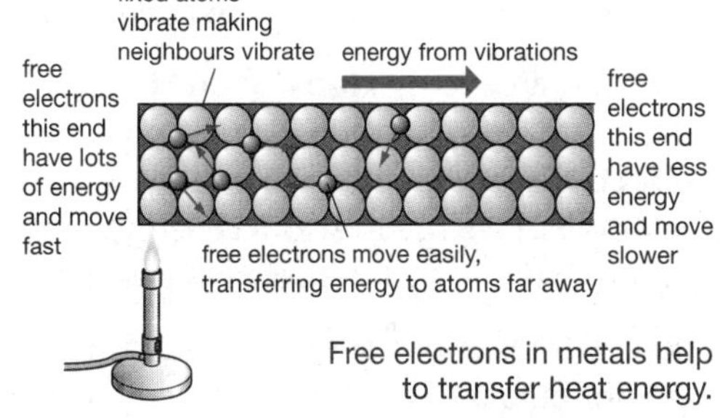
fixed atoms vibrate making neighbours vibrate
energy from vibrations
free electrons this end have lots of energy and move fast
free electrons this end have less energy and move slower
free electrons move easily, transferring energy to atoms far away

Free electrons in metals help to transfer heat energy.

Top Tip!

Do not say that particles in a solid 'start to' vibrate when heated. They are already vibrating! Heat energy makes them vibrate *more vigorously*.

B–A*

- **Copper** is an excellent conductor of thermal energy, but **diamond** is four times better. A diamond's very strongly bonded carbon atoms vibrate in such a way as to conduct heat far more rapidly than any other material. Diamonds are called 'ice' because they feel so cold to the touch.

How Science Works

You should be able to:

- evaluate ways in which heat is transferred in and out of bodies
- evaluate ways in which the rates of these transfers can be reduced.

Questions

Grades D-C

1 Explain how heat energy is conducted along a metal rod.

Grades B-A*

2 Explain why water is a good coolant for engines.

Grades D-C

3 Explain why gases are very poor conductors of heat.

Grades B-A*

4 Explain why diamonds feel cold when you touch them.

Types of energy

Grades

Forms of energy, and energy from the Sun

D–C

- **Energy** is what makes things happen. Energy is measured in **joules (J)**. Raising an apple from floor to table takes about 1 J.
- **Forms** (types) of energy include: **heat**, **light**, **sound**, **chemical**, **microwave**, **radio wave**, **kinetic** and **potential** energy.
- **Stored energy** can be used when required. Energy stores include fuels, batteries, stretched elastic bands and compressed springs.

B–A*

- The Sun's radiation causes most changes that occur on Earth (see diagram). They include the growth of the organisms that formed fossil fuels.
- Solar radiation is also responsible for the Earth's weather systems and ocean currents.

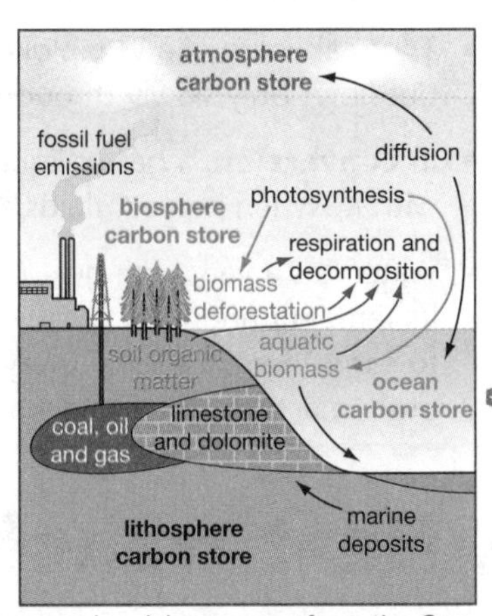

Energy changes involving energy from the Sun.

Energy changes

Transfers, transformations and energy conservation

D–C

- There is an energy **transfer** when the *same* form of energy moves from *one place to another*. For example, a cyclist's **kinetic energy** is transferred to become the bicycle's **kinetic energy**.
- There is an energy **transformation** when *one form of* energy *changes to another form*. For example, a light bulb transforms **electrical** energy to **light** energy, a music system transforms **electrical** energy to **sound** energy, and a battery transforms **chemical** energy to **electrical** energy.

Examples of energy transformations.

- In any transfer or transformation, some energy is always **wasted**: it is not useful. For example, a light bulb also produces heat, and a rocket motor providing kinetic energy also produces heat, light and sound.
- No energy that is transferred or transformed ever 'disappears'. The total amount of energy before and after an event is always the same. The **Law of Conservation of Energy** sums this up:

Energy cannot be created or destroyed.
It can only be transferred or transformed from one form to another.
The total energy always remains constant.

B–A*

- The Sun's immense energy comes from **nuclear fusion** reactions (see also page 130) when the nuclei of atoms combine. Two hydrogen nuclei fuse to form a helium nucleus and produce energy.
- The mass of the helium nucleus is very slightly *less* than the mass of the two hydrogen nuclei because **mass has become energy**: there is *more* energy after the reaction than before it. It seems to contradict the Law of Conservation of Energy. To reconcile this, scientists say that, in nuclear reactions, mass and energy are interchangeable (equivalent), and that mass/energy is conserved.

Questions

Grades D-C

1 If you use 10 million J of energy in a day, how much chemical energy should there be in the food you eat each day?

Grades B-A*

2 Give **two** changes that happen on Earth because of the energy we receive from the Sun.

Grades D-C

3 When you ride a bike, not all the kinetic energy in your legs is transferred to kinetic energy in the bike. Where does the wasted energy go?

Grades B-A*

4 How does the Sun produce energy?

Energy diagrams

Diagrams for energy changes and wastage

D–C

- Only *some* of the energy that is transferred or transformed is **useful energy**. The rest is **wasted energy**.
- For a device (such as a light bulb) that uses energy:

  energy input = energy output = useful energy + wasted energy

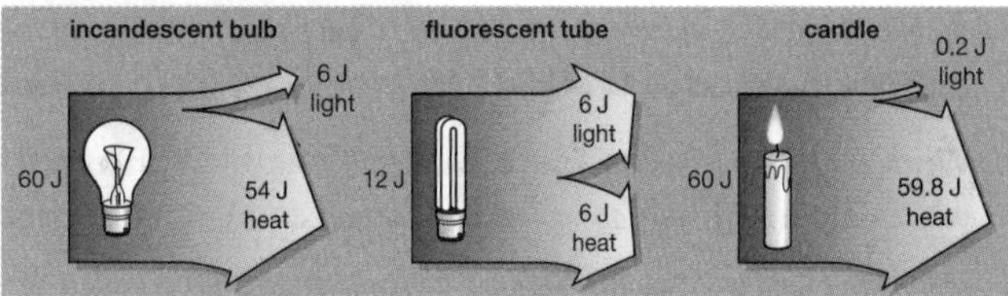

Energy transformations in different light sources.

- A **Sankey diagram** shows the energy input and output of a device and can be used to calculate unknown values.
- A device may have several **energy transfers** or **transformations**. At each one, energy is wasted. The diagram for a petrol engine shows this.

B–A*

- An **electric car** runs on a rechargeable battery and causes no air pollution. A motor, not a combustion engine, drives the wheels. The energy in the battery originates in the energy from either a fossil fuel or a renewable resource (e.g. hydroelectric power) used to generate electricity in a power station.

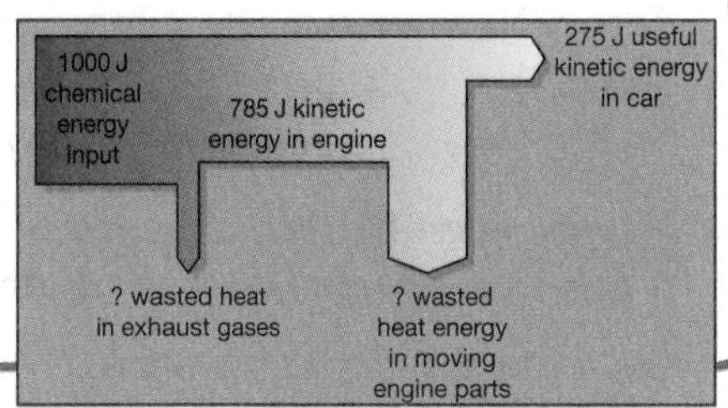

A Sankey diagram for a petrol engine.

Energy and heat

How Science Works

For a range of devices, you should be able to:
- describe the intended energy transfers/transformations
- describe the main energy wastages that occur.

Where wasted heat energy goes

D–C

- In the long run, both wasted energy *and* the energy that is usefully transferred or transformed appear as **heat energy**. This heat energy is transferred to the **surroundings** by conduction, convection and/or thermal radiation, and the surroundings become warmer. For example: electrical energy in a torch is transformed to light and some heat. The light energy hits an object, is absorbed by it, and is transformed into heat energy.
- Energy gradually *spreads out*, and so is less and less useful for further energy transformations. For example: the thermal energy in a pan of hot water can be used for cooking. The water cools as the thermal energy spreads out to warm the surroundings. The energy is no longer useful.
- Energy is more useful when **concentrated**, as in fuels and batteries, which are useful **stores of energy**.
- Inside all objects, the particles are **vibrating**. When an object absorbs any sort of energy, its particles vibrate more vigorously – it heats up. The more energy it absorbs, the more vigorously its particles vibrate and the hotter it gets.

How refrigerators work

B–A*

- A **refrigerator** takes thermal energy spread out amongst items of food, concentrates it and transfers it to the fins at the back of the refrigerator. A refrigerator transfers heat in the opposite direction to normal – from cold food to warm air. It requires electrical energy to do this work.

Questions

Grades D-C

1 Calculate how much heat energy is wasted in the exhaust gases and as heat in the moving parts in the petrol engine in the diagram.

Grades B-A*

2 Explain why electric cars do not cause pollution where they are used, but may indirectly cause pollution elsewhere.

Grades D-C

3 Explain why a bath full of warm water is said to be a 'less concentrated' store of energy than a kettle full of hot water.

Grades B-A*

4 Explain why refrigerators use a lot of energy, even though the food they contain actually *loses* thermal energy.

Energy, work and power

Energy gained, work and power

- When **work** is done on an object, **energy** is transferred to it – it **gains** energy: **energy gained = work done** (both measured in **joules**; 1 joule = 1 newton metre)
- Work is done when a **force** makes something **move** through a **distance**: **work done (joules) = force (newtons) × distance moved (metres)**

Top Tip! Remember to use seconds, not minutes, when calculating power.

- (**1**) A force of 25 N moves a trolley a distance of 3 m. Calculate the work done, and the energy transfered to the trolley.

  work done = 25 × 3 = 75 J = energy transfered to the trolley

D–C

- (**2**) An object gains **potential energy** when lifted to a higher position. A vase of mass 5 kg is lifted 2 m up to a shelf. Calculate the work done and potential energy gained by the vase.

  First, calculate the **force** that lifts the vase's **weight**: weight (N) = mass (kg) × force of gravity (10 N/kg)

  force (N) = weight (N) = mass (kg) × 10 (N/kg)

  force = 5 × 10 = 50 N
  work done = 50 (N) × 2 (m) = 100 J = potential energy gained by the vase

- **Power**, in **watts** (**W**), is the *rate* of doing work, measured as joules of work done in 1 second (J/s). 1 W = 1 J/s: **power (W) = work done (J) ÷ time taken (s)**
- (**3**) A crane lifts 500 kg of planks to a height of 9 m in 3 min. What is its power?

  work done = force × distance moved = weight × distance moved
  = 500 × 10 × 9 = 45 000 J
  power = work done ÷ time taken
  = 45 000 ÷ 180 = 250 W

B–A*

- The Egyptian **pyramids** are steep sided. To lift huge stone blocks up the sides would have taken a very *large* force exerted over a *short* distance. Instead, the ancient Egyptians pulled the blocks up long, shallow ramps built to the sides of the pyramids using a *smaller* force over a *much longer* distance.

How Science Works

You should be able to: calculate the efficiency of a machine using:

$$\text{efficiency} = \frac{\text{useful energy transferred by the machine}}{\text{total energy supplied to the machine}}$$

Efficiency

Calculating efficiency

D–C

- The **efficiency** of a machine is the useful energy **output** (the work a machine does or the energy transformed into useful forms) divided by the energy **input** (all the energy supplied to the machine):

  $$\textbf{efficiency} = \frac{\textbf{useful energy output}}{\textbf{total energy input}}$$

- A crane has an energy input of 2000 J and does 800 J of useful work.

  efficiency = 400 J per 1000 J

B–A*

- The greater the **percentage** of energy that is usefully transformed in a machine, the more efficient a machine is. Referring to the Sankey diagram on page 51 for a petrol engine:

  $$\text{efficiency} = \frac{\text{useful energy output}}{\text{total energy input}} = \frac{300}{1000} = 0.3 \quad \text{percentage efficiency} = 30\%$$

- Many inventors have tried to design a perpetual motion machine – one that is 100% efficient. This is impossible because in all energy transfers and transformations some energy is wasted.

Questions

Grades D–C

1 A girl lifts twenty 2 kg bags of sugar from the ground onto a shelf 3 m above the ground. It took her 1 min to do this. Calculate her power output.

Grades B–A*

2 Pulling a weight up a ramp to a certain height uses the same amount of energy as lifting it straight up. So how does a ramp make it easier?

Grades D–C

3 A machine has an input of 1000 J and a useful energy output of 500 J. Calculate its efficiency.

Grades B–A*

4 Suggest why a machine with moving parts will have a greater efficiency if it is well oiled.

Using energy effectively

Saving energy in the home

- Reducing energy use in the home saves money and benefits the environment. Lights can be turned off and replaced by **energy-saver light bulbs**. The temperature of central heating can be turned down, and old energy-wasting appliances can be replaced.

D–C

- Installing **insulation** reduces the heat energy lost through ceilings, walls, doors and windows. The table compares savings and costs. **Payback time** is how long it takes for savings to equal installation costs.

type of insulation	installation cost (£)	annual saving (£)	payback time (years)
loft insulation	240	60	4
cavity wall insulation	360	60	6
draught-proofing doors and windows	45	15	3
double glazing	2500	25	100

B–A*

- As fossil fuels run low, countries are developing **energy saving policies**. Some local authorities offer grants to help people pay for insulation and draught-proofing.
- Grants are encouraging people to install **solar panels** that collect the Sun's radiant energy to heat domestic water. It takes about 20 years to recoup the cost. Also, energy savings need to be set against the energy cost of manufacturing the materials and equipment required.

Why use electricity?

Using electricity in the home

How Science Works

You should be able to: evaluate the effectiveness and cost effectiveness of methods used to reduce energy consumption.

D–C

- **Mains electricity** provides homes with an instant, convenient, clean, safe and reliable source of energy.
- **Electrical energy** from the mains is **transformed** into heat (e.g. oven), light and sound (e.g. computer) and kinetic (e.g. vacuum cleaner) energy.
- **Batteries** store and supply a *small* amount of portable **chemical energy** that can be transformed into electrical energy, e.g. for a radio or an electric toothbrush.
- In countries where electricity is unavailable or costly, the **chemical** energy stored in either food or fuel is widely used domestically: human, animal and steam power all provide **kinetic energy**; various types of **biomass** (plant material and animal waste) are burnt to supply **heat** energy for cooking; wax candles or gas or oil lamps provide **light**.
- **Solar energy** and the **potential energy** in clockwork springs can charge rechargeable batteries. They enable people to use phones, computers and other devices where there is no electrical supply.

B–A*

- **Light-emitting diodes** (**LEDs**) look like very small light bulbs. They are used on numerous appliances, often to show that they are switched on.
- Instead of a fine coiled-wire filament or a gas, the light-emitting material is a small flat sheet that glows brightly when a tiny current passes through it. The light emerges through a clear plastic bead.
- The material contains atoms whose outer electrons absorb electrical energy, jump to a higher energy level, and emit light when they fall back. Different LED materials give different colours.

Questions

Grades D-C

1 An ordinary light bulb loses 95% of its energy input as heat. A fluorescent bulb is 20% efficient. Which of these two kinds of bulb gives the most light for the same input of electrical energy?

Grades B-A*

2 Suggest why it takes 20 years to recoup the cost of installing solar panels on a roof in England.

Grades D-C

3 Some illuminated signs on country roads are powered by solar and wind energy. Suggest why this is done.

Grades B-A*

4 Describe how electrical energy is transformed to light energy in an LED.

Electricity and heat

The correct fuse would stop a keyboard from catching fire.

Grades

Current, energy and resistance

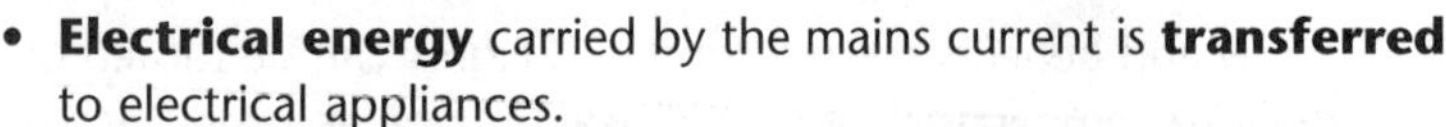

- **Electrical energy** carried by the mains current is **transferred** to electrical appliances.
- Copper is used for circuit wiring because its **resistance** is low compared to other metals. This reduces energy loss as heat. D–C
- Thin wires have a greater resistance than thick wires. An appliance has the correct thickness of wire for the energy it requires. A telephone has a thin wire, while an electric cooker has a very thick wire.
- Some appliances use the heating effect: a bulb filament glows white hot and an electric fire element gives out heat. Too much current through a **fuse** (see page 127) heats it and it melts.

Superconductivity

- At extremely low temperatures, some metals lose all electrical resistance. Loops of superconducting wire can carry currents for years without loss of energy. B–A*
- Expensive cooling materials such as liquid nitrogen keep 'high temperature' superconductors below –150 °C. As an example, MRI scanners use superconducting wires in electromagnet coils.

The cost of electricity

How Science Works

You should be able to:

- compare and contrast the particular advantages and disadvantages of using different electrical devices for a particular application
- calculate the amount of energy transferred from the mains using: energy transferred (kilowatt hour, kWh) = power (kilowatt, kW) × time (hour, h)
- calculate the cost of energy transferred from the mains using: total cost = number of kilowatt hours × cost per kilowatt hour

Measuring energy and calculating costs

- A householder pays for the **electrical energy** that the current *transfers* (passes) to electrical appliances. (The current itself does not change.)
- **Power rating** is the rate at which an appliance *transforms* electrical energy (e.g. to heat, light, etc.):

  1 **joule** (J) of electrical energy per second = 1 **watt** (W) of power

  1000 W = 1 **kilowatt** (kW). (See also page 52.)

- **i** A 1.2 kW microwave oven is used for 5 min. What is the energy transferred? D–C

  total amount of energy transferred (in J) = power rating (in W) × time (in *seconds*)

  Total energy transferred from mains supply to microwave oven = 1200 (W) × 300 (s) = 360 000 J

  ii Electricity bills charge for kilowatt hours (kWh), referred to as Units. How many kWh are used?

  number of kilowatt hours used = power rating (in kW) × time (in *hours*)

  Number of kilowatt hours used = 1.2 (kW) × 5/60 (h) = 0.1 kWh (or 0.1 Unit on the electricity bill)

  iii The cost of a kilowatt hour (1 Unit) is 8p. What is the cost of using the microwave oven?

  total cost = number of kWh used × cost per kWh

  Cost of using the microwave oven = 0.1 × 8 = 0.8p

- Leaving a digital TV set-top box on standby uses up 7 W of power. Two appliances on standby can use as much electricity as an energy-saving light bulb. B–A*

Questions

Grades D-C

1 Explain why a thin wire gets hotter than a thick wire when the same current flows through both of them.

Grades B-A*

2 What is a 'superconductor' and why can superconductors be useful?

Grades D-C

3 A small electric desk fan has a power rating of 50 W. How many joules of electrical energy does it transform (into kinetic energy) every second?

Grades B-A*

4 If electricity costs 8p per kWh, calculate the cost of running the fan for 30 min.

The National Grid

Top Tip!

Very high voltages transmit very large amounts of energy at low current and therefore with relatively small energy losses.

Voltages in the National Grid

D–C

- The **National Grid** is the system that distributes electricity from power stations to consumers. Step-up transformers at the power station increase the voltage so that the electricity is transmitted across the country at a very **high voltage** and hence at very **high energy**.
- To reduce the waste of energy lost as heat, and to enable the use of thinner, cheaper copper wires, electricity is transmitted as a relatively **small current**.
- At **substations**, **step-down transformers** reduce the voltage from 400 000 V to 132 000 V. This is repeated by step-down transformers at smaller substations, so that electricity is at 230 V when it reaches consumers.
- **Power** (in watts) is the energy that an electrical current transfers per second: **power = current × voltage**, or $P = IV$

a.c. and d.c. electricity

B–A*

- The current from a battery is **direct current** (**d.c.**); it flows in one direction.
- The current in mains electricity (see page 126) is **alternating current** (**a.c.**). It constantly changes direction, at a frequency of 50 Hz (50 cycles per second).
- It is cheaper to alter the voltage of a.c. than d.c. at substations.

Generating electricity

Electric motors and dynamos

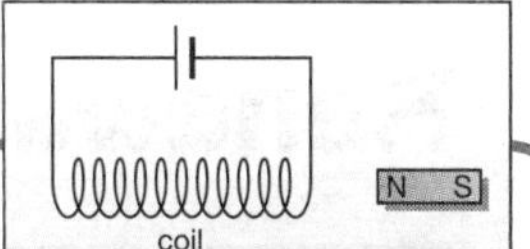

When current flows through the coil, the magnet is attracted to the coil.

D–C

- An electrical current in a wire coil generates a magnetic field that attracts (or repels) a magnet, causing *movement* of coil or magnet.
  - Similarly, in a battery-operated fan, the electrical energy from the current is transformed into **kinetic energy** (energy of movement) by a **motor** which contains a wire coil rotating between magnets.
- When a magnet is moved into or out of a coil, an electrical current is *generated* in the circuit.
  - Similarly, when fan blades are spun by hand (below left), the coil is *moved* relative to the magnets, and an electrical current lights the bulb. This shows that kinetic energy is being transformed into electrical energy. The set-up is an electricity **generator** (or **dynamo**).

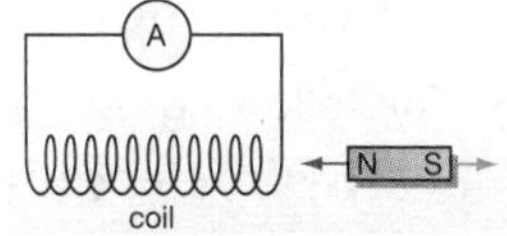

The magnet moving in or out of the coil generates a current in the circuit.

- The diagram on the right shows the arrangement in the generator of a wind turbine where an ammeter can record current flowing when the blades spin.

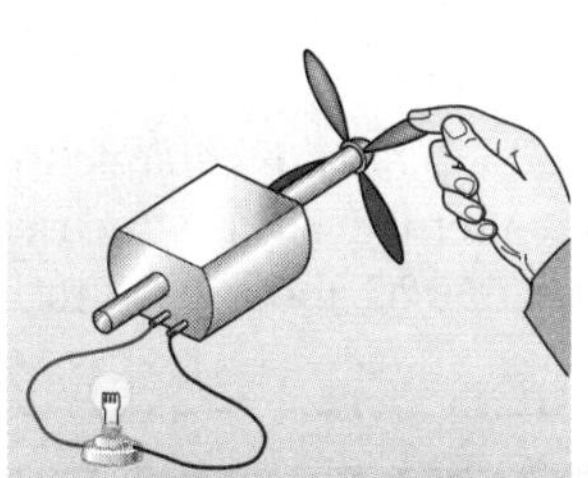

Spinning the blades generates electricity that lights the bulb.

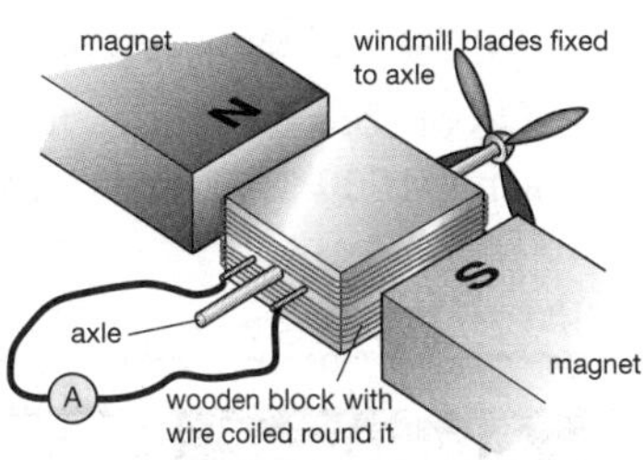

When the blades and coil turn, an electrical current is generated.

Questions

Grades D-C

1 Explain why electricity is transmitted at very high voltages in the National Grid.

Grades B-A*

2 Give **one** reason why electricity is transmitted through the National Grid as a.c. and not as d.c.

Grades D-C

3 Describe how you could use a magnet and a coil of wire to produce electricity.

Grades B-A*

4 How could you alter your 'generator' to make it into a motor?

Power stations

Grades

Fuel-burning and nuclear power stations

D–C

- Many **power stations** use energy to produce **steam** to drive **turbines**. Others use compressed **hot gases**. The turbines turn **generators** which produce the electricity.
- Most British **power stations** burn the fossil fuels **coal, natural gas** or **oil**:
  - **coal-burning** power stations are about 35% efficient
  - **combined cycle gas** power stations use heat left in hot gases to produce steam for a second cycle; efficiency is about 50%
  - a **combined heat and power** station reaches 70–80% efficiency since 'waste' energy is used to heat local buildings.
- **Nuclear** power stations (page 130) use uranium-235 and are 30% efficient. Plutonium-239 is produced as waste.
- Energy transformations in a power station inevitably cause **wasted energy**.

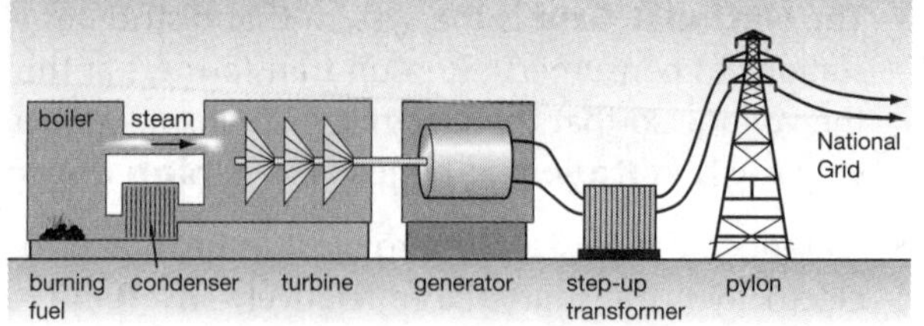

A coal-fired power station.

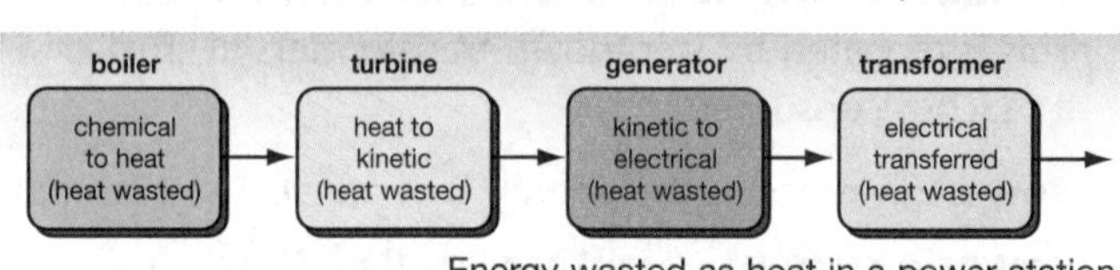

Energy wasted as heat in a power station.

Top Tip!
Make sure you know the sequence of stages in a power station and the energy changes at each stage.

B–A*

- **Landfill** sites containing domestic organic rubbish emit methane, a greenhouse gas with ten times the greenhouse effect of carbon dioxide. Methane is being 'harvested' at over 75 UK landfill sites to generate electricity in new power stations nearby.

Renewable energy

Renewable energy sources

D–C

- **Steam** drives turbines linked to generators in power stations that:
  - burn **biomass** (e.g. willow saplings, grass), and also household rubbish
  - use **geothermal energy**: water pumped down into hot rocks returns as steam.
- The following use the **movement** (kinetic energy) of either **water** or **wind** to drive turbines:
  - a **hydroelectric** power station uses the energy of falling water
  - in a **tidal power** station, the incoming and outgoing tides flow through a turbine
  - a **wave power** generator uses the energy of wave water flowing and ebbing inside a tube; a turbine at the top rotates when the air above the rising and falling water is forced out or sucked in
  - a **wind turbine** uses the energy of moving air; **wind farms** are groups of wind turbines.

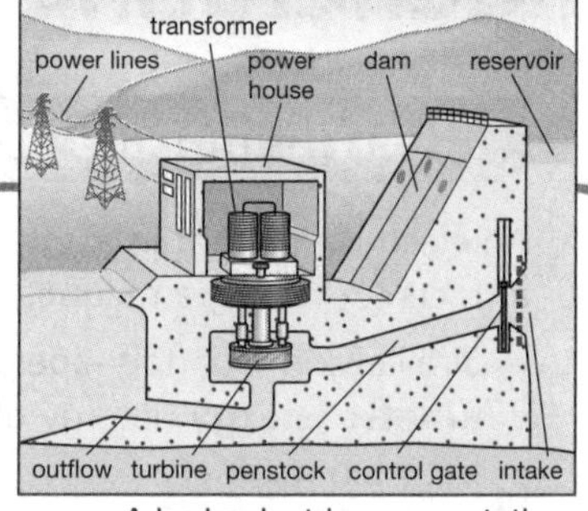
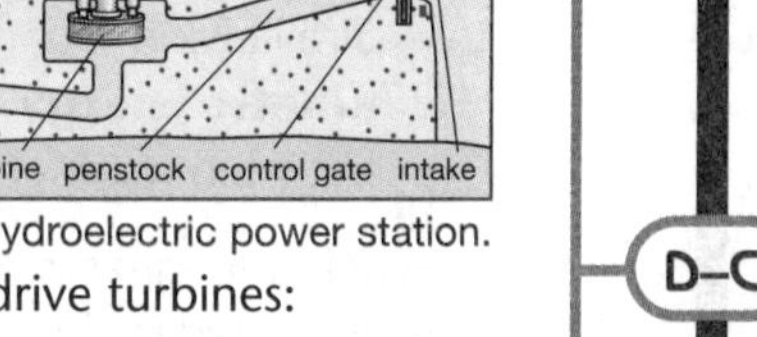

A hydroelectric power station.

B–A*

- **Solar** power stations have been built in hot regions of North America and Europe.
- Banks of concave mirrors focus the Sun's rays onto a tower with a heat-absorbing area of ceramic that reaches 1000 °C. It heats air which, in turn, heats steam that runs a turbine. Alternatively, the focused solar energy heats pipes containing concentrated saline solution that reaches 550 °C and is used to generate steam.

Questions

Grades D-C

1 If 65% of the energy in the fuel a power station uses is wasted, what is the efficiency of the power station?

Grades B-A*

2 Suggest **two** environmental benefits of using methane generated at landfill sites to generate electricity.

Grades D-C

3 In volcanic areas, such as Iceland, hot water and steam come up to the surface of the ground. How can this be used to generate electricity, and what is the name for this kind of energy source?

Grades B-A*

4 Suggest **two** advantages of the new solar power stations over a coal-burning power station.

Electricity and the environment

Comparing means of electricity generation

D–C

power generator	harmful effects	benefits
fossil fuel-burning	gases contribute to climate change	fuel still cheap and abundant
waste-burning	toxic gases may cause cancer or birth defects	reduces need for landfill sites
hydroelectric	dams flood valleys, wildlife habitats, farmland, homes	cheap, clean, renewable energy
tidal and wave	change water flow; disrupt shipping; destroy habitats	very low running costs
geothermal	might release dangerous gases from below Earth's surface	small station, no deliveries required, so low environmental impact
wind turbine	may kill birds; spoil the landscape; noise	cheaper in remote areas than Nat. Grid
nuclear	radioactive pollution; generates nuclear waste	fuel available; no greenhouse gases

- To **limit costs**, power companies: **locate** stations close to consumers; build fossil fuel power stations near roads/rail for fuel access and near rivers for cooling water; site wind turbines in windy areas, hydroelectric power stations amongst hills, and geothermal power stations on easily drilled rock.

Top Tip!

Remember at least one really clear advantage and one really clear disadvantage for each type of power station.

B–A*

- Each hydroelectric power station has a dam. Over the year, its water level fluctuates. Plants grow on the shores during droughts, then are flooded. When submerged, they may rot and give off methane, a potent greenhouse gas.
- A study of one Brazilian hydroelectric dam suggests that it may cause three times the amount of greenhouse gases as an equivalent fossil-fuel power station.

How Science Works

You should be able to: compare and contrast the particular advantages and disadvantages of using different energy sources to generate electricity.

Making comparisons

Which type of power generation?

D–C

- Power stations cannot store electrical energy. They have to match output to **demand**.
- **Solar cells** transform the Sun's light energy *directly* into electrical energy. **Wind turbines** require regular winds over 50 km/h. **Hydroelectric power** requires reliable rainfall, and **solar power** needs near-year-round sunshine.
- Power station planners need to consider **capital** (building) costs and **operating** (fuel and running) costs. **Fossil fuel** power stations are cheaper to build, but fuel costs are rising. Most **renewable** energy power stations currently have high capital costs, lower operating costs and low efficiency, but improved technologies may alter this. New power stations will also need to: use fossil fuels more **efficiently**; use **renewable energy** sources; use **reliable** energy sources; and be **flexible** enough to meet changing demand.
- Certain bacteria in **human sewage** have been found to oxidise organic matter in a process that releases electrons and protons.
- The bacteria cluster round the positive electrode (see diagram) and pass electrons to it, while protons migrate to the central negative electrode. By *separating charge*, a **voltage** is set up.

Generating electricity in sewage.

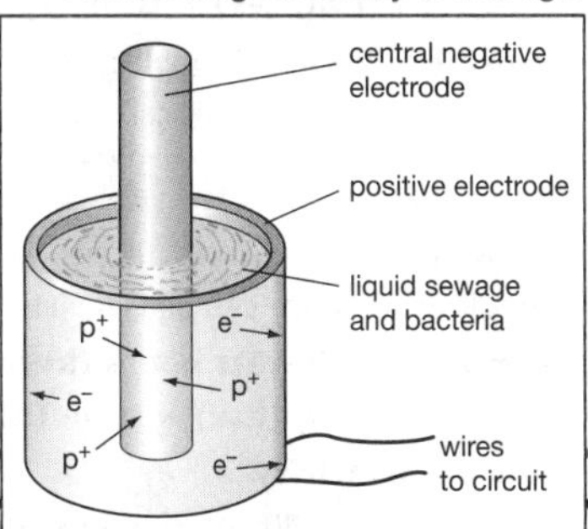

Questions

Grades D-C

1 Why are oil-fuelled power stations often sited near to a major road?

Grades B-A*

2 Explain why the production of methane by a hydroelectric power station could 'cause three times the amount of greenhouse gases as an equivalent fossil-fuel power station'.

Grades D-C

3 Explain why hydroelectric power stations are usually situated in mountainous areas of the country.

Grades B-A*

4 Suggest how sewage could be used to generate electricity in a more conventional way than that shown in the diagram.

P1a summary

Heat energy

Heat is **energy** that matter possesses and that can **transfer** from one material to another.

Temperature is a measure of the heat energy in a material (a solid, liquid or gas).

Heat energy can be transferred by **conduction**, **convection** and **thermal radiation**.

Thermal (**infrared**) **radiation** is the transfer of heat energy by **electromagnetic** waves. All materials **absorb** and **emit** thermal radiation.

The amount of thermal radiation an object **emits** is affected by:

- the object's shape and dimensions
- the type of surface it has.

Energy transfer and transformation

Energy **transfer** occurs when the *same* form of energy *moves* from one place to another. Energy **transformation** occurs when *one* form of energy is *changed* to *another*.

Free electrons in metals help to transfer energy rapidly between the fixed particles.

Thermal conductors (e.g. metals) transfer heat energy easily. **Thermal insulators** (e.g. plastic, air) do not. In **convection**, a gas or liquid (a fluid) moves, carrying thermal energy with it. When heated, fluids **expand**, become **less dense**, and rise.

Energy types and diagrams

Types of energy include: **heat**, **light**, **sound**, **chemical**, **microwaves**, **radio waves**, **kinetic** and **potential**.

The Sun's immense energy comes from **nuclear fusion** reactions; two hydrogen nuclei fuse to form a helium nucleus and produce energy.

The **Law of Conservation of Energy** states: **Energy cannot be created or destroyed. It can only be transferred or transformed from one form to another. The total energy always remains constant.**

A **Sankey diagram** shows the **energy input** and **output** of a device.

Wasted energy is usually heat energy.

energy input = **energy output** = **useful energy** + **wasted energy**

$$\textbf{efficiency} = \frac{\textbf{useful energy output}}{\textbf{total energy input}}$$

Work is measured in joules (J):

work done = force × distance moved

Power is measured in watts (W):

power = work done ÷ time taken

Electricity

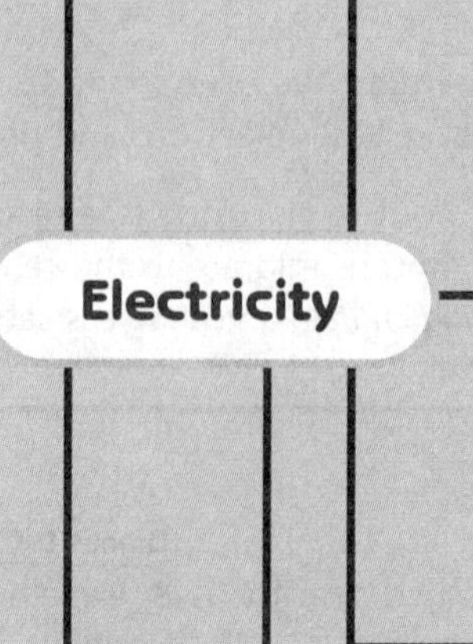

Electrical current from the mains carries **electrical energy** which is **transferred** to **electrical appliances**:

total amount of energy transferred (J) = power rating (W) × time (s)

The **power rating** of an appliance is the rate at which it transforms electrical energy.

The **cost** of the energy that current transfers to an electrical appliance can be calculated:

- **number of kilowatt hours (kWh) used = power rating (kW) × time (h)**
- **cost of electricity = number of kWh used × cost per kWh**

The **National Grid** transmits **electricity** around the country at **high voltage** and **low current** to reduce energy losses.

At **substations**, **step-down transformers** reduce the voltage so that electricity is at 230 V when it reaches consumers.

power = current × voltage, or $P = IV$

The **electrical energy** from a **current** can be transformed into **kinetic energy** (energy of movement) by a **motor**, which contains a wire coil rotating between magnets.

A **dynamo** produces electricity when coils of wire rotate inside a **magnetic field**.

In **power stations**, **fossil fuels** and **biomass** are burned, or **nuclear fuels** react, to release energy as heat.

The heat is used to produce steam: the **steam** turns a **turbine**, a turbine turns a **generator**, and a generator produces **electricity**.

Some power stations use **hot**, **compressed gases** instead of steam.

Renewable energy sources include **wind**, **hydroelectric**, **tidal**, **wave** and **geothermal power**.

All types of **electricity generation** have some **harmful effects** on people or the environment. There are also **limitations** on where they can be used.

Uses of electromagnetic radiation

What is electromagnetic radiation?

D–C

- **Electromagnetic radiation** is **energy** carried by **waves** which: move in **straight lines**; can travel through a **vacuum**; and carry a **range** of energies. All electromagnetic waves travel at the same speed, the speed of light.

The electromagnetic spectrum and some uses of different types of radiation.

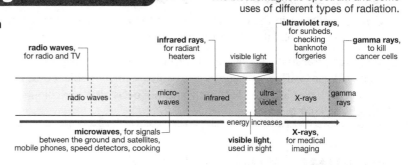

radio waves, for radio and TV
infrared rays, for radiant heaters
visible light
ultraviolet rays, for sunbeds, checking banknote forgeries
gamma rays, to kill cancer cells
radio waves | micro-waves | infrared | ultra-violet | X-rays | gamma rays
energy increases
microwaves, for signals between the ground and satellites, mobile phones, speed detectors, cooking
visible light, used in sight
X-rays, for medical imaging

- The **electromagnetic spectrum** is continuous, but for convenience, different 'types' of radiation are identified. Their effects and uses depend on their energies.

B–A*

- Airport security staff use **X-rays** to form an image of luggage on a screen. It is checked for dangerous and illegal items. X-rays also indicate whether paintings are forgeries.
- **Ultraviolet** radiation shows if passports and banknotes have been forged.
- In burglar alarm systems, an **infrared** beam spans each doorway and window. The alarm goes off if a beam is broken.

Top Tip!

The shorter the wavelength, the higher the frequency. The longer the wavelength, the lower the frequency. The higher the frequency, the higher the energy.

Electromagnetic spectrum 1

Wavelength, frequency and energy

D–C

- A wave has a **wavelength**, measured from one crest to the next.
- The shorter the wavelength, the more **energy** the wave carries.
- The number of waves that pass a point in one second is the **frequency** of the wave, measured in **hertz** (**Hz**). 1 Hz = 1 wave per second

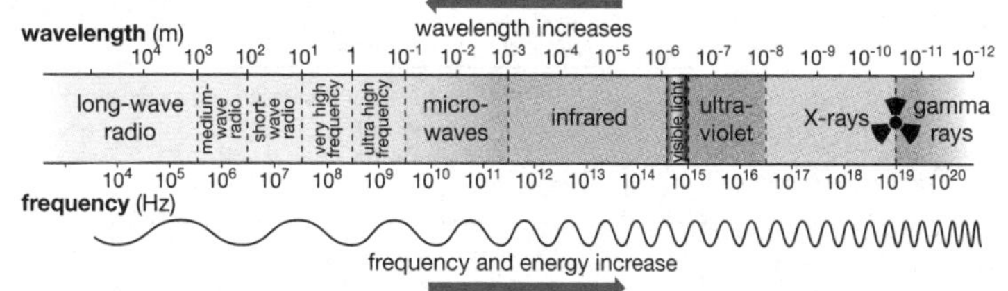
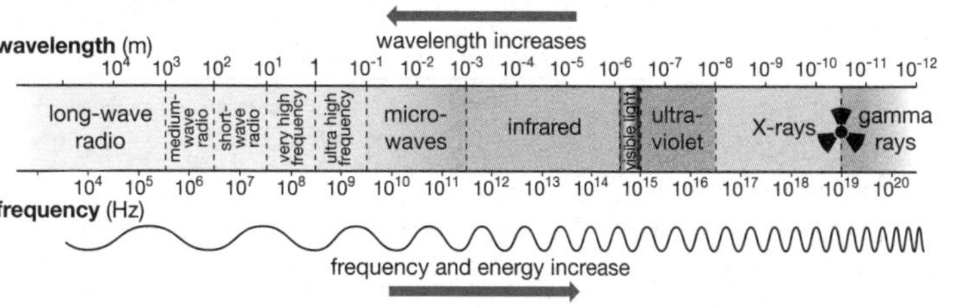

The electromagnetic spectrum with wavelengths and frequencies.

B–A*

- Telecommunications signals are sent out from a transmitter in all directions in the form of electromagnetic radiation.
- When a signal reaches a receiver (e.g. a portable radio), the electromagnetic waves pass through a conducting wire in its circuit.
- Remember that when a magnet moves near a wire in a circuit, an electrical current is generated in the wire (see page 55). This is because the wire *cuts through* a moving magnetic field.

Transmitting and receiving a TV picture.

oscillating electric charge
sound
picture
oscillating electric charge
transmitted signal

- In a similar way, the conducting wire of the radio *cuts through* the travelling electromagnetic waves, and they generate a current in the wire. Being waves, the signal oscillates from plus to minus (see page 126), so the current also oscillates – it is an **alternating current**, and its frequency is the same as the frequency of the incoming waves.

Questions

Grades D–C

1 Give **one** use of microwaves, other than cooking.

Grades B–A*

2 Explain why a burglar may not notice that there is a beam of infrared radiation passing across an entrance.

Grades D–C

3 Which carries more energy – a wave with a short wavelength, or a wave with a long wavelength?

Grades B–A*

4 What is the relationship between the frequency of electromagnetic radiation and the frequency of the oscillation of an alternating current that the radiation produces?

Electromagnetic spectrum 2

Grades

Light radiation and colour

D–C

- **Light** is radiation in the **visible range** of the electromagnetic spectrum. The range consists of the rainbow colours of light which together form white light.
- Matter can transmit or reflect or absorb light. An object that light passes through **transmits** light. An object that **reflects** all light looks white. A *smooth* object that reflects all light is a mirror. An object that reflects red light only and **absorbs** all other colours looks red. One that absorbs all light looks black.

Waves and matter

The mirror reflects the infrared wave and the TV absorbs it.

How waves and matter interact

D–C

- Radiation carries **energy**. The energy carried by different wavelengths of electromagnetic radiation can be absorbed, reflected or transmitted by an object, depending on its substance and surface.
- When matter **absorbs** energy, the energy is often transformed to heat energy:
  - food in a microwave oven absorbs short-wavelength microwave radiation
  - a black object (that absorbs all light radiation) becomes hotter.
- Dark, dull surfaces tend to absorb a large range of radiation.
- Soft tissue is transparent to X-rays, and to the long-wavelength microwave radiation used in mobile phones.
- Particles of a particular substance can absorb the energy of some waves but be transparent to others.

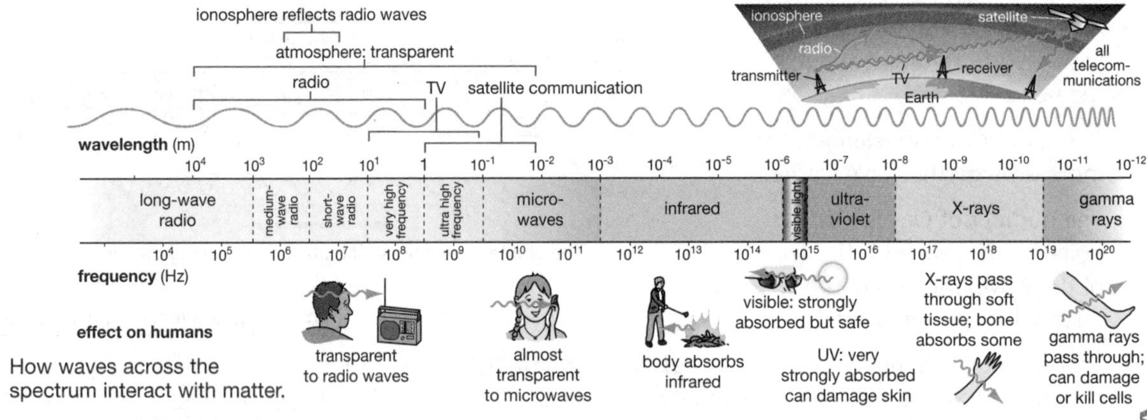

How waves across the spectrum interact with matter.

B–A*

- All electromagnetic radiation travels at the speed of light, 300 000 000 metres per second. Frequency increases as wavelength decreases.
- The **formula** that enables us to find the frequency or wavelength of any wave is:

  wave speed = frequency × wavelength
  m/s — waves/s, Hz — length, m
- So: **frequency = wave speed ÷ wavelength**
  and: **wavelength = wave speed ÷ frequency**
- The wavelength of a signal to a mobile phone is 30 cm (0.3 m). What is its frequency?

frequency = wave speed ÷ wavelength

$$= \frac{300\,000\,000}{0.3} = 1000\,000\,000 \text{ Hz} = 1000 \text{ Mz}$$

Questions

(Grades D-C)

1 List the different types of electromagnetic radiation, in order of their wavelengths (smallest first).

(Grades B-A*)

2 Your mobile phone produces microwaves. Are they absorbed in your head?

(Grades D-C)

3 The wavelength of an electromagnetic wave is 10^4 metres. What type of wave is it?

(Grades B-A*)

4 Calculate the frequency of the wave in question 3.

Grades

Dangers of radiation

Effects of radiation on living cells, and safeguards

D–C

- Electromagnetic radiation does no harm to living things if it is reflected or passes through living matter without energy being absorbed. Radiation that *transfers energy* to cells can cause damage. Effects depend on the type of radiation and size of dose (which regulations exist to control). In order of increasing frequency:
  - radio waves transfer virtually no energy and are safe; the low frequency microwaves of mobile phones transfer almost no energy and are safe
  - higher frequency microwaves in a cooker heat up water molecules and make the molecules vibrate more vigorously; for safety, a microwave oven operates only when the door is closed
  - human bodies absorb (and transmit) infrared radiation (from e.g. Sun, fires); small doses and/or short exposures are safe
  - molecules in the retina absorb visible light; if too intense, light can damage these molecules
  - ultraviolet radiation can cause skin cancer; sun-block protects skin and underlying tissues if it blocks long *and* short (UVA and UVB) ultraviolet radiation
  - X-rays are used for imaging and gamma rays for killing cancer cells; lead aprons and lead-glass screens protect hospital staff.

How Science Works

You should be able to:
- evaluate the possible hazards associated with the use of different types of electromagnetic radiation
- evaluate methods to reduce exposure to different types of electromagnetic radiation.

Ionising radiations can cause cancer

B–A*

- Ultraviolet radiation, X-rays and gamma radiation all **ionise** atoms and molecules in living cells.
- Damage to DNA may cause a mutation (change) in a gene that prompts the cell to divide out of control and grow into a cancerous **tumour**. A cancer that spreads is described as malignant.

Telecommunications

Satellites, fibre optics and digital signals

D–C

- Radiations, from long-wave radio to microwave, used in **telecommunications** pass through the atmosphere (see diagram on page 60) to and from **geostationary satellites**. The satellites receive and transmit ultra-high frequency and microwave radiations.
- Infrared and visible (**laser**) rays transmit information along **fibre optic cables** (see page 62). The fibre optic cables are cheaper to make than copper electrical cables, they carry many more messages at one moment than wires, the signal travels at the speed of light and the signal quality remains high.
- Rather than as a continuous signal, messages can be transmitted as pulses of radiation, in a **digital signal** compatible with computers and digital TVs.

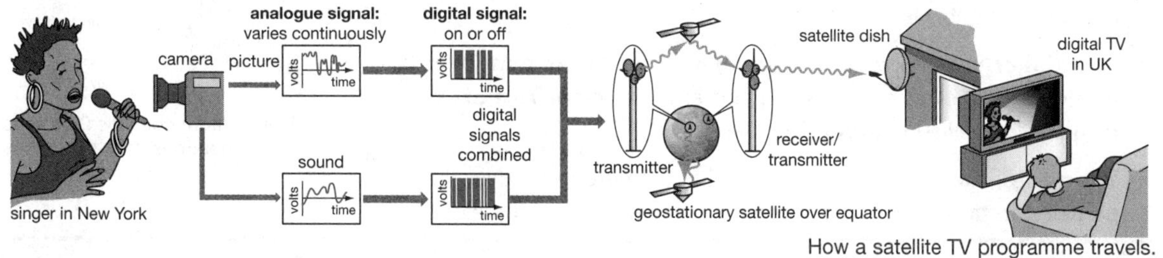

How a satellite TV programme travels.

Questions

Grades D-C
1 What type of electromagnetic radiation causes skin cancer?

Grades B-A*
2 Explain how this type of radiation causes cancer.

Grades D-C
3 Explain what is meant by a 'geostationary satellite'.

Grades B-A*
4 What types of electromagnetic radiation are transmitted and received by communications satellites?

Fibre optics: digital signals

Grades

How light travels along optical fibres

- An optical fibre is a very fine strand of pure silica glass. Black cladding prevents light from escaping.

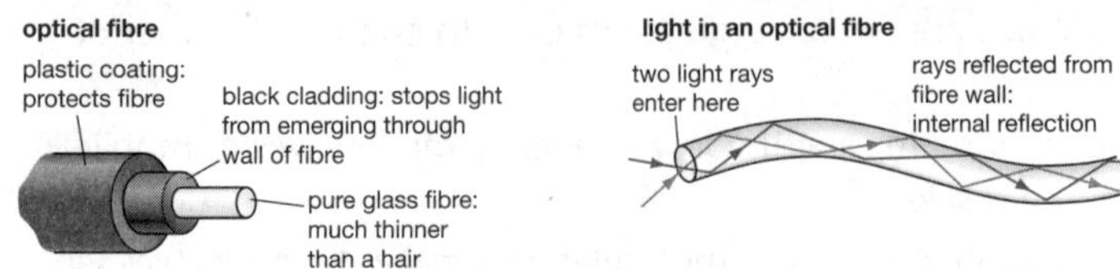

Optical fibres carry information for phones, cable TV and computers.

D–C

- A beam of laser light (infrared to visible) travels inside a glass fibre. The beam hits the glass boundary at an **angle of incidence** over 42° and undergoes **total internal reflection**.
- One fibre can carry several wavelengths of light simultaneously.
- The light is **pulsed** (turned on/off) to produce a **digital signal**. Up to 10 billion pulses travel per second, enabling one fibre to carry over 10 000 phone calls simultaneously.
- Between 12 and 200 optical fibres are bundled together in a cable. The light does not heat the cable, and signals in adjacent fibres do not interfere.

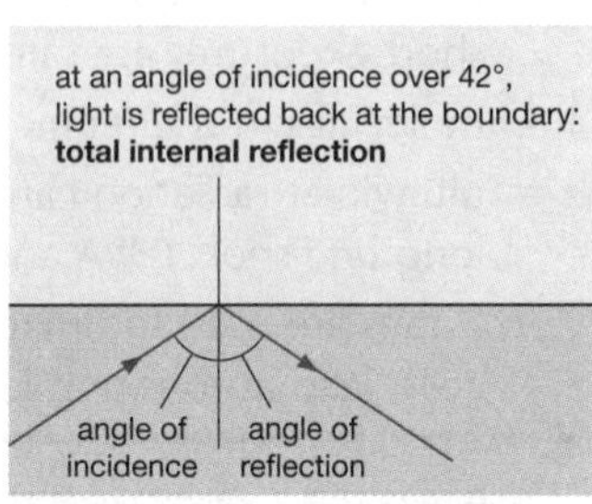

Total internal reflection enables the light to remain inside the fibre.

Digital technology

B–A*

- Steadily, equipment using analogue (continuous signal) technology is being replaced by digital equipment (using a pulsed signal). It is more compact and versatile and includes cameras and digital audio players.
- Card driving licences contain digitised information about drivers. In the future, similar identification (ID) cards could hold extensive digital information including unique biometric data such as an eye retina scan or a voice pattern.

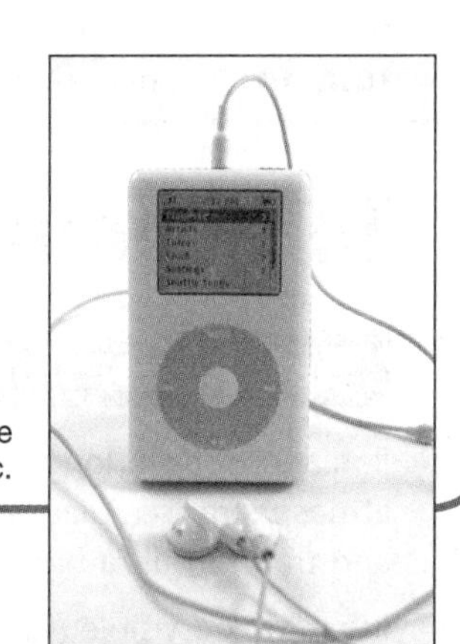

An iPod is used to store and play digital music.

Radioactivity

Unstable isotopes

B–A*

- All atoms of an **element** have the same number of protons. **Isotopes** are different forms of an element, each with a different number of neutrons.
- A **radioactive** isotope (radioisotope) is an **unstable** form of an element. Its nucleus breaks down and emits **nuclear radiation** by a process called **radioactive decay**.
- The nuclear radiation carries energy and is emitted as either alpha particles, beta particles or gamma rays, depending on the element.

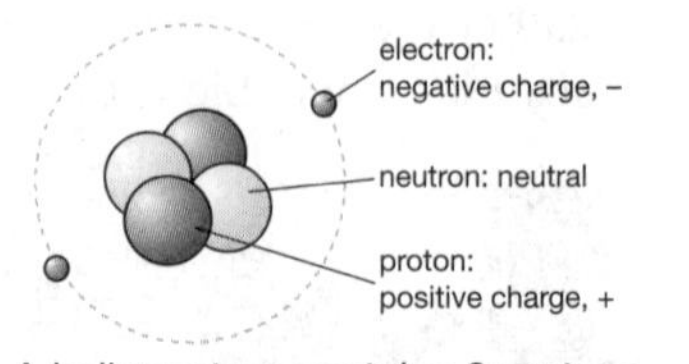

A helium atom contains 2 protons, 2 neutrons and 2 electrons.

Questions

Grades D-C

1 Explain why light rays do not escape from optical fibres.

Grades B-A*

2 List **two** advantages of sending information in a digital rather than an analogue form.

Grades D-C

3 Explain why 'nuclear radiation' is given that name.

Grades B-A*

4 List the **three** types of nuclear radiation.

Alpha, beta and gamma rays 1

Three types of nuclear radiation

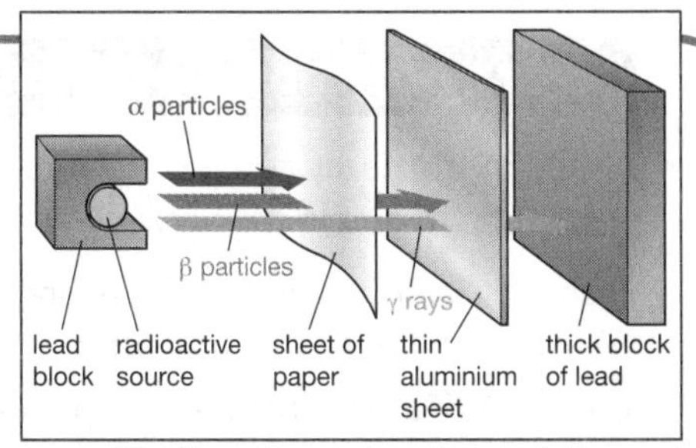

The three types of nuclear radiation and their ability to penetrate materials.

D–C

- When the nucleus in an atom of a radioisotope decays (breaks up), it emits one of three types of **nuclear radiation** (see diagram and table for characteristics):
  - an **alpha particle** is a helium nucleus, ^4_2He, of charge 2+
  - a **beta particle** is an electron, e^-, formed in the nucleus when a neutron changes to a proton: $n \rightarrow p^+ + e^-$
  - **gamma rays** are electromagnetic radiation (see page 59).
- Nuclear radiation carries energy that depends on the radiation source. Values in the table are averages.

radiation	alpha (α) particle	beta (β) particle	gamma (γ) rays
range in air (approx.)	1–5 cm	10–100 cm	very long range
ionising effect	very high	high	low
skin penetration	fraction of a millimetre	0.1 mm	centimetres
damage to cells	severe, if it's inside body	moderate	low doses: little damage (weakly ionising)

B–A*

- Because **alpha** and **beta** particles have a charge, electric fields and magnetic fields (see diagrams) deflect them.
- As a type of electromagnetic radiation, **gamma** rays have no charge (or mass), so electric and magnetic fields do not affect them.

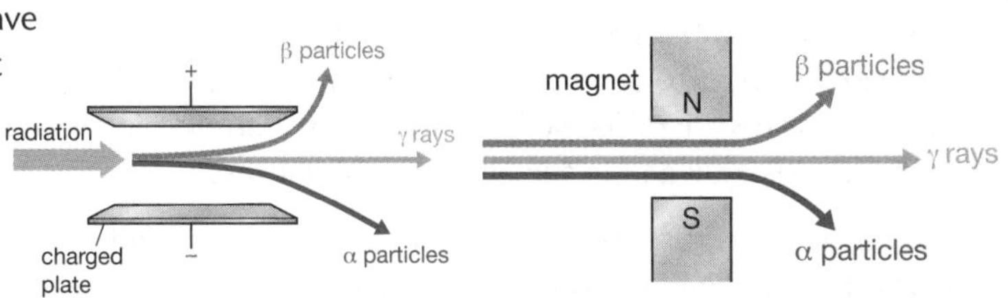

Effect of electric and magnetic fields on type of radiation.

Background radiation 1

Sources of background radiation

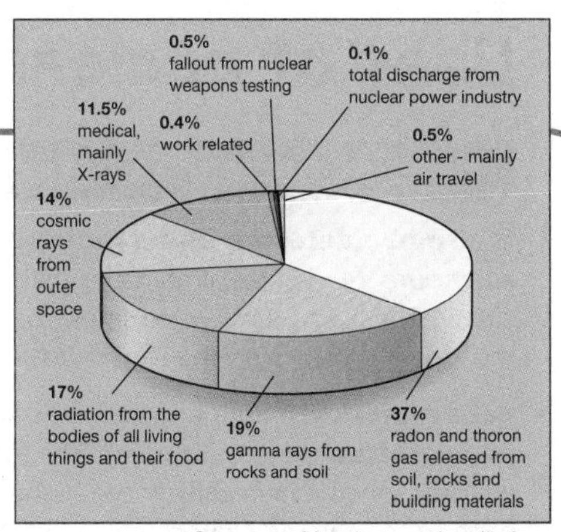

Sources of background radiation.

D–C

- There is a natural low level of **background radiation** in the environment. It includes man-made radiation.
- The Earth's magnetic field and atmosphere shield us from most harmful solar **cosmic rays**. Planes fly in a thin atmosphere and a weaker magnetic field, so aircrew monitor their flying time to maintain safe radiation levels.
- Radioisotopes in granite rocks emit **gamma rays** and form **radon** gas, also radioactive. This raises background radiation levels in granite areas of the UK.
- Cells can repair damage caused by average levels of background radiation.
- Workers using radiation sources wear a **radiation badge** to monitor exposure (see also page 65).

B–A*

- For an experiment using a radioactive source, the background radiation count is first recorded and then subtracted. A Geiger counter, used to monitor radiation, emits a click at each decay.

How Science Works

You should be able to:
- evaluate the possible hazards associated with the use of different types of nuclear radiation
- evaluate measures that can be taken to reduce exposure to nuclear radiations.

Questions

Grades D-C

1 Explain why an isotope emitting alpha radiation is very dangerous if you swallow it, but not if you just touch it.

Grades B-A*

2 Explain why gamma radiation is not affected by an electric field.

Grades D-C

3 Explain why people who work near radiation must wear radiation badges.

Grades B-A*

4 A Geiger counter recorded 9 counts per second when held in the middle of a room. It was then placed in front of a radioactive source, and gave a reading of 80 counts per second. What was the count for the radiation emitted from the radioactive source?

Grades

Half-life

Decay of radioisotopes

- Although the decay of a particular radioactive atom is random, we do know how many atoms in a radioisotope sample will decay in a particular time. This is the **activity (count) rate**.
- Each radioisotope has its unique rate, and scientists use this to identify unknown radioisotopes. We measure the rate as **half-life**: D–C

The half-life of a radioisotope is the average time it takes for half of its atoms to decay.

- For a radioisotope sample, the **half-life** is the average time taken:
  - for half the radioactive nuclei to decay, and therefore
  - for the activity (count rate) to decrease to half its starting rate.
- The graph shows the decay of strontium-93, an artificial (manufactured) radioisotope.
- Several artificial radioisotopes with short half-lives are used in **medicine** to monitor body processes. The radioisotopes are safe, give results in a short time and can be used in small doses.

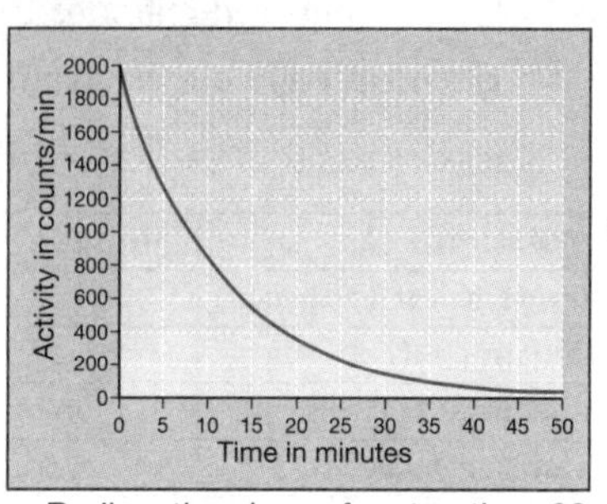

Radioactive decay for strontium-93.

- Cosmic rays bombarding nitrogen-14 in the atmosphere convert it to the radioisotope carbon-14. It combines with oxygen giving carbon-14 dioxide which stays at a steady proportion of atmospheric carbon dioxide – 1 molecule in 12 million million. It is in the same proportion in plants which take up carbon dioxide and in the animals that eat them. B–A*
- The half-life of carbon-14 is 5700 years. From the amount of carbon-14 left in ancient organic material, scientists use its half-life in **radiocarbon dating** to work out the age of anything that lived up to 60 000 years ago (see also page 128).

Uses of nuclear radiation

How Science Works

You should be able to: evaluate the appropriateness of radioactive sources for particular uses, including as tracers, in terms of the types of radiation emitted and their half-lives.

Uses of nuclear radiation

- In a **smoke detector**, **alpha radiation** from americium-241 ionises air particles which complete the circuit. The alarm starts when smoke particles prevent the ions from reaching the electrodes.
- **Beta radiation** is used in paper mills to control **paper thickness** (see diagram). The rollers are adjusted when a high activity count shows that the paper is too thin or a low count shows that it is too thick. D–C
- **Gamma rays** are used to: **sterilise** surgical instruments, killing bacteria; kill cancer cells in **radiotherapy**; trace leaks in water pipes – a radioisotope added to the water leaks out into the soil and is detected by a Geiger counter.

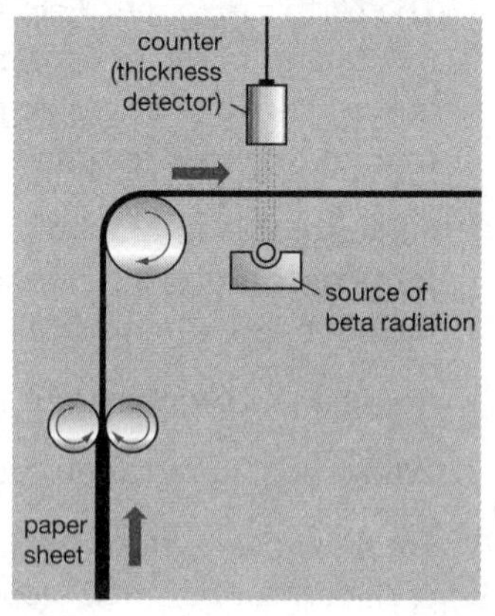

Controlling paper thickness.

Questions

Grades D-C

1 A piece of a radioactive isotope contains 10 million atoms. After 12 years, 5 million of these atoms have decayed. What can you say about the count rates you could detect from the isotope, using a Geiger counter, at the start and then 12 years later?

Grades B-A*

2 Name the isotope of carbon that is used in radiocarbon dating.

Grades D-C

3 Explain why the amount of this radioactive carbon isotope in a piece of wood steadily decreases over time.

Grades B-A*

4 Suggest why beta radiation, not alpha or gamma, is the best to use for checking the thickness of paper.

Grades

Safety first

Radiation hazards and precautions

Top Tip!

Do not say that radiation ionises 'cells'. You cannot ionise cells, only atoms and molecules.

D–C

- The **energy** of nuclear radiation can knock electrons from atoms in cells: the **ionised** atoms then disrupt the chemical processes of cells. If damage to DNA causes a gene **mutation**, cells may divide uncontrollably and form a cancerous tumour.
- Long or intense radiation exposure causes **radiation sickness** and often death, because more cells die than the body can replace.
- Workers using radiation sources, such as staff in hospitals and nuclear power stations, follow strict **health and safety** procedures. They handle radiation sources carefully, not touching them directly. They wear protective clothing and a badge that monitors the radiation they receive.
- The disposal of **spent nuclear fuel** causes problems of safety.

Searching space

Looking into space

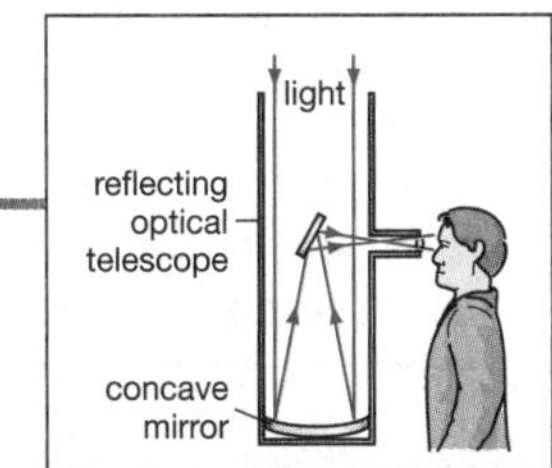

A reflecting optical telescope.

D–C

- **Optical telescopes** form images of objects in space using energy in the visible spectrum (light). A simple optical telescope forms poor images because it is too short and narrow to gather much light. **Reflecting** optical telescopes, of large diameter and with mirrors to extend their length, give far better images.
- Varying air density, humidity and dust pollution distort optical images taken at ground level. Large ground-based telescopes are therefore sited high up in remote, dry places.

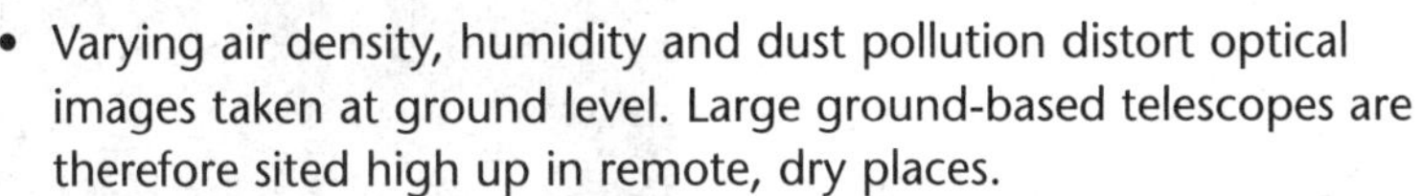

- **Radio telescopes** use a huge dish or an array of aerials to collect radio waves from gas clouds collapsing to form stars, supernovae and pulsars (see page 67).
- The orbiting **Hubble space telescope** has formed optical images (see page 66) further and deeper into space than ever before.

The Arecibo radio observatory in Puerto Rico.

Seeing the invisible

How Science Works

You should be able to: compare and contrast the particular advantages and disadvantages of using different types of telescope on Earth and in space to make observations on and deductions about the Universe.

B–A*

- Objects in space transmit radiation in all parts of the **electromagnetic spectrum** (see page 59). Gases in the atmosphere absorb a wide range of electromagnetic radiation including ultraviolet, X-rays and infrared. So our use of ground-based telescopes is limited.
- Telescopes on orbiting satellites now form images from radio waves, infrared, ultraviolet, X-rays and gamma rays of objects in space (see table and pages 66 and 67).

wavelength	objects 'seen' in space
gamma rays	neutron stars
X-rays	neutron stars
ultraviolet	hot stars, quasars
visible	stars
infrared	red giants
far infrared	forming stars, planets
radio	supernovae, pulsars

Questions

Grades D-C

1 Explain why disposing safely of spent nuclear fuel is extremely important.

Grades D-C

2 Why did the invention of the reflecting telescope allow better images to be seen?

Grades D-C

3 Why can the Hubble telescope produce better images of objects in space than ground-based telescopes?

Grades B-A*

4 Suggest why using telescopes that 'see' different wavelengths of the electromagnetic spectrum gives us more information about objects in space than just using light.

Gravity

Grades

Gravity, movement and orbits

Voyager 2's route across the Solar System.

D–C

- Owing to its **mass**, an object exerts a force of *attraction* – **gravity** – on other objects. The larger the mass, and the closer the objects, the stronger the force of gravity.
- Objects in steady **orbit** do not fall together. The force due to their **movement** *balances* the force of gravity. Gravity and movement keep all the bodies of the **Solar System** in their orbits. Without gravity, these objects would drift apart. Without orbital movement, they would fall together.
- Launched in 1977, the spacecraft Voyagers 1 and 2 have investigated the giant planets and are travelling in space beyond the Sun's gravity. To reduce the fuel load, the spacecraft flew close to Jupiter and Saturn and accelerated past them, gaining momentum from their huge gravitational pull.

B–A*

- As a spacecraft travels from the Earth to the Moon, the Earth's gravitational force diminishes and the Moon's gravitational force increases. At one point in the journey, the forces balance precisely – there is a net zero force – and the spacecraft and its crew feel **weightless**.
- Astronauts orbiting the Earth also feel weightless but the feeling continues. This is because, for as long as the spacecraft remains in that orbit, the force due to the spacecraft's orbital velocity (see page 111) balances the force of the Earth's gravity.

Birth of a star

Stars from dust

D–C

- Immense gas and dust clouds form in space. Over millions of years, **gravity** pulls the particles close together, they heat up to millions of degrees and atomic nuclei undergo **nuclear fusion**, and a **star** is formed.

$$\textbf{hydrogen} \xrightarrow{\text{nuclear fusion}} \textbf{helium} + \textbf{ENERGY} \text{ (including light)}$$

Spectacular gas clouds in the Eagle nebula, taken by the Hubble space telescope.

B–A*

- In nuclear fusion, the hydrogen in a new star forms helium. When hydrogen runs out, helium becomes the fuel, giving heavier elements, and so on. The elements up to iron are formed.
- When a very large ageing star collapses, it becomes a much hotter, smaller **supernova** (see page 67). This explodes with such energy that the elements heavier than iron are formed, again by the nuclear fusion of lighter elements.
- Since the Earth contains about 100 elements, we know that the gas and dust that formed the Solar System must have included the remnants of a much bigger star than our Sun.

Steps in the formation of a star.

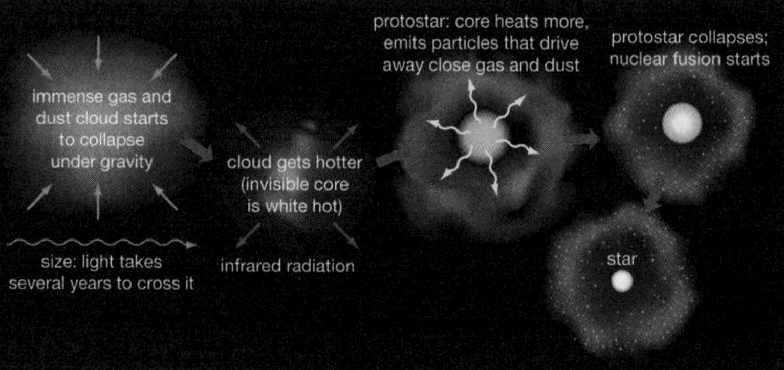

Questions

Grades D-C

1 How did gravity help the two Voyager craft to save fuel?

Grades B-A*

2 Why do astronauts orbiting Earth in the space shuttle feel weightless?

Grades D-C

3 What happens to hydrogen atoms when nuclear fusion takes place?

Grades B-A*

4 You are made of many different elements – mostly hydrogen, but also a lot of carbon, oxygen and nitrogen, as well as small amounts of iron and other elements. Where were these elements made?

Formation of the Solar System

Birth and shape of the Solar System

D–C

- The Solar System began as a huge cloud of gas and dust. **Gravity** drew the material together in a disc. Particles clumped to form rocks. A central star formed.
- The star became the Sun. Its intense radiation drove the lighter gaseous material outwards, while heavier material including rocks stayed closer.
  - The rocks became the inner **rocky planets**, Mercury, Venus, Earth and Mars. Beyond them, the **asteroid belt** of smaller rocks formed.
  - The outer **gas giants**, Jupiter, Saturn, Uranus and Neptune, formed from gases. Beyond them, ice and dust particles became the Oort cloud, from where **meteorites** come.

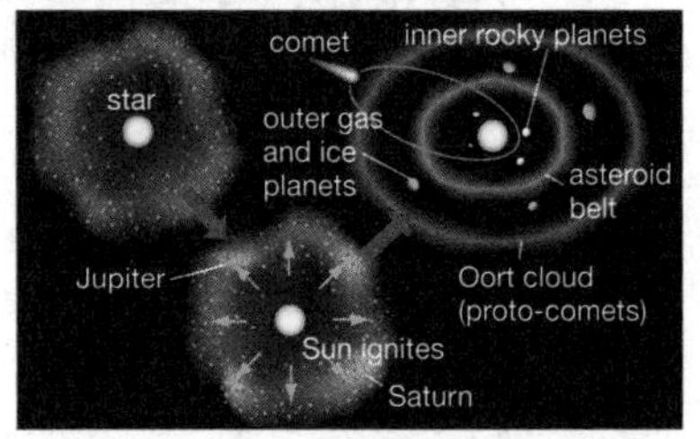

How the Solar System was formed.

B–A*

- As well as emitting electromagnetic radiation, every second the Sun emits 1 million tonnes of electrically charged **particles** in all directions. This **solar wind** is not steady: it has massive bursts called **magnetic storms**.
- The solar wind takes two days to reach Earth. The particles not deflected by the Earth's magnetic field interact with particles in the atmosphere:
  - the **aurora borealis** or northern lights are a spectacular display of glowing atmospheric particles
  - magnetic storms disrupt telecommunication systems including satellites.

The aurora borealis.

Life and death of a star

Events in a star's life and death

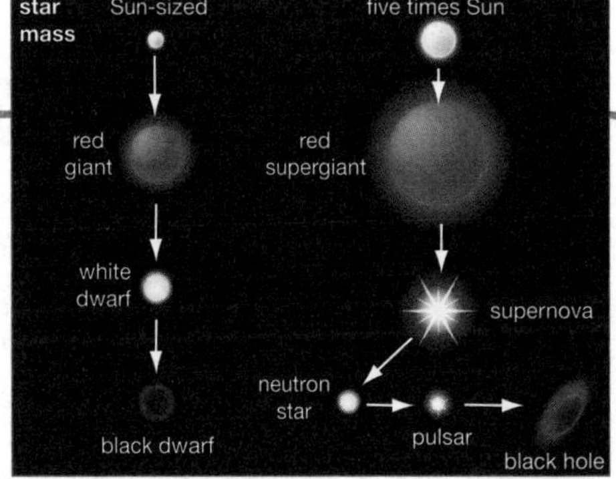

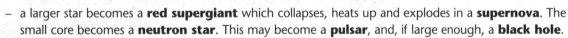

Events in the life and death of Sun-sized and larger stars.

D–C

- Fuel burns in a star and it gradually loses **mass**, as energy (see page 50) and particles in **solar wind**. Its gravity lessens and so it expands. Then, as hydrogen fuel runs out, it cools and collapses under the force of its own gravity.
- A star's **mass** (see diagram) determines the **speed** of collapse and hence the objects that then form:
  - a Sun-sized star eventually becomes a **red giant** which collapses under gravity to form a **white dwarf**, then a cold **black dwarf**
  - a larger star becomes a **red supergiant** which collapses, heats up and explodes in a **supernova**. The small core becomes a **neutron star**. This may become a **pulsar**, and, if large enough, a **black hole**.

B–A*

- In a stable star, two factors are in balance:
  - huge energy and streams of particles from **nuclear fusion** reactions in the core produce an *outward* force called **radiation pressure**
  - the mass of gas exerts an *inward* **gravitational force**.
- These stay balanced until hydrogen in the core runs out. Then, hydrogen fusion continues in the outer parts of the star and the star expands. It continues to lose mass. Then the cooling core contracts, and the star collapses.

Balanced forces in a stable star.

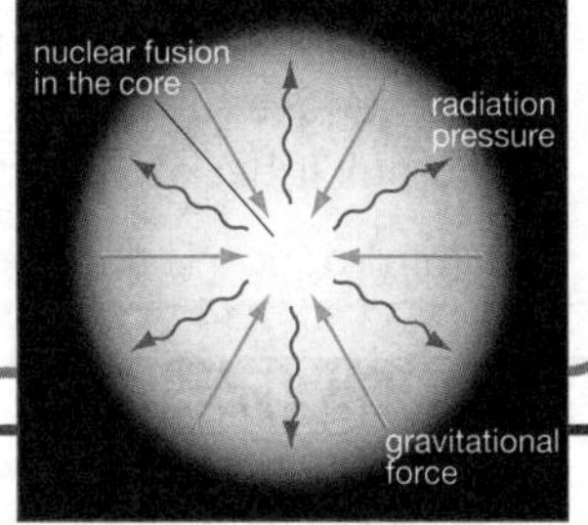

Questions

Grades D-C

1 Explain why the gas giants are the planets furthest from the Sun.

Grades B-A*

2 What causes the aurora borealis?

Grades D-C

3 Why do stars like the Sun slowly lose mass?

Grades B-A*

4 Explain why a star eventually collapses when its hydrogen fuel runs out.

In the beginning

The Big Bang: when energy became matter

- The **Big Bang** theory states that: about 14 billion years ago there was a huge amount of energy at a single point; the energy expanded violently in the Big Bang; within seconds, energy began changing into **protons**, **neutrons** and **electrons**. The diagram includes later events.

The different stages of the Big Bang.

a fraction of a second after
three minutes after
300 000 years after
3.8 billion years after
quarks, electrons
protons, neutrons, nuclei
hydrogen

- The first element to form was hydrogen, the main element in the first gas clouds. These were the birthplace of stars in which nuclear fusion produced other elements.
- Because of the Big Bang explosion, everything in the Universe continues to expand outwards.

D–C

Looking into the origins of the Universe

- Light from the Sun takes 8 minutes to reach Earth. Light from the furthest galaxy yet observed in the Universe took over 13 billion years. So we are seeing it as it was soon after the Big Bang.
- Wherever astronomers look in the Universe they detect **background radiation**, microwaves which are the remnant of the heat (thermal radiation) generated at the Big Bang.

B–A*

The expanding Universe

Red shift and the expanding Universe

The spectral pattern for hydrogen.

- In the visible spectrum, wavelengths at the red end are longest. A **spectroscope** is an instrument that splits light from any source into its wavelengths, forming a **spectral pattern** of bands. When heated, every element emits light with a unique spectral pattern (see the pattern for hydrogen).
- If a light source is moving *away* from an observer, the wavelengths *appear* longer, shifting towards the red end of the spectrum. The difference between actual and apparent wavelengths is the **red shift**.
- Using the red shift values for hydrogen from different galaxies, Edwin Hubble calculated that they are moving away from each other. The further away, the faster they are receding.
- George Gamow worked back to a single position as the place where the galaxies started from. This position and red shifts support the Big Bang theory.

D–C

The redder the galaxy appears, the faster it is receding, and therefore the greater its distance from Earth.

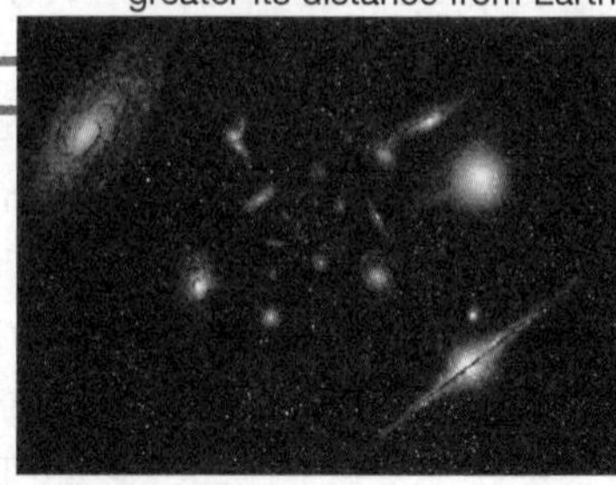

The possible fates of the Universe

- Some astronomers think that the Universe will expand forever.
- Others think that, at some point, the **gravity** of all its matter will (like bungee rubber) be strong enough to slow everything down and pull it back to collide in a **Big Crunch**.
- A third option is that the Universe will slow down and then stop in a steady state.

B–A*

Questions

Grades D-C

1 Which was the first element that was formed by the Big Bang?

Grades B-A*

2 Explain why we see the furthest galaxies as they were 13 billion years ago.

Grades D-C

3 What is 'red shift' and why do we see a red shift in the light that reaches us from distant galaxies?

Grades B-A*

4 Explain how the red shift provides evidence for the Big Bang.

P1b summary

Electromagnetic radiation

Electromagnetic radiation is **energy** carried by **waves** that:

- travel in **straight lines**
- can travel through a **vacuum**
- have a range of energies
- travel at the **speed of light** (300 000 000 m/s).

Electromagnetic radiation can be **reflected**, **absorbed** or **transmitted** by an object depending on its substance and surface.

Wavelength is measured from one **crest** to another. **Frequency** is the number of waves passing a point in a second, measured in **hertz** (**Hz**).

Wave speed is given by:

wave speed = frequency × wavelength

A **fibre optic cable** contains **optical fibres**, which carry **digital** signals in the form of pulsed **laser** light. **Total internal reflection** enables the light to remain inside the fibre.

The **electromagnetic spectrum** is divided into regions according to the energies and uses of each type of wave.

From low frequency and energy (longest wavelength) to high frequency and energy (shortest wavelength) the types are: radio waves, microwaves, infrared, visible light, ultraviolet, X-rays, gamma rays.

Radiation that **transfers energy** to living **cells** can **damage** them. Effects depend on the type of radiation and size of dose. **Ionising** radiation (ultraviolet, X-rays and gamma rays) can ionise atoms, notably in **DNA**, and can cause **cancer**. Long or intense radiation exposure causes **radiation sickness** and death.

Electromagnetic radiation has many uses in **telecommunication** systems including: radio, TV, satellites, cable and mobile phone networks. **Geostationary satellites** are used to receive and transmit signals.

Radioactivity

When the **nucleus** in an atom of a **radioisotope decays**, it emits one type of **nuclear radiation**: either an **alpha particle** or a **beta particle** or **gamma rays** (electromagnetic radiation).

Each type has its own uses and hazards depending on its: range; ability to penetrate material; ionising effect; and reaction to electric and magnetic fields.

Background radiation is all around us. **Granite** rocks can emit **gamma rays** and form radioactive **radon** gas.

The **activity** (**count**) **rate** of a **radioisotope** is measured as its **half-life**, which is the average time it takes for half of its atoms to **decay**.

The Universe

Reflecting optical telescopes built on high ground in dry places and **space telescopes** (e.g. the **Hubble** telescope) form the best images of objects in space as they reduce or avoid **atmospheric distortion** of the image. Telescopes mainly on orbiting satellites can form images from radio waves, infrared, ultraviolet, X-rays and gamma rays of objects in space.

Gravity and **movement** keep all planets and other bodies in the **Solar System** in orbit. At one point on a journey in a spacecraft travelling from the Earth to the Moon, the **gravitational forces** of the Earth and Moon balance – there is a **net zero force** – and the spacecraft and its crew seem **weightless**.

A **star** develops when the particles in a **nebula** are pulled together by **gravity**, heat up and form a **protostar**. This collapses and **nuclear fusion** starts in the star:

$$\textbf{hydrogen} \xrightarrow{\textbf{nuclear fusion}} \textbf{helium + ENERGY}$$

A star's **mass** determines the **speed** of its **collapse** and the objects that may form.

A **sun-sized star** becomes: red giant → white dwarf → black dwarf.

A **larger star** becomes: red supergiant → supernova → neutron star → pulsar → black hole.

A **supernova** explodes with such energy that the elements heavier than iron are formed by the nuclear fusion of lighter elements.

Current evidence suggests that: the **Universe** is **expanding**; a huge amount of **energy** at a single point expanded violently in the **Big Bang**; energy became **matter**; **hydrogen** was formed into gas clouds; **stars** were born; and **nuclear fusion** produced the other **elements**.

The Universe may: expand forever; undergo a **Big Crunch**; or slow down and then stop in a steady state.

Red shift indicates that: **galaxies** are moving apart; the **further** away a galaxy is, the **faster** it recedes.

Cells

Animal and plant cells and their organelles

D–C

- The upper diagram shows a typical **animal cell**, its **organelles** and their **functions**. A typical **plant cell** (lower diagram) also has a cell wall and may contain chloroplasts and a permanent vacuole.

cell membrane, which controls the passage of substances in and out of the cell

cytoplasm, where metabolic reactions controlled by enzymes take place

mitochondrion, where energy is released in respiration

ribosomes, where protein synthesis takes place

nucleus, which controls the activities of the cell

An animal cell and its organelles.

- An **electron microscope** shows far more detail in cells than a light microscope can. A beam of electrons passes through the specimen to form an image on a photographic film.
- The structural detail shown in electron micrographs helps scientists to understand how cells function.

B–A*

cell wall, which strengthens the cell

ribosome

mitochondrion

cytoplasm, where metabolic reactions take place

cell membrane, which controls the passage of substances in and out of the cell

chloroplast, which absorbs light energy to make food

permanent vacuole, filled with cell sap

nucleus, which controls the activities of the cell

A plant cell and its organelles.

Be very careful to use the terms cell *wall* and cell *membrane* correctly.

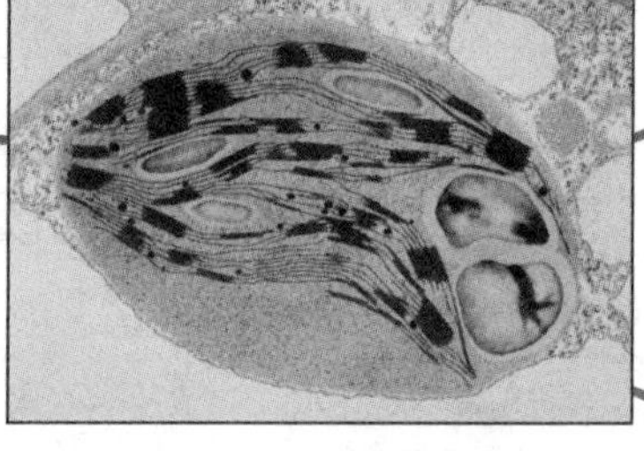

An electron microscope image of a chloroplast: stacks of tiny membranes containing chlorophyll appear black.

Specialised cells

Different cells perform different functions

D–C

- A single-celled organism carries out all its functions in the one **cell**.
- In a multicelled organism, **specialised** tissues and organs carry out different functions. The diagrams show some specialised cells in humans.

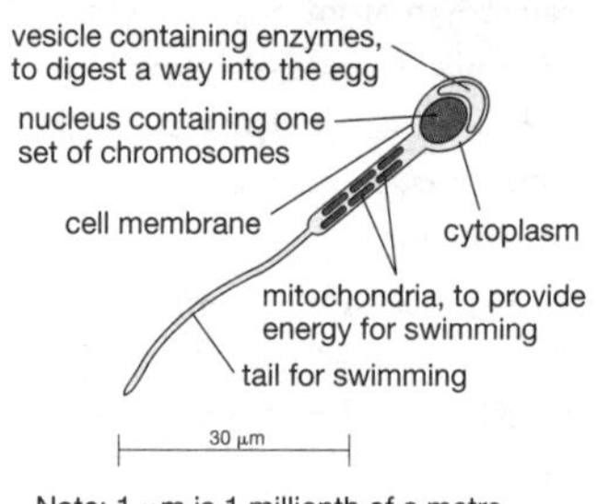

Note: 1 µm is 1 millionth of a metre

A sperm cell.

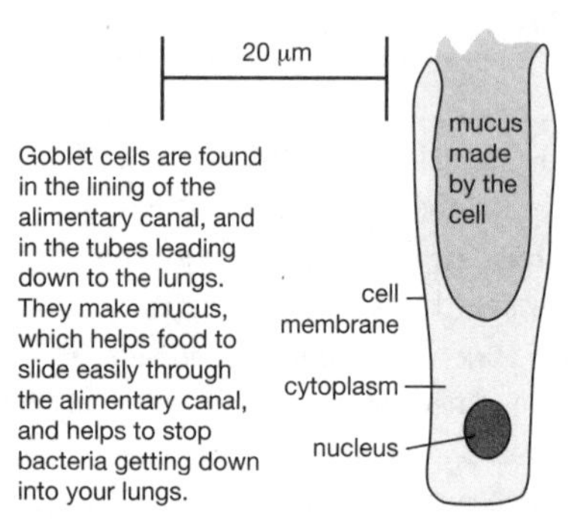

Goblet cells are found in the lining of the alimentary canal, and in the tubes leading down to the lungs. They make mucus, which helps food to slide easily through the alimentary canal, and helps to stop bacteria getting down into your lungs.

A goblet cell.

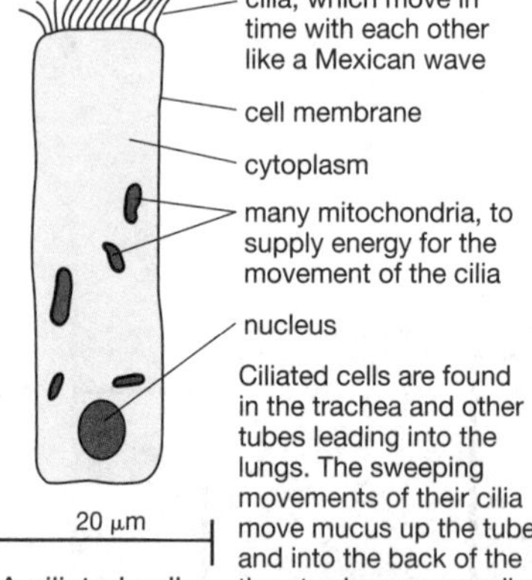

Ciliated cells are found in the trachea and other tubes leading into the lungs. The sweeping movements of their cilia move mucus up the tubes and into the back of the throat, where you swallow it.

A ciliated cell.

How Science Works

You should be able to: relate the structure of different types of cells to their function in a tissue or an organ.

B–A*

- In plant roots, cells are similar in the region behind the root tip where they are rapidly dividing, and in the region of growing cells. Older cells further up are starting to specialise into phloem and xylem tissue and root hairs.

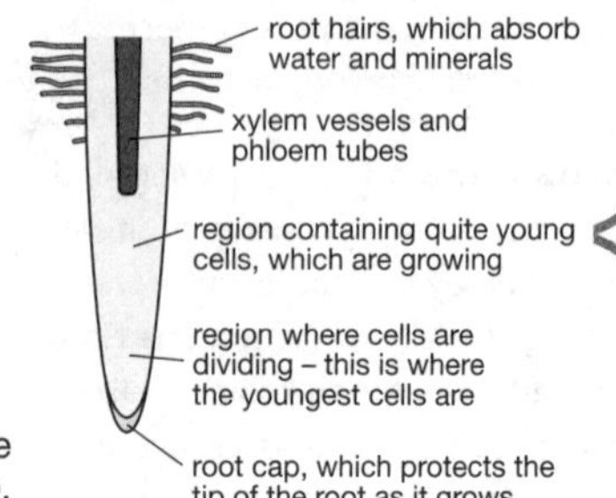

A section of a root. Cells start to specialise above the regions of division and growth in a root tip.

Questions

Grades D-C

1 Name the organelle where protein synthesis takes place.

Grades B-A*

2 Describe what can be seen inside a chloroplast, using an electron microscope.

Grades D-C

3 Where are goblet cells found and what do they do?

Grades B-A*

4 Name **three** types of specialised cells in plant roots.

Diffusion 1

What is diffusion?

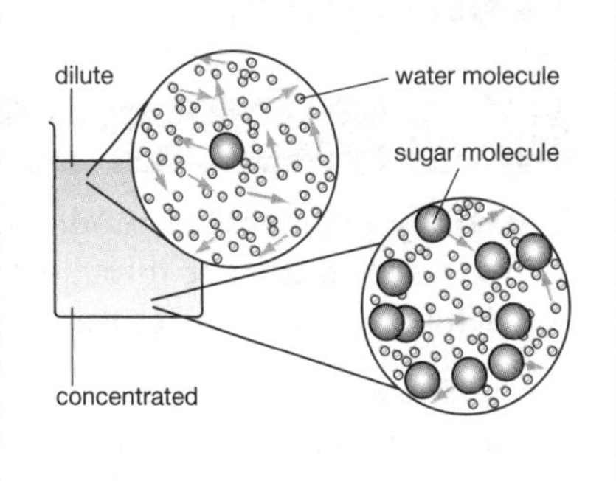

Diffusion in a beaker of sugar solution. Sugar is most concentrated at the bottom.

- **Diffusion** is the **net movement** of particles of a gas, or of a solute in solution, from a region where the particles are at a higher **concentration** to a region of lower concentration. This results from the random movement of particles.

D–C
- In the beaker, there are more sugar molecules to move by diffusion up towards the less concentrated region than in the other direction. Similarly, there are more water molecules at the top to move downwards. Eventually, sugar and water molecules will be evenly spread throughout the solution.
- Particles move faster at higher **temperatures**. Therefore, a temperature rise speeds up diffusion.

Top Tip!

Particles do not go in a particular direction on purpose. They just move around randomly.

Illustrating diffusion in gases

B–A*
- As soon as the apparatus is set up, molecules of ammonia gas start to diffuse from one end of the tube, and molecules of hydrogen chloride gas diffuse from the other end.
- Eventually, the molecules meet along the tube, they react and form a white cloud of ammonium chloride molecules.

Diffusion of two gases.

Diffusion 2

Diffusion of oxygen into cells

D–C
- Cells continually use up **oxygen** in **respiration** according to the equation:
  glucose + oxygen → carbon dioxide + water (+ energy)
- As oxygen concentration in the cell drops, oxygen diffuses in through the cell membrane from the blood in which oxygen is at a higher concentration.
- At a higher respiration rate, the difference in oxygen concentration inside and outside the cell increases, so the rate of diffusion of oxygen into the cell increases.

Villi in the intestine and glucose absorption

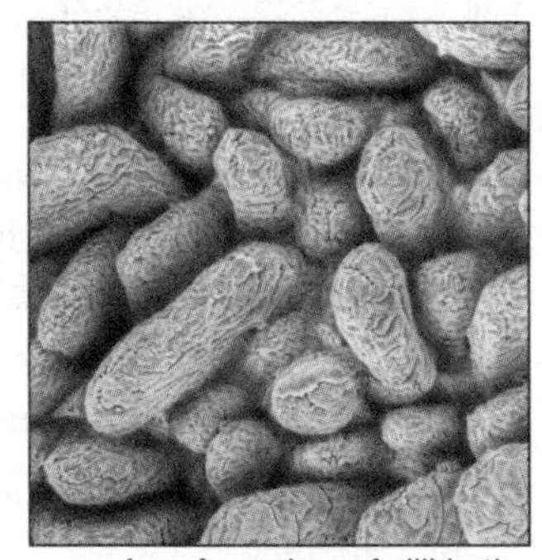
A surface view of villi in the small intestine.

B–A*
- The **small intestine** is lined by millions of 1 mm high finger-like projections called **villi**. These greatly increase the **surface area** in contact with the food. Each villus contains a network of blood capillaries.
- Digested food in the intestine contains **glucose**, needed for respiration. Glucose passes easily into the capillaries of the villi and dissolves in the blood which carries it to respiring cells all over the body.
- The greater the surface area, the faster the rate of diffusion.

Questions

Grades D-C

1 How does a decrease in temperature affect the rate of diffusion?

Grades B-A*

2 Explain how hydrogen chloride and ammonia gas spread along the tube in the diffusion experiment.

Grades D-C

3 How does difference in concentration affect the rate of diffusion?

Grades B-A*

4 Explain how villi speed up the rate of diffusion across the wall of the small intestine.

Osmosis 1

Demonstrating osmosis

D–C

- A **partially permeable membrane** allows small molecules such as water to **diffuse** through it, but blocks large molecules. The diffusion of small molecules only is called **osmosis**.
- In osmosis, water diffuses from a dilute to a more concentrated solution, through a partially permeable membrane that allows the passage of water molecules only.

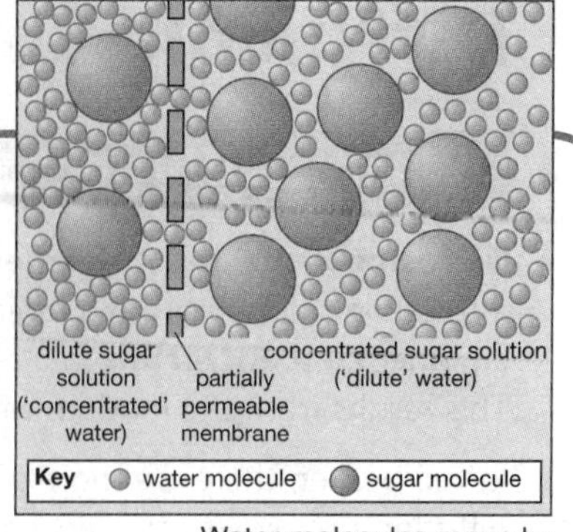

Water molecules move by osmosis through a partially permeable membrane.

Top Tip!

Osmosis is just a special kind of diffusion – the diffusion of water molecules (or small molecules) through a partially permeable membrane.

Which particles diffuse?

B–A*

- In the first diagram, the concentrations of the iodine solution outside the membrane, and the starch solution inside the partially permeable membrane, are the same. Water therefore has the *same* concentration inside and out. Iodine turns blue in the presence of starch.
- The second diagram shows the set-up 10 minutes later.

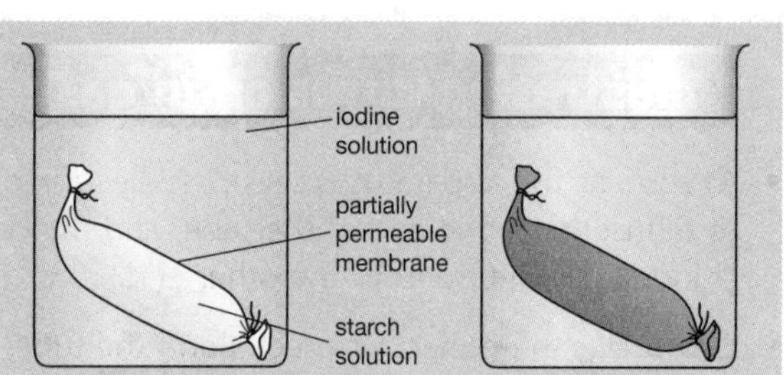

Left: Set-up to test which particles diffuse by osmosis. Right: After 10 minutes.

Osmosis 2

Osmosis in animal and plant cells

D–C

concentrated solution

cell membrane is pulled away from the cell wall

strong cell wall stays the same

the cell shrinks, and pulls away from the cell wall

The cell membrane of a plant cell is partially permeable, while the cell wall is fully permeable.

- The **cytoplasm** of an **animal** cell is a fairly concentrated solution. The **cell membrane** is **partially permeable**. It allows for osmosis: water can diffuse through it in either direction, but most other substances cannot.
- An animal cell in distilled water will take in water until it bursts. If placed in a highly concentrated solution, the cell loses water and collapses.
- A plant cell also has a strong, fully permeable, **cell wall**. It prevents the plant cell from bursting when the cell has taken up a lot of water. If placed in a highly concentrated solution, the cell loses water by osmosis and the **cell membrane** may pull away from the cell wall.

The effect of osmosis on potato cells

B–A*

- Three potato chips are cut and placed in different solutions for 30 minutes.
- The table shows how the length of each chip changes.

solution	change in length (mm)
water	+6
dilute sugar solution	+1
concentrated sugar solution	–4

Questions

Grades D-C

1 Which contains more water molecules – a dilute solution or a concentrated solution?

Grades B-A*

2 Explain the results of the starch and iodine experiment in terms of the diffusion of: **a** starch molecules; **b** iodine molecules; **c** water molecules.

Grades D-C

3 Name the structure that stops a plant cell bursting when it absorbs a lot of water.

Grades B-A*

4 Using the terms 'osmosis' and 'partially permeable membrane', explain why the potato chip in concentrated sugar solution got shorter.

Photosynthesis

Using energy and making food

D–C

- In **photosynthesis**, the green plant pigment, **chlorophyll**, traps energy from sunlight. The energy is used to convert **carbon dioxide** and water into **glucose** and the by-product **oxygen**:

carbon dioxide + water (+ light energy) → glucose + oxygen

- Plants can convert and store excess glucose in the form of **starch** which is insoluble, and in the form of fats and proteins.
- The chemical energy stored in glucose molecules is unlocked when organisms use glucose in **respiration**:

glucose + oxygen → carbon dioxide + water + energy

Top Tip!

Plants respire *and* photosynthesise during the day, and just respire at night.

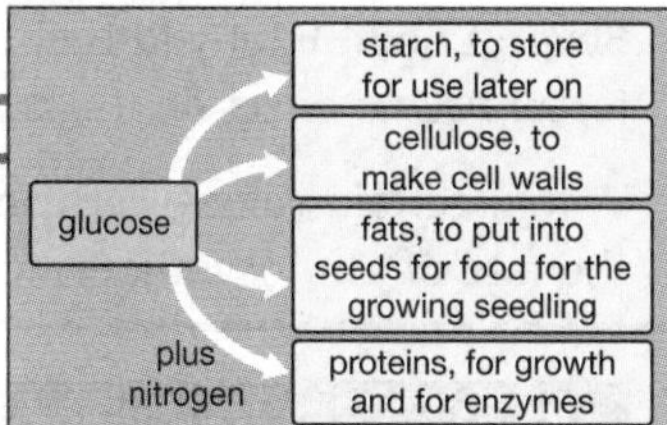

Substances that plants make from glucose.

Substances made from glucose

B–A*

- **Starch** is a good **storage** product for cells. Being insoluble, it is isolated from the cell's metabolic processes and, unlike a soluble substance, it has no effect on osmosis in the cell.
- Many seeds contain starch as food for the growing seedling before it can make its own food in photosynthesis. Seeds often also contain fats and oils as energy sources, and proteins for growth.

Leaves

Leaves: adaptations for photosynthesis

D–C

- A **leaf** is **adapted** for **photosynthesis**. Loosely packed cells of the spongy mesophyll inside allow gases to diffuse.
- In photosynthesis, the concentration of **carbon dioxide** drops inside the leaf, and so carbon dioxide **diffuses** in through **stomata**. The concentration of oxygen rises in the leaf, so this gas diffuses out into the air.

How a leaf is adapted for photosynthesis.

there are tiny pores (stomata) on the underside of the leaf which allow carbon dioxide to diffuse into the leaf

the broad, flat surface of the leaf gives it a very large surface area, so lots of sunlight and carbon dioxide reach it

the leaf is very thin, so sunlight and carbon dioxide can easily reach the cells where photosynthesis happens

the veins in the leaf contain tubes which run all the way up from the roots, carrying water and minerals

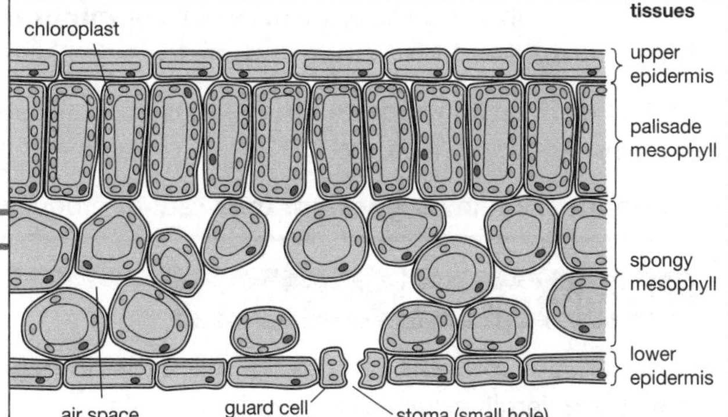

The tissues of a leaf in section.

A highly adapted leaf

B–A*

- Some of the leaves of the carnivorous **pitcher plant** are not the usual thin, flat shape but are funnel shaped. As well as photosynthesising, they are adapted to trap insects.
- Sugary nectar produced at the base of a deep funnel with slippery sides, lures insects in. Those that fall in cannot crawl out. Instead, they are digested by enzymes and provide the plant with **nutrients**.

Questions

Grades D-C

1 Name the process in which cells get energy from glucose.

Grades B-A*

2 Explain why starch makes a good storage substance in plant leaves.

Grades D-C

3 How does the position of palisade cells in a leaf help them to photosynthesise?

Grades B-A*

4 Plants make glucose from photosynthesis in their leaves. What other nutrients do you think pitcher plant leaves provide to the plant?

Limiting factors

Factors that limit photosynthesis

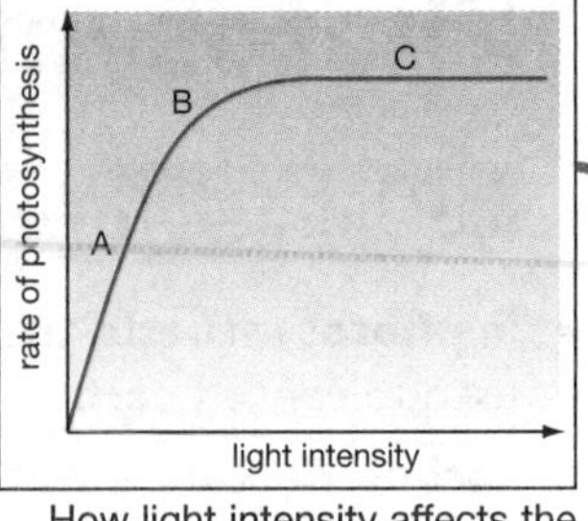

How light intensity affects the rate of photosynthesis.

- The rate of **photosynthesis** increases when there is an increase in: **light intensity**; **carbon dioxide concentration**; or **temperature**.
- Any of these can be a **limiting factor** for the rate of photosynthesis. They can affect photosynthetic rate separately, as the graphs for light and carbon dioxide show, or they can interact.

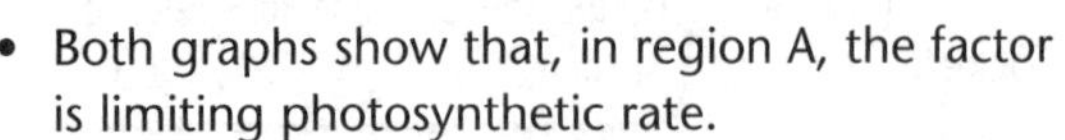

- Both graphs show that, in region A, the factor is limiting photosynthetic rate.
- Beyond point B, any further increase in the factor has no effect on the rate of photosynthesis.
- Within certain values, **temperature** also limits the rate of photosynthesis.

rate of photosynthesis
carbon dioxide concentration (%)
0 5 10 15 20 25
A B C

How carbon dioxide concentration affects the rate of photosynthesis.

D–C

How Science Works

You should be able to:
- interpret data showing how factors affect the rate of photosynthesis
- evaluate the benefits of artificially manipulating the environment in which plants are grown.

Temperature and carbon dioxide in glasshouses

- A tomato grower may consider burning fuel in his glasshouses. This speeds up growth rate in two ways: it raises the temperature and it increases carbon dioxide concentration, which can be a limiting factor in glasshouses.
- His tomatoes would then be ready earlier and attract a higher price.
- The grower's decision depends on whether the increased price is more than the cost of the fuel.

B–A*

Healthy plants

Healthy plants need mineral salts

- To make proteins and chlorophyll, plants use **carbohydrates** (containing C, H and O), and other elements.
- Proteins contain nitrogen, and chlorophyll contains nitrogen and magnesium. Through their roots, plants absorb these elements in ions of **mineral salts** in the soil – **nitrogen** as **nitrate ions** and **magnesium** as **magnesium ions**.
- A plant **deficient** in **nitrate ions** cannot make enough amino acids and therefore not enough proteins. Its growth is stunted.
- A plant deficient in **magnesium ions** cannot produce enough chlorophyll, so its leaves are yellow.

Using fertiliser carefully

- A farmer adds **mineral salts** to the soil, either in **organic fertiliser** such as cattle manure, or as **inorganic** fertiliser such as ammonium nitrate.
- Fertilisers are costly. To avoid over-use, a farmer tests soil samples for deficiency in minerals. The farmer can use a global positioning system to help map the mineral content in different parts of a field.
- The farmer also considers the differing mineral requirement of different crops.

B–A*

Questions

Grades D-C

1 In the light intensity graph, why does the line stay flat in region C?

Grades B-A*

2 Explain the advantages and disadvantages of burning a fuel such as paraffin in a glasshouse used for growing tomatoes.

Grades D-C

3 How do nitrate ions help to make proteins?

Grades B-A*

4 Explain why it is useful for a farmer to test the soil before adding fertiliser to it.

Food chains

A lion with the carcass of an antelope, and vultures in the background.

Energy flow, food chains and food webs

D–C

- A **food chain** represents the **energy** originating from sunlight that passes from one organism to another.
- In the food chain: grass → antelope → lion grass is the **producer**, the antelope is the **primary** (first) **consumer** (and a **herbivore**), and the lion is a **secondary consumer** (and a **carnivore**).
- Food chains can interact to form a **food web**.

- A leaf uses only a little of the solar energy reaching it to store in cell substances. The rest is wasted.

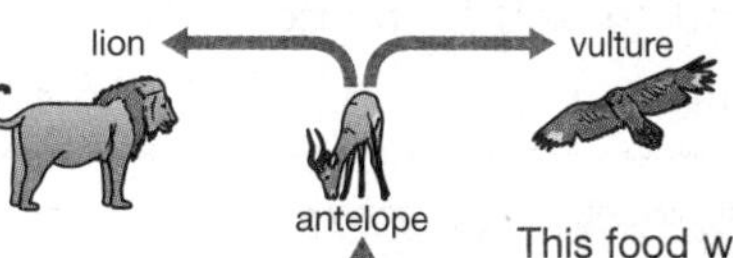

This food web shows that both vultures and lions feed on antelope.

Efficiency of energy transfer

B–A*

- The efficiency of energy transfer between stages in a food chain can be calculated from:

$$\textbf{efficiency} = \frac{\textbf{useful energy transferred}}{\textbf{original amount of energy}} \times \textbf{100\%}$$

For a leaf receiving 200 units of solar energy, and using 40 units in **photosynthesis**:

$$\text{efficiency} = \frac{40}{200} \times 100 = 20\%$$

- A plant may use 20% of the energy that it received from sunlight to make carbohydrates (or other substances). The energy stored in these substances is available to the animal that eats the plant.

Biomass

How Science Works

You should be able to: interpret pyramids of biomass and construct them from appropriate information.

Pyramids of biomass

D–C

- A mass of living material is called **biomass**.
- A **pyramid of biomass** is a scale diagram representing the mass of organisms at each step in a food chain. Biomass is related to the energy transferred, which also diminishes up the pyramid.
- The diagram below shows that energy:
  - is transferred to tissues and organs as organisms grow
  - is lost in respiration that supplies energy for living processes including movement, and for heating the body, especially in mammals and birds
  - is lost as waste matter.

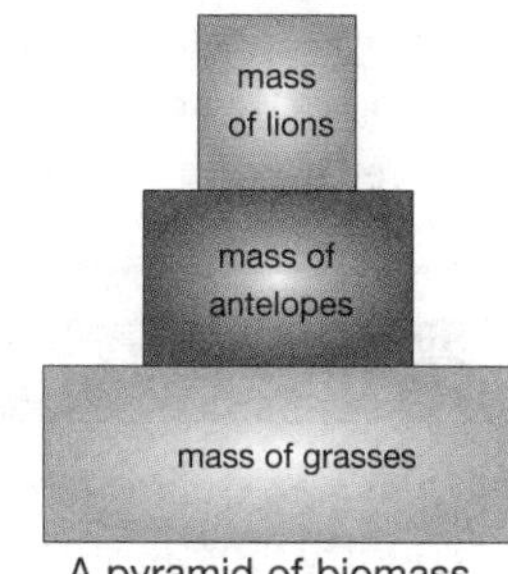

A pyramid of biomass.

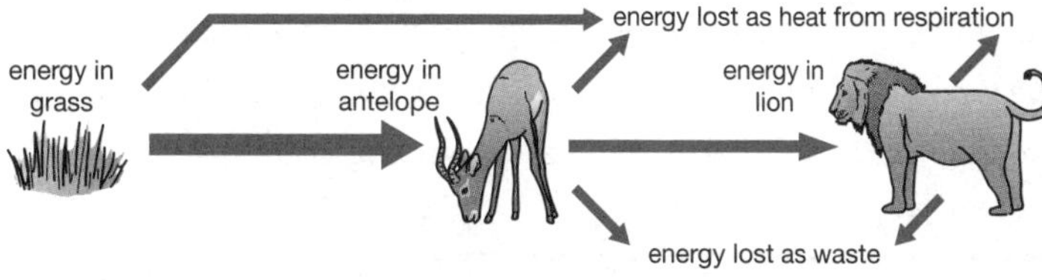

How energy is lost in this food chain.

Questions

Grades D–C

1 Suggest why plants are not able to use all the energy in the sunlight where they grow.

Grades B–A*

2 A leaf receives 500 units of light energy. The leaf uses 90 units of energy to make carbohydrates. Calculate the efficiency of this energy transfer.

Grades D–C

3 Why do mammals need more food than reptiles of the same size?

Grades B–A*

4 Suggest why food chains rarely have more than five links in them.

Food production

Grades

Efficiency in food production

D–C

- **Energy** is lost at each step along a **food chain**. The shorter the food chain, the smaller the energy loss.
- Farmers find ways to reduce energy losses to increase the **efficiency** of **food production**. To reduce energy losses in rearing poultry and livestock:
  - birds and animals are housed in barns to reduce their movement
  - barns are warmed to reduce loss of the heat generated when birds and animals respire.

Food for the world?

B–A*

- In the search to feed the world's population, scientists look to growing **microorganisms** for food. This can be done in huge vats rather than on land.
- Quorn, made from **fungi**, was developed in the UK. The fungal material is grown in 40 m high steel containers called fermenters. It is fed on glucose syrup and other plant products.
- Quorn contains protein and fibre and is low in fat. It is formed into differently flavoured shapes.

Quorn is made from a fungus.

The cost of good food

Food miles and animal welfare

D–C

- Supermarkets save transport costs by moving large quantities of food in container lorries from huge central distribution warehouses.
- There may be a conflict between the need for cheap food and standards of **animal welfare**. Standards vary between countries and are often lower than in the UK.

Battery hens.

Imported food

B–A*

- Over half the food eaten in the UK is grown in another country. Some, such as bananas and citrus fruit, cannot be grown here.
- We buy imported food that we could produce because:
  - we want it at times of the year when it is not available locally
  - we want to support farmers in a developing country
  - it costs less.

How Science Works

You should be able to:

- evaluate the positive and negative effects of managing food production and distribution
- recognise that practical solutions to human needs may require compromise between competing priorities.

Questions

Grades D-C

1 How can keeping animals warm reduce a farmer's costs?

Grades B-A*

2 Explain why growing Quorn for food can be said to be energy efficient.

Grades D-C

3 Suggest how the demand for cheaply produced eggs has resulted in poor welfare standards for hens kept for egg production.

Grades B-A*

4 Give **three** reasons why supermarkets in the UK import food, rather than sourcing it from UK growers.

Death and decay

Decay

D–C

- In the **decay** process, **microorganisms** (e.g. bacteria and fungi) use **enzymes** to break down (**digest**) the materials of dead plants and animals.
- Microorganisms function most rapidly at **warm temperatures**, typically between 25 and 45 °C.
- To grow and cause decay, microorganisms need **moisture**.
- To be active, many microorganisms need **oxygen** for **aerobic respiration**.

freezer (–4 to –10 °C) no growth
fridge (0 to 4 °C) slow growth
fast growth
boiling water (100 °C) no growth – cells killed in 10 minutes (but not spores)
pressure cooker (120 °C) cells and spores killed in 15 minutes

How temperature affects the activity of microorganisms.

Foods that have been treated to slow down decay.

Preventing decay

B–A*

- Microorganisms cannot cause decay in dried or vacuum-packed food.
- Other food processing methods slow down or stop the activity of microorganisms. Methods include boiling, then sealing in tins or jars, and pickling (preserving in vinegar or concentrated sugar solution).

Cycles

Recycling

D–C

- If **decay** did not happen, dead and waste plant and animal materials would build up, and nutrients would be locked up in dead and waste matter.
- Within a **community** of organisms, materials are constantly removed from the environment and returned to it – they are **cycled**. Where removal processes and return processes are balanced, the community is stable.
- **Detritus feeders** (e.g. earthworms) eat dead bodies and waste materials.

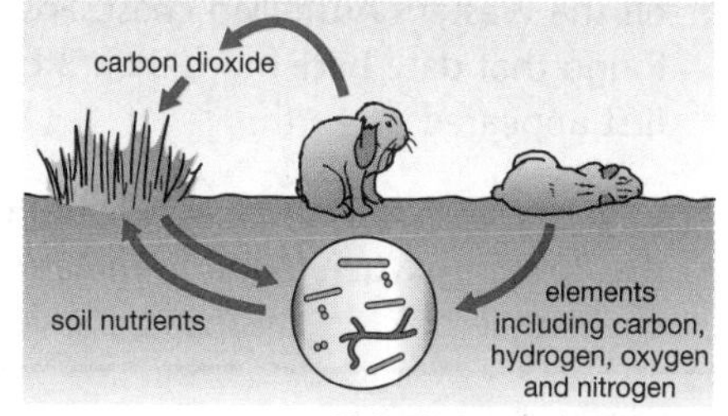

This diagram shows how materials are recycled in an ecosystem.

The fate of a dead whale

B–A*

- A whale that dies at sea will probably sink to the darkness of the ocean floor.
- There, whole communities of crabs, worms and fish will eat the whale's body. Microorganisms gradually decay the whale's tissues. The bones are last to decay.
- The entire process can take decades.

It takes years for a dead whale to decay.

Questions

Grades D-C

1 Why does decay happen more quickly when it is warmer?

Grades B-A*

2 Using what you know about osmosis and decay organisms, suggest why keeping fruit in a concentrated sugar solution could slow down the rate at which it decays.

Grades D-C

3 What kind of organisms are fed on by decay microorganisms?

Grades B-A*

4 Explain why food chains at the bottom of the ocean are usually based on detritus feeders rather than on plants.

The carbon cycle 1

Grades

The importance of carbon

- In **photosynthesis**, green plants use **carbon dioxide** to make glucose:

carbon dioxide + water $\xrightarrow{\text{light energy}}$ glucose (a carbohydrate) + oxygen

- For growth, plant cells make fats, proteins and carbohydrates from glucose. All these contain carbon.
- Plant and animal cells also use glucose in **respiration**:
  - to produce carbon dioxide which enters the air
  - to release energy used for growth and cell functions:

  glucose + oxygen → carbon dioxide + water (+ energy)
- Plants are eaten by animals, and some animals eat each other.

D–C

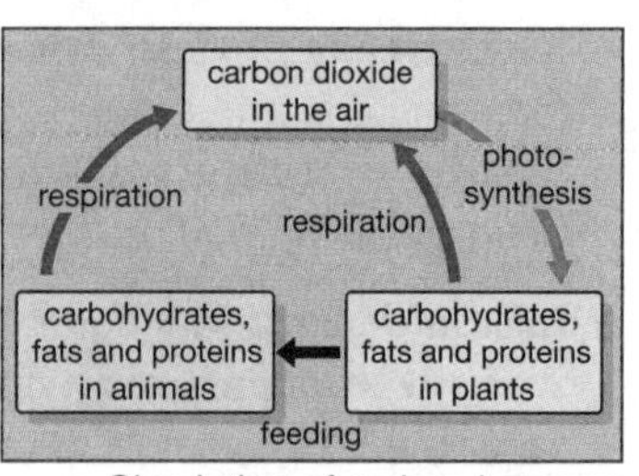

Circulation of carbon between the air, plants and animals.

An oxygen-free Earth

- The atmosphere of the early Earth contained no oxygen. The first organisms to evolve were able to live **anaerobically** – without oxygen.
- Then organisms that could photosynthesise appeared and produced oxygen as a waste product.
- Among the first photosynthesising organisms are thought to be the **cyanobacteria**, which are blue-green algae. Some cyanobacteria make structures called stromatolites which are found in the sea off the Western Australian coast. Fossil stromatolites have been found that date back more than 3 billion years, soon after life first appeared on Earth.
- Organisms such as these are thought to have produced much of the 20% of oxygen found in the atmosphere today. The oxygen enabled aerobic organisms to evolve.

B–A*

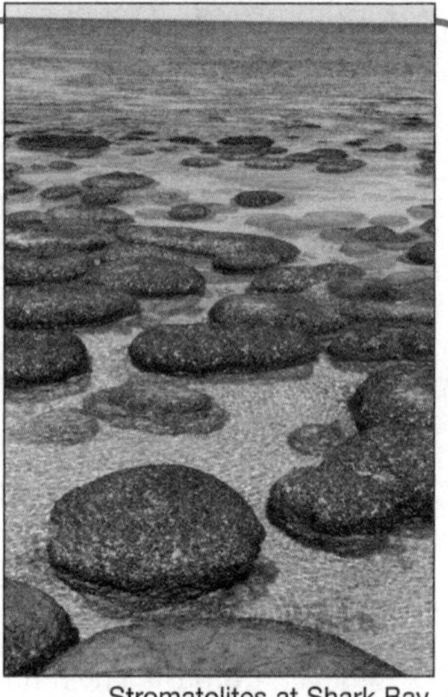

Stromatolites at Shark Bay, Western Australia.

The carbon cycle 2

The carbon cycle

- Microorganisms that are **decomposers**, and the **detritus feeders**, return to the soil the plant nutrients from dead organisms and plant and animal waste.
- When decomposers respire, they release carbon dioxide.
- The diagram shows how animals, plants and decomposers all interact in the carbon cycle.

D–C

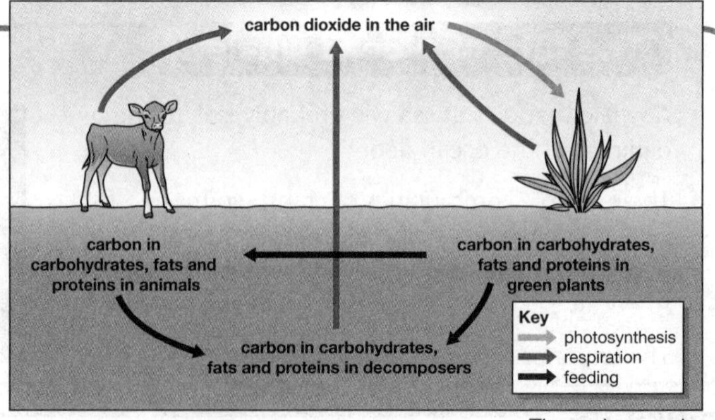

The carbon cycle.

Questions

Grades D-C

1 List **three** types of substances in plants and animals that contain carbon atoms.

Grades B-A*

2 Where did the oxygen in our atmosphere come from?

Grades D-C

3 How do decomposers return carbon dioxide to the air?

Grades B-A*

4 Describe the role of detritus feeders and decomposers in the carbon cycle.

B2a summary

Cells

Animal cells have a cell membrane, cytoplasm and a nucleus.

In addition, **plant cells** may also have a permanent vacuole, chloroplasts and a cell wall.

In multicellular organisms, different cells are **specialised** to perform different functions.

In **plant roots**, rapidly dividing and growing cells in the **root tip** differentiate to form **phloem** and **xylem** tissue and **root hairs**.

Diffusion is the **net movement** of particles of a gas, or a solute in a solution, from a region where particles are at a higher **concentration** to a region of lower concentration resulting from their **random movement**.

Osmosis is the special diffusion of water molecules through a **partially permeable membrane**.

The **difference** in **oxygen concentration** inside and outside **respiring cells** allows oxygen to diffuse into them from the **blood**.

Photosynthesis

Green plants use **chlorophyll** arranged on tiny membranes to maximise **absorption of sunlight** to perform **photosynthesis**:
carbon dioxide + water → glucose + oxygen

Leaves are **adapted** for photosynthesis by being very thin and having stomata, a broad, flat surface and veins.

Glucose can be used in **respiration**:
glucose + oxygen → carbon dioxide + water + ENERGY

The **rate** of photosynthesis is affected by:
- light intensity
- carbon dioxide concentration
- temperature.

These can all be **limiting factors**.

Ions of **mineral salts** in the soil can be used by plants to make proteins and chlorophyll.

Lack of a particular mineral ion results in a **deficiency symptom** in a plant.

Food chains and cycles

Energy passes along **food chains**, but some energy is lost at each transfer.

The shorter the food chain, the smaller the energy loss.

Reducing energy loss increases the **efficiency of food production**.

Food chains interact to form **food webs**.

In **decay**, **microorganisms** use **enzymes** to break down waste materials.

Decay is affected by the levels of:
- temperature
- moisture
- oxygen.

Decomposers and **detritus feeders** feed on the organisms in the food chain, recycling materials, e.g. carbon.

In the **carbon cycle**, carbon is removed from the air as carbon dioxide during photosynthesis and returned to it by respiration.

The **atmosphere** of early Earth had **no oxygen** and the first organisms lived **anaerobically**.

Cyanobacteria were the first organisms to **photosynthesise** and are thought to have produced much of the **oxygen** present in the atmosphere today.

The presence of oxygen enabled **aerobic organisms** to evolve.

Enzymes – biological catalysts

Grades

Enzyme action, heat and pH

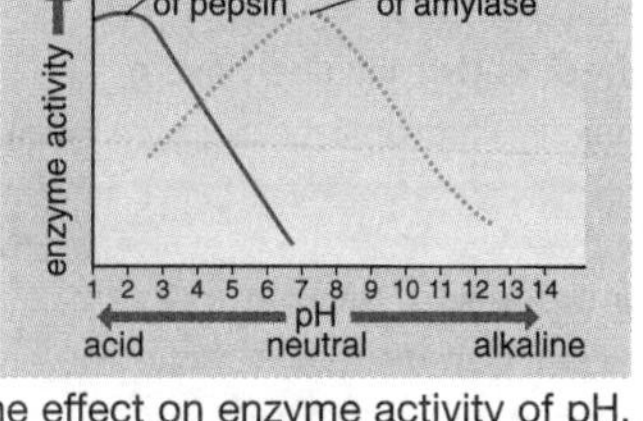

The effect on enzyme activity of pH.

- **Enzymes** are **biological catalysts** that speed up the **rate** of chemical reactions in living organisms without being used up themselves. Each enzyme controls a particular reaction.
- An enzyme is a **protein** molecule made of a long, folded chain of **amino acids**. The **substrate** molecule reacts in the **active site**. **Temperature** and **pH** affect the shape of the active site.
- An enzyme works best at or near its **optimum** temperature and pH. High temperatures or extremes of pH **denature** (destroy) an enzyme. Cold inactivates enzymes.
- Many enzymes catalyse the cell reactions of **respiration**, **protein synthesis** and **photosynthesis**. Several of these enzymes build up larger molecules (e.g. amino acids and proteins) from smaller ones.

D–C

Top Tip!

Do not say that enzymes are 'killed' by high temperatures. They are just chemicals, not living things. They are *denatured*.

- At temperatures below their optimum, enzymes *double* their reaction rate for every 10 °C increase in temperature.
- The enzyme **catalase** catalyses the breakdown of hydrogen peroxide, a poisonous by-product of metabolism in some plant and animal cells:

$$\textbf{hydrogen peroxide} \xrightarrow{\textbf{catalase}} \textbf{water + oxygen}$$

- In an experiment, raw potato is liquidised and an equal amount put into eight test tubes, each kept at a different temperature. The same volume of hydrogen peroxide is added to each test tube, and the oxygen released in 1 min is collected from each. These are the results:

temperature (°C)	0	10	20	30	40	50	60	70
volume of O_2 (cm^3)	2	4	9	17	32	15	1	0

B–A*

Enzymes and digestion

Digestive enzymes

- In the digestive system, glands secrete **digestive enzymes** onto food to split large insoluble molecules into smaller ones that can be absorbed into the bloodstream.
- **Amylase** catalyses the breakdown of starch into sugars in the mouth and small intestine.
- **Proteases** catalyse the breakdown of proteins into amino acids in the stomach and small intestine.
- **Lipases** catalyse the breakdown of fats and oils into fatty acids and glycerol in the small intestine.
- Stomach cells produce **hydrochloric acid** to provide stomach enzymes with their optimum pH.
- **Liver** cells produce **bile**, stored in the **gall bladder**. Released into the small intestine, bile neutralises the acidic food from the stomach and provides an alkaline pH for the enzymes of the small intestine.

D–C

- Food takes about 48 h to pass through the 7-m long human digestive system.
- To reduce their appetite, obese people can have their stomach temporarily made smaller by stapling off a large part of it.

B–A*

Questions

Grades D-C

1 Explain why an enzyme does not work at an extreme pH.

Grades B-A*

2 From the table of results, give an estimate of the optimum temperature for catalase, and explain why you cannot give the exact optimum temperature.

Grades D-C

3 How does bile help in digestion?

Grades B-A*

4 Name the type of nutrient that would be likely to be digested less efficiently in a person with a stapled stomach.

Enzymes at home

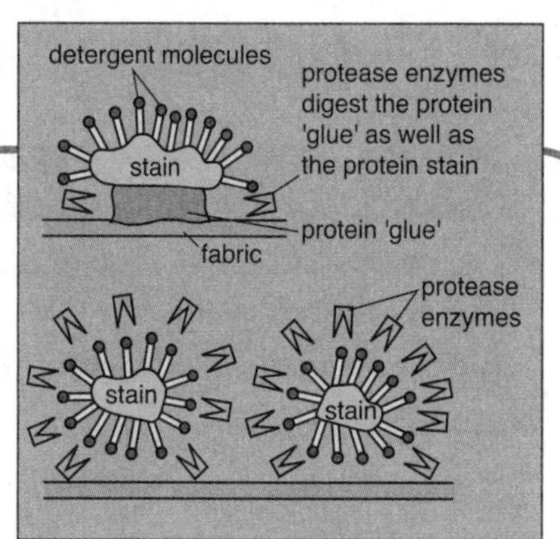

Removing a protein stain.

Biological washing powders

D–C

- Insoluble protein and fat stains are difficult to remove from fabrics. The cells of some microorganisms secrete protein- and fat-digesting enzymes. These enzymes are purified and added to **biological** washing powders that also contain **detergent**.
- The **detergent** molecules make greasy stains **soluble**, lifting them from the fabric, and the **enzymes** break them down:
  - **protease** enzymes digest protein (e.g. in blood and egg)
  - **lipase** enzymes digest fat droplets (e.g. butter and oil).

These enzymes often work best at about 35–40 °C.

Enzyme toothpaste

B–A*

- Saliva contains several enzymes, and is slightly alkaline. Some enzymes start digestion before food is swallowed. Others prevent the build-up of bacteria that produce acid which decays teeth.
- People who suffer from 'xerostomia' (dry mouth) do not produce enough saliva. Their teeth rot and they find food difficult to swallow.
- A manufacturer has produced a toothpaste for xerostomia sufferers. It contains three enzymes that reduce the numbers of harmful bacteria but leave beneficial bacteria.

Enzymes and industry

How Science Works

You should be able to: evaluate the advantages and disadvantages of using enzymes in the home and in industry.

Enzymes used by the food industry

D–C

- Small babies do not digest proteins easily. **Proteases** in some baby foods pre-digest the protein into amino acids. Babies can absorb amino acids into their bloodstream for transport to all their body tissues.
- **Carbohydrases** are used to convert **starch** from potatoes and maize into **sugar syrups** (saturated sugar solutions) that are used in sports drinks.
- To reduce the sugar content of sugars in slimming foods, the enzyme **isomerase** is used to convert **glucose** into the sugar **fructose**, which tastes much sweeter.

Soft-centred chocolates

B–A*

- The fillings of chocolates with soft centres contain sucrose, fruit flavourings, colourings and a little water. They also include the enzyme **sucrase**, added to catalyse the conversion of less soluble sucrose into glucose and fructose which are much more soluble and become runny in water:

$$\textbf{sucrose} \xrightarrow{\text{sucrase}} \textbf{glucose + fructose}$$

- The fillings are enclosed in chocolate and then warmed to 37 °C, the best temperature for sucrase activity but below the melting point of chocolate. As a result, the chocolates have soft, runny centres.

Questions

(Grades D-C)

1 The enzymes in many biological detergents are made by bacteria that normally live in hot springs. Suggest why this may make these detergents easier to use and more effective than other biological detergents.

(Grades B-A*)

2 Suggest why people with xerostomia may suffer from decaying teeth.

(Grades D-C)

3 Why are proteases sometimes used in the production of baby foods?

(Grades B-A*)

4 Explain how sucrase can help chocolate manufacturers to make chocolates with soft, runny centres.

Respiration and energy

Grades

Energy: getting it and using it

D–C

- Energy that living cells require is supplied by **aerobic respiration**. This process takes place in **mitochondria**, and the reactions are summarised in the equation:

glucose + oxygen → carbon dioxide + water (+ energy)

- The energy released is used:
  - to build large molecules from small ones
  - by plants to build sugars, nitrates and other nutrients into amino acids, then proteins (see diagram)
  - by mammals and birds to maintain a steady body temperature in colder surroundings
  - by animals to make muscles **contract**.

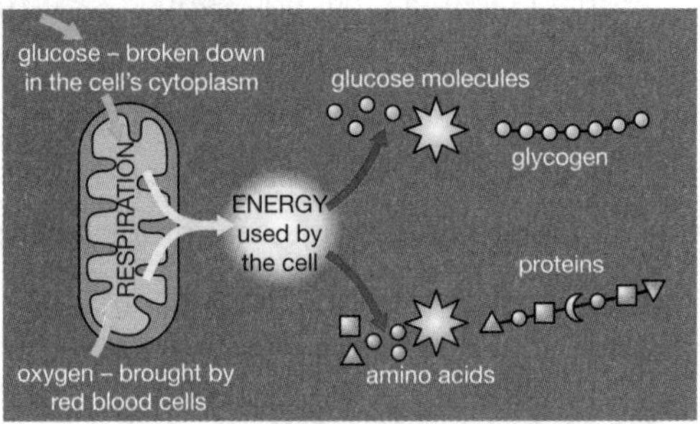

Small molecules join to make big ones.

Most sperms do not have the energy

B–A*

- A sperm at the top of a woman's vagina still has 10–15 cm to swim before it reaches an egg. This distance is about 2000 times its own length.
- Of 200 million sperms released in sexual intercourse, most die, and only a few hundred get near the egg.
- Though their tails are packed with mitochondria, they have no food reserves for respiration, and have to stop and start during their journey.

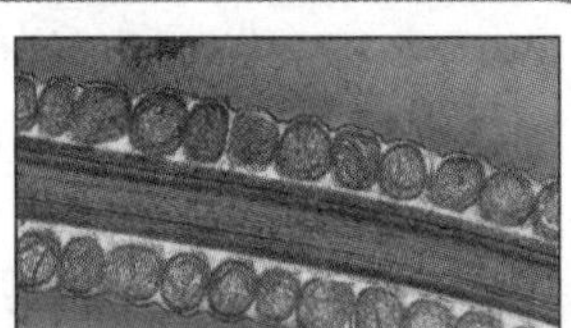

Tightly packed rows of mitochondria are seen in this section through the tail of a sperm.

Top Tip!

Do not confuse respiration and breathing. Respiration happens inside cells. Breathing is movement of air in and out of the lungs.

Removing waste: lungs

The lungs and gas exchange

D–C

- **Carbon dioxide** is a waste product of **respiration** in cells and must be removed. It **diffuses** from cells into the blood, and then from blood into the air sacs (alveoli) of the lungs.

Alveoli have thin walls to allow rapid diffusion of gases.

SCUBA diving

B–A*

- **S**elf-**c**ontained **u**nderwater **b**reathing **a**pparatus (SCUBA) allows swimmers to dive and stay underwater. The gas tank contains compressed air which is 20% oxygen and 79% nitrogen.
- On land, the amount of nitrogen that goes into solution in the blood is negligible. But at the pressures in deep water, nitrogen is absorbed, and stays in solution since the body has no use for it.
- A diver who comes up too rapidly (moving from high pressure to low pressure), does not allow time for the nitrogen to diffuse out of the blood and be expelled in exhaled air. Bubbles of nitrogen form in the blood vessels, causing the 'bends' which can be very painful and life-threatening.

Questions

Grades D–C

1 Why do mammals and birds need more energy than fish?

Grades B–A*

2 Explain why a sperm's tail is packed with mitochondria.

Grades D–C

3 Describe how carbon dioxide produced by a respiring muscle cell is removed from the body.

Grades B–A*

4 Suggest why carbon dioxide diffuses faster from the blood into the alveoli when a person has been exercising hard.

Removing waste: liver and kidneys

Producing urea from amino acids

D–C

- **Protein** in the diet is broken down into **amino acids**. The **liver** converts any surplus amino acids into **urea**, a soluble substance which the blood transports to the **kidneys**.
- The kidneys filter waste substances from the blood, including urea, producing **urine** which flows to the **bladder** for temporary storage before being excreted.

Top Tip!

Remember: *urea* is made in the liver, not in the kidneys. The kidneys make *urine*, which contains urea dissolved in water.

The liver
food in the small intestines
amino acids used to make proteins
proteins digested here
waste amino acids converted into urea
amino acids transported by the blood
urea transported to the kidneys in solution in blood plasma
kidney
blood is filtered here and urea is removed in urine
urine passes down this tube - the ureter - to be stored in the bladder

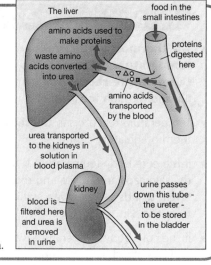

How our bodies remove urea.

Homeostasis

Blood sugar, temperature, ions and water

D–C

- **Homeostasis** means keeping the body's **internal conditions** constant. (See also page 5.)
- **Insulin**, secreted by the pancreas, enables cells to take in **blood sugar** (glucose) for respiration. Excess blood sugar can be fatal, and insulin also causes it to be converted into insoluble glycogen (see page 84).
- Enzymes in cells need a **temperature** of about 37 °C to work optimally. When the body overheats, **sweating** cools it. When it is too cold, blood is re-routed deeper in the body where it loses less heat to the surroundings.
- Cells gain **water** and **mineral ions** from food. Mineral ions draw water into cells (see osmosis, page 72). Excess water could burst cells, and excess ions could be toxic. The diagram shows how the body balances its **water** gains and losses to avoid cell dehydration or damage.

body attempts to return factor to optimum
body functioning at its optimum
body attempts to return factor to optimum
factor rises
factor falls

How homeostasis works.

Daily water in, water out – it must balance.

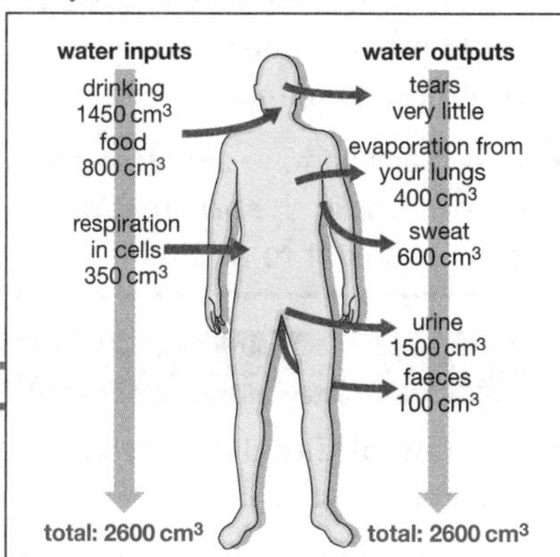

Water balance in freshwater fish

B–A*

- The body fluids of freshwater fish are more concentrated than the surrounding water. They use their gills to obtain mineral salts and oxygen from the water, but more water and minerals than they need also enter through the gills.
- To regulate their internal conditions, freshwater fish:
  - are covered with skin, scales and mucus that are impermeable to water and mineral salts
  - constantly lose minerals (and carbon dioxide, the respiratory waste product) through their gills
  - lose very large quantities of dilute urine produced by their kidneys.

Questions

Grades D-C

1 Explain how the liver and kidneys help to remove excess amino acids from the body.

Grades B-A*

2 Urea is sometimes called a 'nitrogenous waste product'. Suggest why it is given this name.

Grades D-C

3 Why is less urine often produced on a hot day than on a cold day?

Grades B-A*

4 Explain why freshwater fish constantly excrete dilute urine.

Grades

Keeping warm, staying cool

Keeping close to 37 °C

D–C

- The **thermoregulatory centre** in the brain maintains **core body temperature** at between 35.5 and 37.5 °C. The centre contains **temperature receptors** that monitor the temperature of blood flowing through the brain.
- Temperature receptors in the **skin** also send nerve impulses to the thermoregulatory centre.

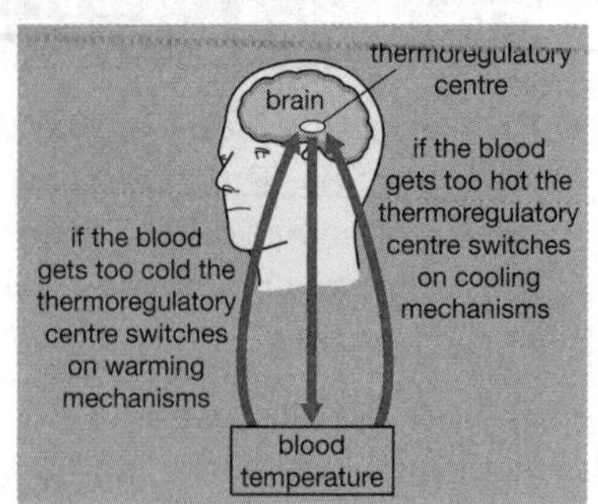

Temperature control is a homeostatic mechanism.

B–A*

- If the core body temperature is too *high*, blood vessels supplying the skin capillaries dilate and more blood flows through capillaries. Sweat glands secrete sweat, which evaporates and takes heat from the skin.
- If the core body temperature is too *low*, blood vessels supplying the skin capillaries constrict and less blood flows through capillaries. Muscles may shiver – their contraction needs respiration, which releases some heat energy.

Treating diabetes

Glucose balance and insulin

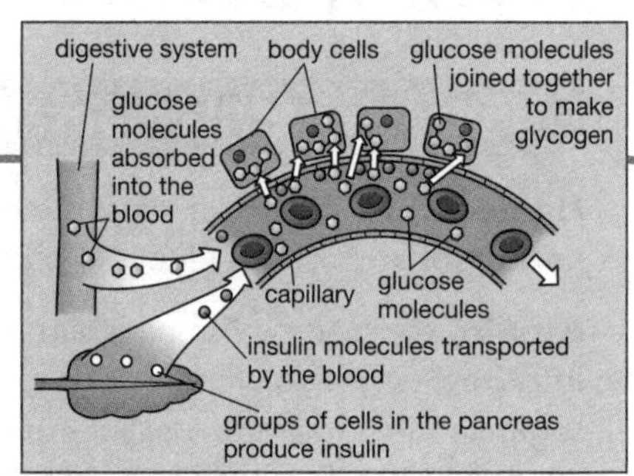

Blood glucose, insulin and cell membranes.

D–C

- The **pancreas** monitors blood **glucose** and controls its level by producing the **hormone insulin**, which reduces blood glucose level.
- Insulin allows glucose to move from the blood into cells (e.g. of muscle and liver), which convert soluble glucose to the insoluble storage material **glycogen**.
- People with some forms of **diabetes** produce too little or no insulin, and so their blood glucose could rise to a fatally high level. They may have to inject insulin regularly and follow a diet to control sugar intake.
- Canadians Banting and Best discovered insulin in 1921. They collected fluid from the islets of Langerhans in the pancreas of a healthy dog. They injected the fluid into a dog with diabetes, and restored it to health.

New diabetes treatments

B–A*

- Scientists at City University in London are developing an 'artificial pancreas' that will:
  - measure blood glucose concentrations
  - automatically inject the right amount of insulin when it is needed.
- Other teams are working on injecting groups of the pancreas cells that produce insulin into the blood of a person with diabetes. In trials, it is found that, instead of the pancreas, the cells end up in the liver, yet they still produce insulin.

How Science Works

You should be able to:

- evaluate the data from the experiments by Banting and Best which led to the discovery of insulin
- evaluate modern methods of treating diabetes.

Questions

Grades D-C

1 Where is the body's thermoregulatory centre, and what does it do?

Grades B-A*

2 Describe **two** ways in which changes in the skin help your body to lose more heat when you are too hot.

Grades D-C

3 How did we discover what insulin does?

Grades B-A*

4 Suggest how an artificial pancreas could keep blood glucose levels of a person with diabetes more constant than if they relied on insulin injections and care over their diet.

Grades

Cell division – mitosis

How new body cells are formed

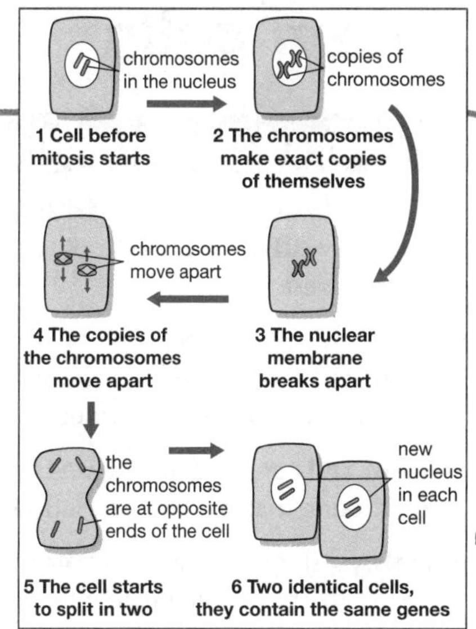

The main stages of mitosis.

D–C

- A human **body cell** has two sets of 23 **chromosomes**, one set from each parent, therefore 23 pairs totalling 46. (Sex cells, gametes, have only one set.) Chromosomes contain **genes** that control inherited characteristics.
- Body cells divide by **mitosis** – for growth or to replace cells.
- Some organisms reproduce **asexually**. The parent produces cells by **mitosis**, and each grows into a new organism, genetically identical to its parent.

B–A*

- **Special** genes ensure that mitosis happens successfully. One type tells a cell when to divide; another type tells it when not to. **Repair** genes mend damaged DNA and **suicide** genes tell a cell to self destruct.
- A **mutation** is a change in a gene's structure, hence in its function. A **mutagen** (e.g. tobacco smoke chemicals, strong UV in sunlight, X-rays) can cause a mutation.
- Some mutagens cause **cancers**, when mitosis is not regulated successfully and a tumour develops.

Gametes and fertilisation

Importance of sexual reproduction and gamete formation

D–C

- Two **gametes** – a **sperm** and an **ovum** – fuse in **fertilisation**, giving a single body cell that divides repeatedly by mitosis to form the new individual. Gametes have only 23 chromosomes, and each of the body cells in the new individual has 46.
- **Alleles** are different varieties of a gene. Different alleles determine different variations of an inherited characteristic.
- An individual has two copies of each gene, one from each parent. Sexual reproduction gives the new individual genes from both parents in a mixture that is different from either parent. This is a cause of **variation** between individuals.

B–A*

- In the **reproductive organs** (**ovaries** and **testes**) cells divide by **meiosis** to form **gametes**. Copies of the chromosomes are made, and the cell divides twice to form four gametes, each with a single set of chromosomes (see diagram).

nucleus with two chromosomes
four chromosomes
1 Cell before meiosis starts
2 Each chromosome makes a copy of itself
chromosomes move apart
4 The pairs of chromosomes move apart
3 The nuclear membrane disappears
four chromosomes move apart
one chromosome in each cell
5 A chromosome from each pair is now separate. The cell starts to split into four
6 Four new cells (gametes) made

The main stages of meiosis.

The differences between mitosis and meiosis.

mitosis	**meiosis**
two cells are made	four cells are made
cells made can go on and divide again	cells made cannot divide
all cells have the same number of chromosomes	cells made have half the number of chromosomes
body cells are made	gametes are made

Questions

Grades D-C

1 Name the kind of cell division that makes new cells for growth, repair and asexual reproduction.

Grades B-A*

2 Explain how a mutagen can cause cancer.

Grades D-C

3 Why are offspring produced by sexual reproduction not exactly like their parents?

Grades B-A*

4 A cell with two sets of chromosomes divides by meiosis. How many cells will be produced, and how many sets of chromosomes will they each have?

Grades

Stem cells

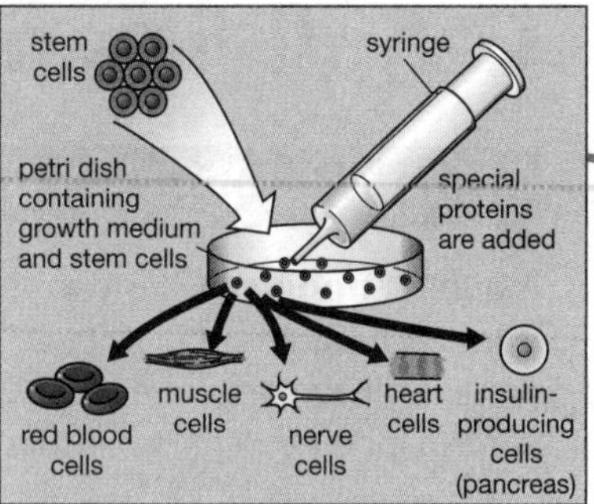

Embryo stem cells can become many types of cells depending on the proteins added.

Unspecialised, specialised and stem cells

- Some cells are **unspecialised** and can develop into any type of cell. They **differentiate** to do just one job – they become **specialised**.
- In mature animals, cell division is mainly of differentiated cells for repair and replacement. In **plants**, many cells can differentiate into any type of cell throughout their life.
- **Stem cells** are undifferentiated cells. They are found in embryos and adult bone marrow and can form many different types of cell.
- **Embryo** stem cells can be treated with particular proteins to turn genes on or off. They differentiate into different tissues which perhaps can be used to repair damaged body parts or restore function (e.g. after paralysis).
- **Bone marrow** stem cells can develop into red or white blood cells, or into cells to repair bone, cartilage, tendons or fat tissue.

D–C

How Science Works

You should be able to: make informed judgements about the social and ethical issues concerning the use of stem cells from embryos in medical research and treatments.

Treating Parkinson's disease with stem cells

B–A*

- Parkinson's disease affects the nervous system of some older people. Their hands shake and they cannot move easily because special neurones fail to produce a vital chemical **dopamine**. In the search for a cure:
  - scientists injected embryonic stem cells into the brains of rats with a form of Parkinson's disease, and their mobility improved
  - embryonic stem cells are being used to make neurones that secrete dopamine.

Chromosomes, genes and DNA

Chromosomes, DNA and genes

- A **chromosome** is a very long molecule of **deoxyribonucleic acid** (**DNA**).
- Chromosomes contain **genes** which together form the **genetic code**, the instructions for joining amino acids to make proteins (e.g. enzymes) that determine our characteristics. Each gene codes for just one protein.

A cell
cell membrane
nucleus
gene
chromosome

Expanding diagram to illustrate DNA.

Genetic information and its ownership

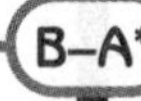

- In the **Human Genome Project**, geneticists worked out the sequence of all the bases in human chromosomes. It is hoped that this information will enable researchers to develop treatments for inherited diseases in particular.
- A question that arises is: Who owns this information about genes? Pharmaceutical companies spending money to develop treatments would like to own the copyright of specific genes. This would prevent other companies from using the same genetic information to develop competitive treatments.

Questions

Grades D-C

1 Name **two** different kinds of cells that stem cells in bone marrow can produce.

Grades B-A*

2 Explain why injecting stem cells into the brain could relieve some of the symptoms of Parkinson's disease.

Grades D-C

3 Why must DNA duplicate itself before mitosis begins?

Grades B-A*

4 How do genes determine the kind of proteins made in a cell?

Inheritance

Top Tip!

Genetic diagrams do not tell you how many offspring two parents will have. They just tell you the chances of getting different combinations of alleles.

Homozygous and heterozygous crosses

D–C

- Gregor Mendel's experiments on pea plants provided evidence for the rules of genetic **inheritance**.
- In the pea plants, a single **gene** controls flower colour. Its two **alleles** (see page 85) give red (R) and white (r) flowers. A plant with the same alleles (RR or rr) is **homozygous** (pure breeding). A plant with different alleles (Rr) is **heterozygous**.
- When a homozygous red plant (RR) is **crossed** with a homozygous white plant (rr), all offspring (Rr) have red flowers. Red (R) is **dominant**; white (r) is **recessive** and shows only when red is absent (rr).
- The grid shows that a cross of heterozygous plants (Rr) gives three red-flowered plants to one white.

A heterozygous cross.

	(R)	(r)
(R)	**RR** red flowers	**Rr** red flowers
(r)	**Rr** red flowers	**rr** white flowers

Genetics of the fruit fly

B–A*

- In the fruit fly, the allele for normal wings is **dominant** to the allele for vestigial (very small) wings.
- A pure breeding (homozygous) normal winged fly is crossed with a pure breeding vestigial winged fly. A cross of the resulting (F1) generation gives an F2 generation of which about one-quarter have vestigial wings.

How Science Works

You should be able to:
- explain why Mendel proposed the idea of separately inherited factors and why the importance of this discovery was not recognised until after his death
- interpret genetic diagrams
- predict and/or explain the outcome of crosses between individuals for each possible combination of dominant and recessive alleles of the same gene
- construct genetic diagrams.

How is sex inherited?

Sex inheritance and sex ratio

D–C

- In human **body cells**, one pair of the 23 chromosome pairs carries the genes which determine **sex**. In females, the sex chromosomes are the same, **XX**. In males they are different, **XY**.
- An egg or sperm contains one sex chromosome only.
- There are about the same number of females as males (see table).

The sex chromosome in sperms determine a child's sex.

sex chromosome		sex chromosomes	
in egg	**in sperm**	**in embryo**	**sex of embryo**
X	X	XX	female
X	Y	XY	male

Temperature-dependent sex determination

B–A*

- The sex of many reptiles depends on the **temperature** at which eggs are incubated, not on sex chromosomes.
- Sea turtles incubated at 30 °C or higher are female, at 28 °C they are male and female, and below 28 °C they are male.
- In breeding programmes for endangered species, scientists apply the effect of temperature on sex to produce appropriate numbers of males and females.

Questions

Grades D-C

1 What does 'pure-breeding' mean?

Grades B-A*

2 In a cross between many fruit flies, all heterozygous for wing length, 2300 flies were produced. Calculate how many of these you would expect to have normal wings.

Grades D-C

3 Why are roughly equal numbers of boy babies and girl babies born?

Grades B-A*

4 What evidence is there that sex in turtles is not determined by their chromosomes?

Inherited disorders

Grades

Inherited disorders

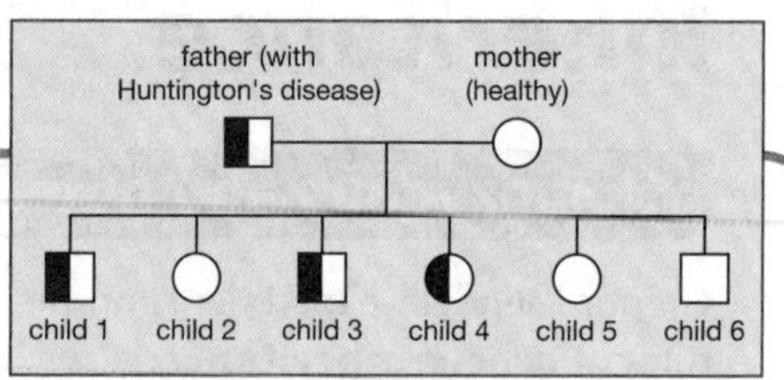

Inheritance of Huntington's disease.

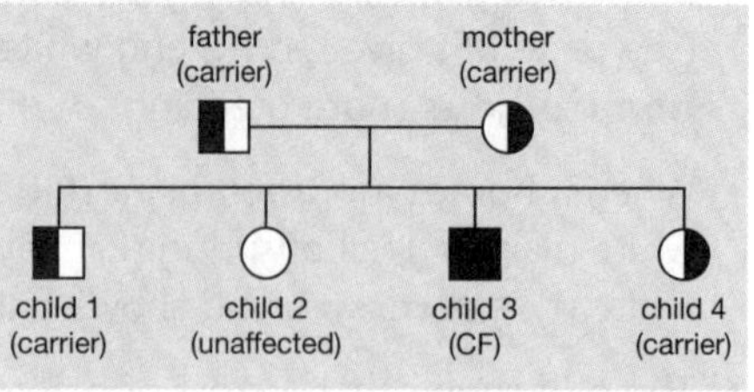

Inheritance of cystic fibrosis.

D–C

- An **inherited disorder** lasts for life, since all body cells carry the faulty genetic material.
- A **dominant allele** causes **Huntington's disease** which affects the nervous system. A child inherits Huntington's disease if it receives this allele from just one parent.
- A **recessive allele** causes **cystic fibrosis** (CF) which affects cell membranes. Two healthy parents may be **carriers**. A child with CF will have inherited the recessive allele from both parents.

Screening for inherited diseases

B–A*

- Embryos can be **screened** for the Huntington's disease allele (on chromosome 4) and the cystic fibrosis allele (on chromosome 7), and parents can receive genetic counselling.
- Someone with Huntington's disease may not show symptoms until middle age. Then, they develop uncontrollable muscle and eye movements, poor memory and slowed thinking, depression, anger and loss of appetite. Death may follow after 15–20 years.
- Someone with CF cannot digest food properly, has frequent chest infections, fertility problems and a shortened life. There is no cure yet, but treatments can minimise the symptoms and tackle infections.

How Science Works

You should be able to: make informed judgements about the economic, social and ethical issues concerning embryo screening that you have studied, or from information you have received.

DNA fingerprinting

What is a DNA fingerprint?

D–C

- Each person's **DNA** differs from everyone else's except an identical twin's. DNA cannot be changed.
- A **DNA** (or **genetic**) **fingerprint** or a **DNA profile** can be used to identify a single individual, alive or dead.
- DNA fingerprinting requires just a tiny sample of DNA (from e.g. saliva, hair, blood, semen). It is cut into different lengths by enzymes. The bits are separated into bands which are treated with radioactive chemicals to make an image of them on X-ray film.

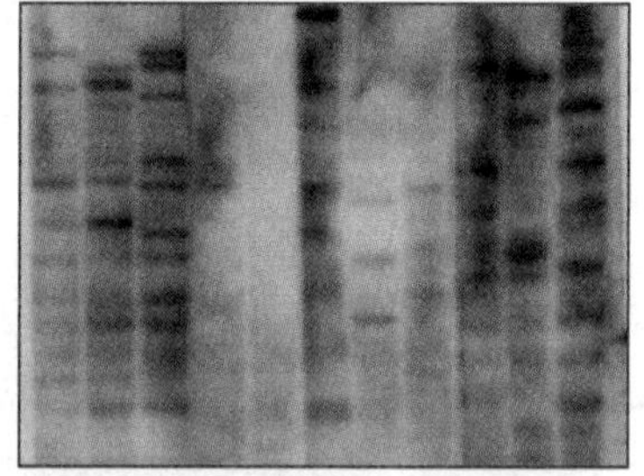
A good DNA fingerprint identifies just one person.

- In **solving crime**, DNA from the crime scene can be matched to DNA from a suspect.
- In **paternity disputes**, the bands shared by a man and a child show whether he is the child's father.

B–A*

- Identity cards are issued to all citizens of some countries, and their use in the UK is being discussed.
- Some propose that, in addition to a person's name and bank details, an ID card in the UK could include blood group, medical records and their DNA profile. There is strong disagreement about some of these suggestions.

Questions

Grades D-C

1 How can two parents without CF have a child with CF?

Grades B-A*

2 What decision would parents have to make if genetic screening shows that their unborn embryo has an allele for Huntington's disease?

Grades D-C

3 Name **two** uses of DNA fingerprinting.

Grades B-A*

4 How could DNA fingerprinting help to eliminate a suspect from a murder enquiry?

B2b summary

Enzymes

Enzymes are **biological catalysts** that speed up chemical reactions and are not used up.
An enzyme is a **protein molecule** made of a long, folded chain of **amino acids**.

Enzymes are involved in **respiration**, **photosynthesis** and **protein synthesis**.

Enzymes are used in industry in **food production** and in the home in **biological detergents**.

The digestive enzymes **amylases**, **lipases** and **proteases** break down food.
Catalase catalyses the breakdown of hydrogen peroxide, a poisonous by-product of metabolism:
hydrogen peroxide → water + oxygen

Each enzyme works at an **optimum pH** and **temperature**.
High temperatures or extremes of pH **denature** enzymes by affecting the **shape** of their **active sites**.

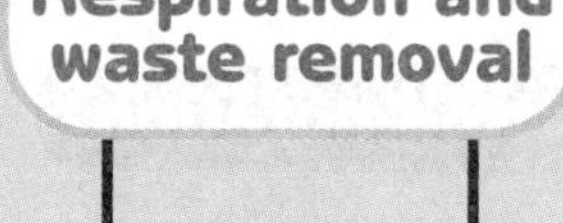

Respiration and waste removal

Aerobic respiration is carried out in **mitochondria** in our cells to provide energy:
glucose + oxygen → carbon dioxide + water + ENERGY
The **tail** of a **sperm** cell is packed with mitochondria to enable it to swim the long distance to the egg.

Breathing allows the **diffusion** of oxygen into the body for respiration and removal of waste carbon dioxide out of the body via the **alveoli** in the **lungs**.
The dissolved gases are carried round the body in the blood.

Released **energy** is used:
- to **build** large molecules from small ones
- by plants to build sugars, nitrates and other nutrients into amino acids, then **proteins**
- by mammals and birds to maintain a **steady body temperature**
- by animals to make **muscles contract**.

Excess amino acids are broken down by the **liver** into **urea**.
The **kidneys** make **urine**, which contains dissolved urea.
The urine is stored in the **bladder**, then excreted.

Homeostasis

Blood sugar levels are controlled by the **pancreas** which produces **insulin**. People with **diabetes** produce too little insulin.

Human core body **temperature** is kept constant at about 37 °C.
The **thermoregulatory** centre in the brain contains **temperature receptors**.

Sweating cools the body down as it evaporates.
Re-routing blood deeper in the body reduces heat loss.

Water inputs from drinking and cell respiration are **balanced** with **water outputs** from tears, evaporation from the lungs, sweat, urine and faeces. Excess water could burst cells, and excess ions could be toxic.

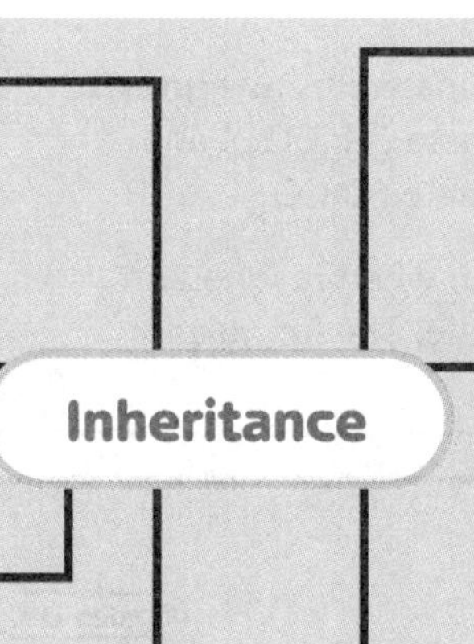

Inheritance

Chromosomes are long molecules of **DNA** made up of **genes**.
Genes contain the **genetic code**, which contains the instructions for making proteins, including **enzymes**.

Body cells divide by **mitosis**. Each new cell has the same number of identical chromosomes as the original cell.

Undifferentiated **stem cells** can **specialise** into many types of cells.

Sex chromosomes determine the sex of the offspring (male XY, female XX).

In **reproductive organs**, the **ovaries** and **testes**, cells divide by **meiosis** to form **gametes** (**sperms** and **ova**) that have half the number of chromosomes of body cells.

Gametes fuse at **fertilisation** to produce a new organism with the same number of chromosomes in their body cells as the parents.
The mixture of **alleles** from the parents causes **variation** in the offspring.

Some inherited **characteristics** are controlled by a single gene.
Different forms of a gene are called **alleles**.
In **homozygous** individuals, the alleles inherited from each parent are the same.
In **heterozygous** individuals, they are different.
Alleles can be **recessive** or **dominant**.

Atomic structure 2

Top Tip!

Remember the relative mass and charge of a proton, a neutron and an electron (see page 26).

The structure of an atom

D–C

- The **nucleus** of an **atom** contains **protons** and **neutrons**. Round the nucleus are **electrons**.
- All atoms of an **element** have the same number of protons and electrons. An atom has no overall electrical charge.
- See page 26 for: **relative masses** and **charges** of sub-atomic particles; elements arranged in the **periodic table** in order of atomic number; **symbol**, **mass number** and **atomic number** of each element.

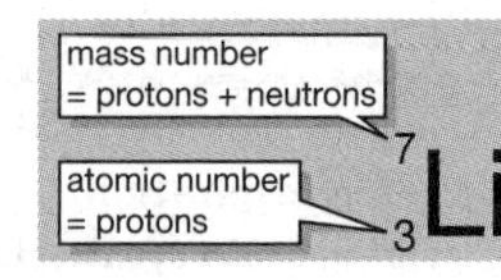

B–A*

- During a **chemical** reaction, the elements remain the same. Only *electrons* are involved. They are transferred or are shared between atoms. A small amount of energy is either taken in or given out.
- In **nuclear** reactions, *nuclei* of atoms either split or join together. In both cases, huge amounts of energy are released:
  - in nuclear **fission**, the nucleus of one atom of a large element splits to give two nuclei of smaller elements
  - in nuclear **fusion**, the nuclei of two small atoms join together to form a new element.

Electronic structure

How Science Works

You should be able to: represent the electronic structure of the first 20 elements of the periodic table in the following forms.

For sodium: and 2, 8, 1

Electrons in shells and electron notation

D–C

- Electrons are arranged in **shells**. The innermost shell holds up to two electrons, the second up to eight. Each shell represents a different **energy level**. The innermost shell has the lowest energy level. Electrons occupy the lowest available energy levels.
- Atoms with a full outer electron shell are stable and unreactive. When an atom reacts, it gains, loses or shares one or more electrons to achieve a full outer electron shell.
- **Groups** in the periodic table contain elements with the same number of outer electrons, so elements in a group undergo similar reactions. In Group 2, elements react by losing **two** outer electrons:
  - a **magnesium** atom has 12 electrons. Its **electron notation** is 2, 8, 2
  - a **calcium** atom has 20 electrons. Its electron notation is 2, 8, 8, 2.

Top Tip!

Electrons fill the electron shells in order, starting with the shell closest to the nucleus.

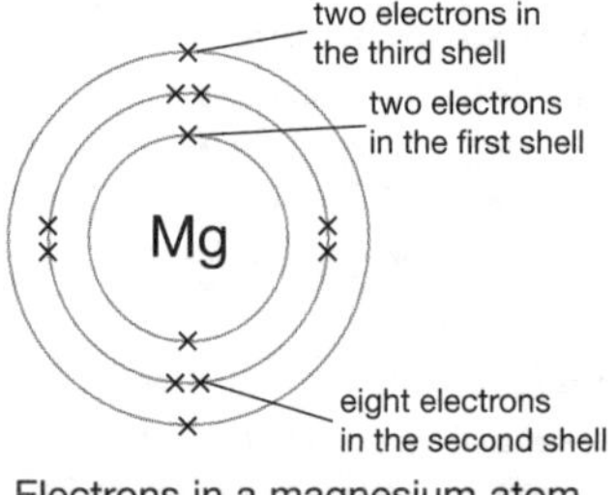

Electrons in a magnesium atom.

B–A*

- Magnesium and calcium both have carbonate ores: **magnesite** is the magnesium ore magnesium carbonate ($MgCO_3$), and **calcite** is the calcium ore calcium carbonate ($CaCO_3$).
- A third ore, **dolomite**, is thought to form when magnesium ions swap with some calcium ions in calcite. The formula for dolomite is $CaMg(CO_3)_2$.

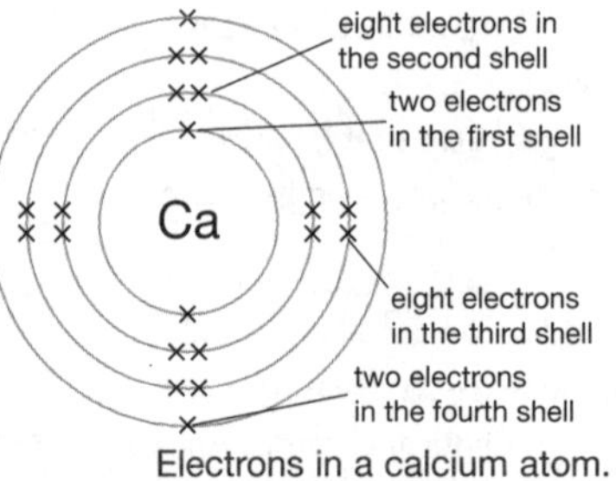

Electrons in a calcium atom.

Questions

Grades D-C

1 Look at the periodic table on page 26. What is the mass number of potassium? What does this tell you about the structure of a potassium atom?

Grades B-A*

2 Explain the difference between a chemical reaction and a nuclear reaction.

Grades D-C

3 How can a potassium atom become stable?

Grades B-A*

4 Suggest why magnesium and calcium can easily interchange in carbonates.

Mass number and isotopes

Elements, groups and relative atomic mass

D–C

- A **group** in the **periodic table** (page 26) contains elements with the same number of electrons in their outermost shell, so elements in a group have similar chemical properties.
- The **mass number** of an element is the sum of protons and neutrons in one of its atoms.
- Atoms of an element may have different numbers of neutrons. Such atoms are **isotopes** of the element (see also page 127), and have different mass numbers.

B–A*

- The **relative atomic mass**, A_r, of an element is its mass compared with the mass of a carbon-12 atom, where A_r for carbon = 12. To find the relative atomic mass of an element with *several isotopes*, the average of the mass numbers is calculated (taking into account the proportion of each isotope).

Ionic bonding

Ion formation and ionic compounds

- To form **ionic compounds**, the atoms of reacting elements either lose electrons to become **positively** charged **ions**, or gain electrons to become **negatively** charged ions. Ions achieve a full outermost electron shell, with the electronic structure of a noble gas (Group 0).
- **Ionic bonds** are strong forces of attraction between oppositely charged ions. Ionic compounds are solids at room temperature.

D–C

- **Metal** atoms *lose* electrons to form positive ions. **Non-metal** atoms *gain* electrons to form negative ions.

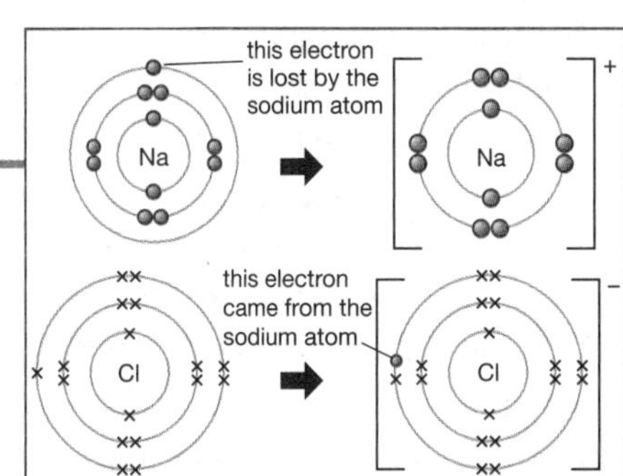

Sodium metal and chlorine gas form ions when they react. Electron arrangements: sodium atom 2, 8, 1 → sodium ion [2, 8]$^+$; chlorine atom 2, 8, 7 → chloride ion [2, 8, 8]$^-$.

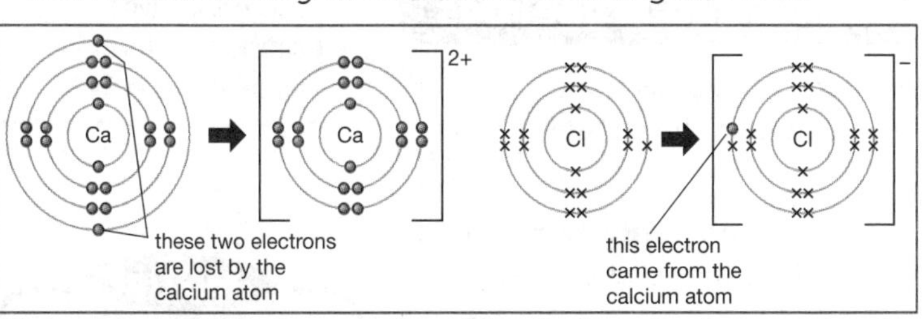

When calcium and chlorine react, the calcium ion has a 2+ charge, and *each of two* chlorine atoms has a 1– charge. $Ca + Cl_2 \rightarrow Ca(Cl)_2$.

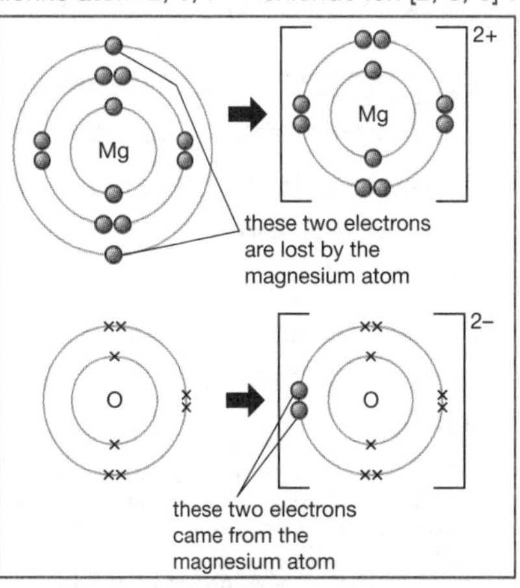

Magnesium metal and oxygen gas form ions when they react. A magnesium ion has a 2+ charge. An oxygen ion has a 2– charge. $2Mg + O_2 \rightarrow 2MgO$

How Science Works

You should be able to: represent the electronic structure of the ions in sodium chloride, magnesium oxide and calcium chloride in the following forms.

For sodium ion (Na^+): 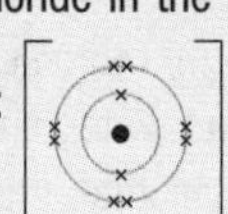 and $[2, 8]^+$

B–A*

- Salt, NaCl, contains oppositely charged Na^+ and Cl^- ions. These ions separate and move freely in an aqueous solution.
- With water only, the bulb does not light. With salt, the bulb lights because electricity flows: charge is carried by the ions moving in the water.

An aqueous solution of ions conducts electricity.

Questions

Grades D-C

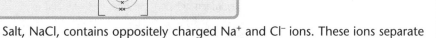

1 What do all the elements in the same group in the periodic table have in common?

Grades B-A*

2 The relative atomic mass of chlorine is 35.5. Why is this not a whole number?

Grades D-C

3 Explain why the *overall* charge of the ions in an ionic compound is zero.

Grades B-A*

4 How does charge move through the salty water in the diagram?

Ionic compounds

Ionic compounds: bonding and properties

D–C

- In an **ionic compound**, **positive** metal ions and **negative** non-metal ions are arranged in a regular **lattice** to form a **giant ionic structure**. Strong electrostatic **forces of attraction** acting in all directions form ionic bonds between oppositely charged ions.
- Ionic compounds are solids, with high melting points and boiling points. The *solids* do not conduct electricity but when *molten* or in *aqueous* solution, they do conduct electricity (can be electrolysed) because the ions are free to move.

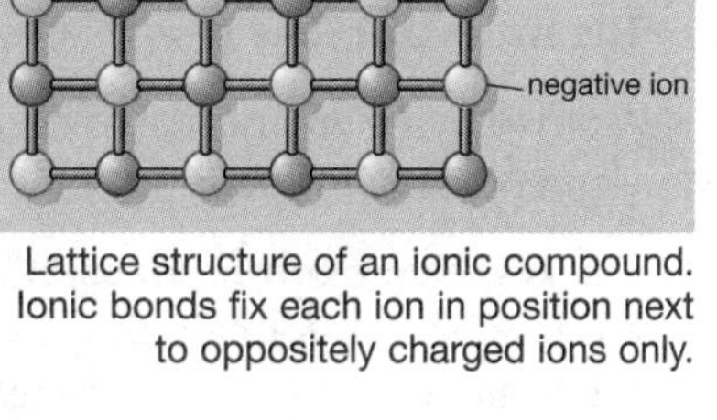

Lattice structure of an ionic compound. Ionic bonds fix each ion in position next to oppositely charged ions only.

B–A*

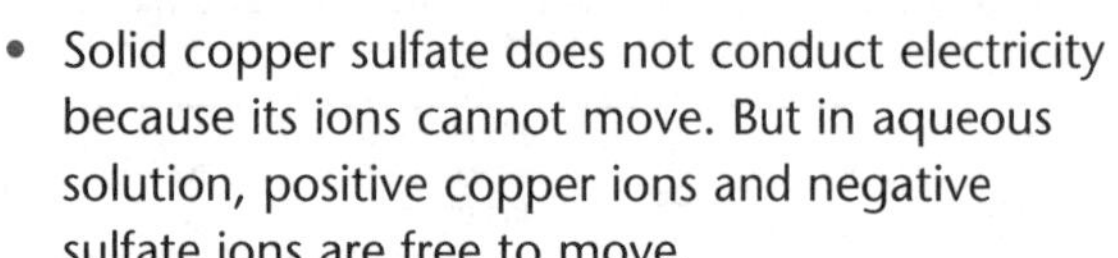

- Solid copper sulfate does not conduct electricity because its ions cannot move. But in aqueous solution, positive copper ions and negative sulfate ions are free to move.

Top Tip!

Ionic compounds have lots of very strong bonds.

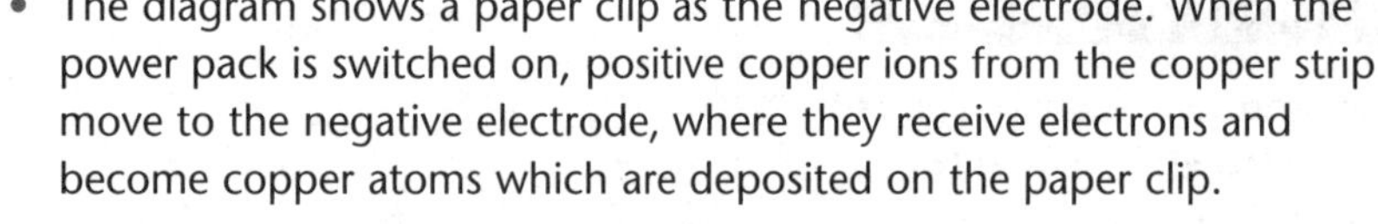

- The diagram shows a paper clip as the negative electrode. When the power pack is switched on, positive copper ions from the copper strip move to the negative electrode, where they receive electrons and become copper atoms which are deposited on the paper clip.
- At the positive electrode, copper atoms lose electrons and go into solution, replenishing the supply of copper ions.

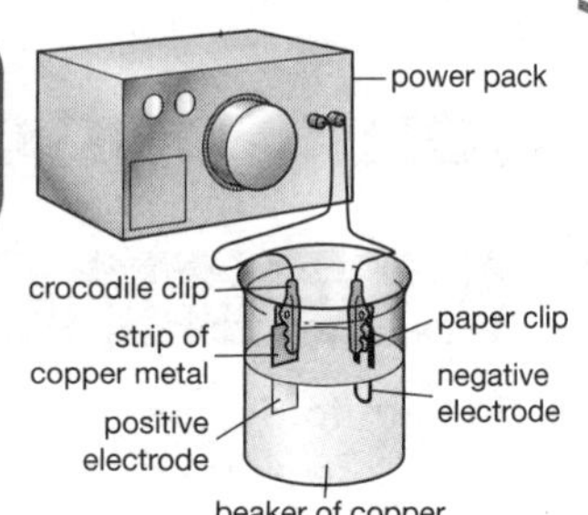

When ionic compounds are molten or in solution, they can conduct electricity.

Covalent bonding

How Science Works

You should be able to: suggest the type of structure of a substance, given its properties.

Covalent compounds: bonding and properties

D–C

- A **covalent bond** is formed when two atoms **share** electrons so that each achieves a **stable** outermost shell of electrons (2 in the first shell, 8 in the second).
- Covalent bonds are strong, holding atoms together in **simple molecules** that include hydrogen, chlorine, oxygen, hydrogen chloride, water and ammonia.

An oxygen atom has 6 electrons in its outermost shell. To make 8, each oxygen atom shares 2 electrons with the other oxygen atom, making a **double** bond.

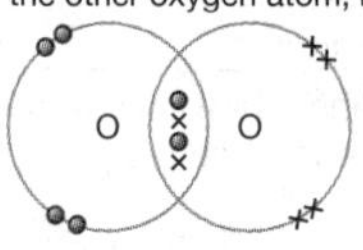

A molecule of oxygen, O_2, O=O.

An oxygen atom shares 1 electron from each hydrogen atom to make 8, and each hydrogen atom shares 1 electron from oxygen to make 2.

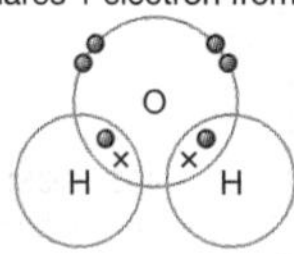

A molecule of water, H_2O, H–O–H.

Top Tip!

A covalent bond is a shared pair of electrons.

B–A*

- Bonding *within* **covalent** iodine molecules is strong. However, forces of attraction *between* molecules are weak.
- Iodine is a dark grey solid at room temperature, but at just 30 °C it **sublimes**, changing into a purple vapour. Many covalently bonded simple molecules are gases.

How Science Works

You should be able to: represent the covalent bonds in molecules such as water, ammonia, hydrogen, hydrogen chloride, chlorine, methane and oxygen.

For ammonia (NH_3):

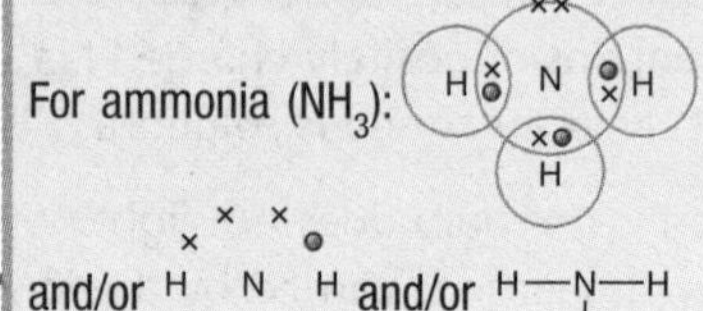

Questions

Grades D-C

1 Explain what is meant by a 'giant ionic structure'.

Grades B-A*

2 Explain what happens to copper ions at the negative electrode, during electrolysis of copper sulfate.

Grades D-C

3 Explain why two oxygen atoms can both become stable if they share two pairs of electrons.

Grades B-A*

4 Explain, in terms of forces of attraction, why many covalently bonded simple molecules, such as O_2 and N_2, are gases at room temperature.

Simple molecules

> **Top Tip!**
> Simple molecules consist of small numbers of atoms held together by shared pairs of electrons forming covalent bonds.

Simple molecules and their properties

D–C

- Strong **covalent bonds** join the atoms in **simple molecules**.
- Simple molecular substances are either gases, liquids or solids with relatively low melting and boiling points. They do not conduct electricity as they have no overall electrical charge.
- Simple molecules that are:
  - **elements** include hydrogen (H_2), H–H; oxygen (O_2), O=O; and chlorine (Cl_2), Cl–Cl
  - **compounds** include water (H_2O), H–O–H; hydrogen chloride (HCl), H–Cl; methane (CH_4), H–C(H)(H)–H and ammonia (NH_3), H–N(H)–H

B–A*

- Though covalent bonds *within* simple molecules are strong, the forces *between* molecules are weak. When a substance formed of simple molecules either melts or boils, it is these weak **intermolecular** forces that are overcome.

Fragrances evaporate at room temperature

B–A*

- **Fragrant substances** in perfumes include oils from plants (e.g. violet or almond) or animals (e.g. deer musk). Their covalently bonded molecules evaporate readily (they are volatile).
- In perfumes, these substances are dissolved in solvents, such as ethanol, which are even more volatile. On the skin, the solvent evaporates quickly, while the fragrance molecules evaporate more gradually.

Giant covalent structures

> **How Science Works**
> You should be able to: represent the covalent bonds in giant structures, e.g. diamond and silicon dioxide.

Giant covalent structures and their properties

D–C

- In **giant covalent structures**, very strong covalent bonds join many atoms arranged in a 3-D **lattice**, forming **macromolecules**.
- They have high melting points. Many are very hard. They do not conduct electricity or dissolve in water.
- Diamond and graphite, two forms of **carbon**, are giant covalent structures.

- In **diamond**, covalent bonds join each carbon to four others, making diamond the hardest known natural **mineral** (used in diamond-tipped drills for rock and metal drilling).
- In **graphite**, each carbon atom bonds to three others, forming layers which are free to slide over each other, so graphite is soft and slippery (used as a lubricant).

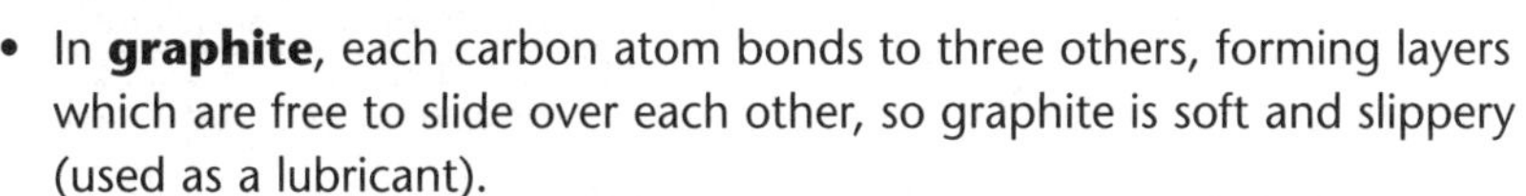

B–A*

- In graphite, one electron from each carbon atom is **delocalised**. The delocalised electrons enable graphite to conduct electricity and heat.

D–C

- The **compound silicon dioxide** is silica (sand). It has strong silicon-oxygen covalent bonds in a giant structure, so it is very hard and has a high melting point.

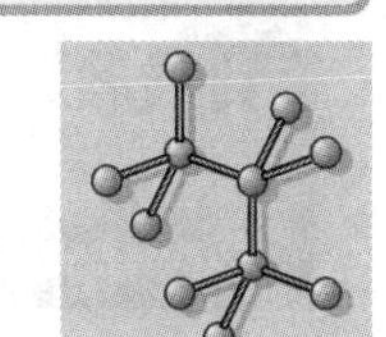

A small part of the diamond lattice.

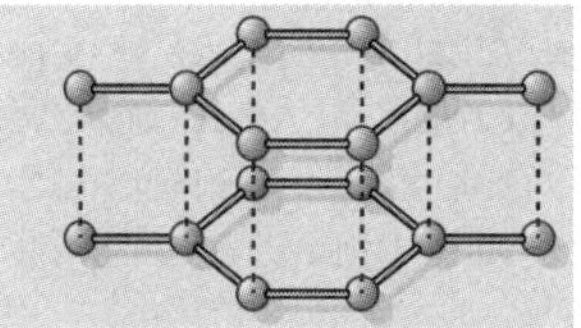

Graphite has free electrons between layers.

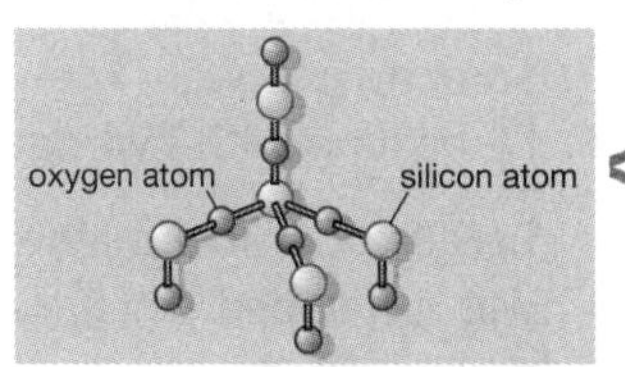

A small part of the silicon dioxide lattice.

> **Top Tip!**
> In general, covalent bonds are just as strong as ionic bonds.

Questions

(Grades D-C)
1 Explain why compounds made of molecules do not conduct electricity.

(Grades B-A*)
2 Explain why fragrant oils from plants evaporate readily.

(Grades D-C)
3 Explain why graphite is soft.

(Grades B-A*)
4 Explain why graphite can conduct electricity, even though it is a covalent compound.

Metals

How the structure of metals relates to their properties

D–C

- **Metals** are reactive and so in ores they occur in combination with other elements. In chemical reactions, metals lose the electrons in their outermost shell and form **positive** ions.
- The layers of atoms in metals can slide over each other, allowing metals to be bent and shaped. Heavier metals have high densities and strength, and high melting and boiling points.

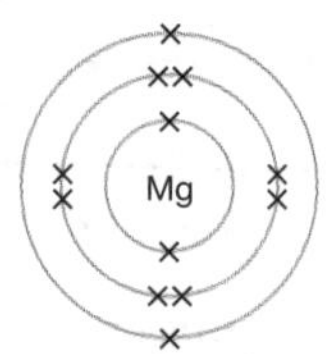

An atom of the metal element magnesium. It loses the two outer electrons when it reacts.

B–A*

- Metals consist of **giant lattice** structures of regularly arranged atoms. Electrons in the outer shell of the atoms are delocalised – free to move. We can picture the structure as fixed positive ions with mobile electrons between them. The structure is held together by strong **electrostatic forces** – attractions between oppositely charged particles.
- Metals conduct **heat** well. Heating increases the kinetic energy of the delocalised electrons, they move faster and transfer **thermal energy** to the fixed positive ions. In this way, heat moves rapidly by conduction through the metal.
- The delocalised electrons also make metals good conductors of **electricity**. In a metal wire in a circuit, delocalised electrons carry negative electrical charge through the wire.
- Metal **alloys** contain atoms of different sizes. This prevents atoms from being regularly arranged and the delocalised electrons cannot move so freely. Such alloys have an increased electrical resistance, so they heat up in a circuit.
- Nichrome, containing nickel, iron and chromium, is such an alloy, and is used as the heating element in toasters.

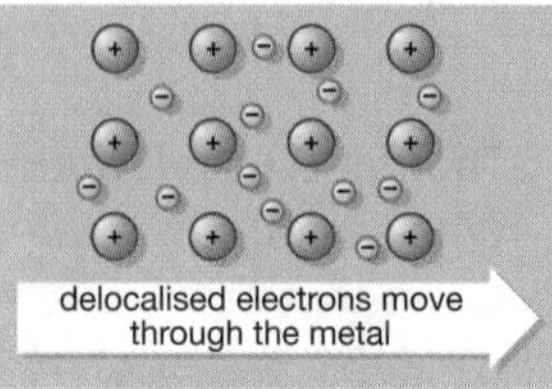

In a circuit, delocalised electrons move in one direction to carry negative charge to the positive terminal of a cell.

How Science Works

You should be able to: represent the bonding in metals in the following form:

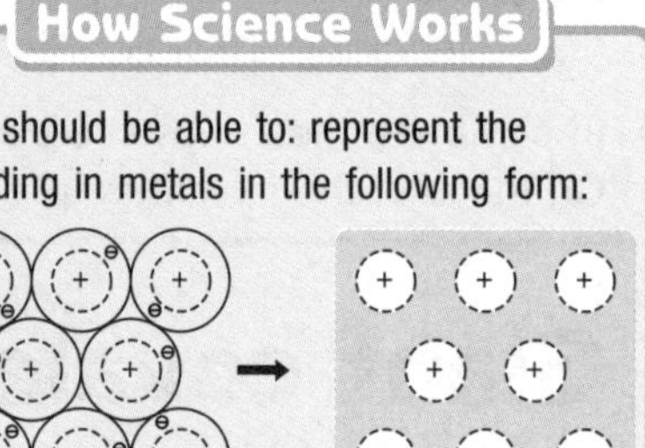

Alkali metals

Group 1 – the alkali metals

D–C

- The elements of **Group 1**, in column 1 of the periodic table, are the **alkali metals**.
- Alkali metals are very reactive, with reactivity increasing down the group. In reactions, they form **ions** with a 1+ charge by losing the single electron in their outermost shell and leaving them with the full outer shell of a noble gas.
- All alkali metals react vigorously with water. With non-metals they form **salts** which are **ionic compounds**.

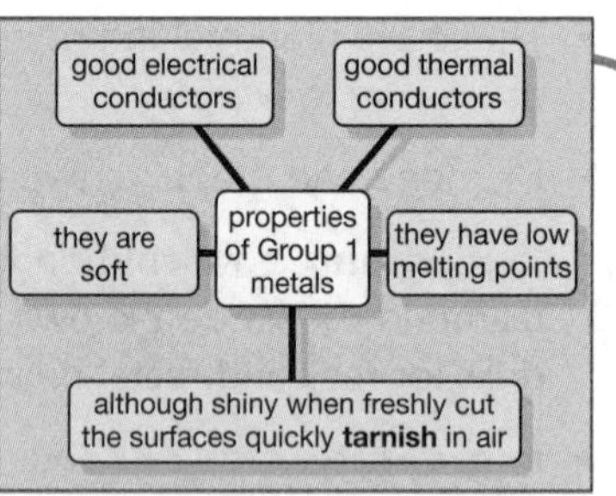

Properties of Group 1 metals.

B–A*

- Some metals were known in ancient times because they could be extracted from their compounds by heating with hydrogen or carbon.
- Because Group 1 metals are so reactive, they were only discovered 200 years ago, when electrolysis was developed. Then, since the metals react so readily with water, the molten salts of Group 1 metals, rather than aqueous solutions, were electrolysed to produce the metal.

Questions

Grades D-C

1 Explain why metals can be bent and shaped.

Grades B-A*

2 Explain why alloys do not usually conduct electricity as well as pure metals.

Grades D-C

3 Which is more reactive – sodium or rubidium? Explain your answer.

Grades B-A*

4 Explain why people learned how to extract iron from its ore before discovering how to extract magnesium from its ore.

Halogens

Group 7 – the halogens

D–C

- The elements in Group 7 of the periodic table are the halogens. All exist as molecules containing two covalently bonded atoms.
- The halogens are: **fluorine**, a pale yellow gas, the halogen with the smallest atoms and the most reactive; **chlorine**, a green gas; **bromine**, a dark red liquid; **iodine**, a dark grey crystalline solid that sublimes on heating to give a purple vapour; and **astatine**, a highly radioactive solid.

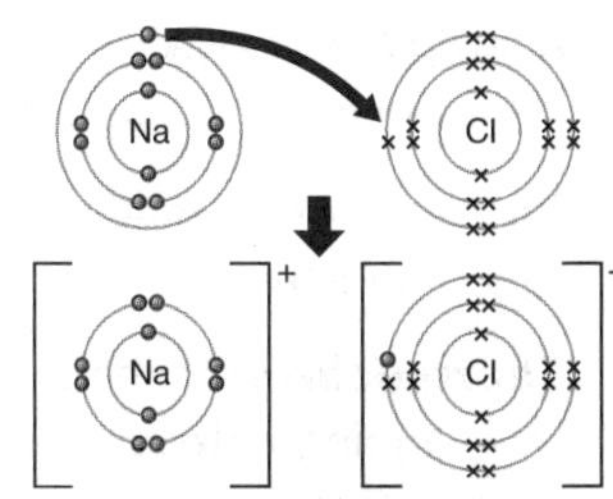

Chlorine reacting with sodium. When a chlorine atom reacts it gains an electron to form a chloride ion: this has a 1– charge and the electron structure of noble gas argon with a full outer electron shell.

- The halogens have similar chemical properties. Reactivity decreases down the group. With alkali metals they form **salts – ionic compounds**; the **halide** ion has a single negative charge.
- Sodium metal burnt in chlorine gas forms the **ionic compound** sodium chloride:

  sodium + chlorine → sodium chloride

B–A*

- Reactivity decreases down Group 7, so the most reactive halogen is fluorine. A more reactive halogen can **displace** a less reactive halogen from a compound.
- When chlorine gas is bubbled through a solution of (colourless) potassium bromide, chlorine displaces the bromide to produce bromine that colours the solution orange:

  chlorine + potassium bromide → bromine + potassium chloride

Nanoparticles

What is a nanoparticle?

D–C

- A **nanoparticle** is a tiny manufactured structure of a few hundred atoms, from 1 to about 100 nanometres long. (A human hair is 100 000 nanometres wide.) Nanoparticles are made from carbon, metals and metal compounds.
- The properties of nanoparticles are different from the properties of the materials in bulk.
- The atoms in nanoparticles are regularly arranged in hollow structures such as tubes and spheres one atom thick. Being so small, nanoparticles have a very high **surface area to volume ratio**.

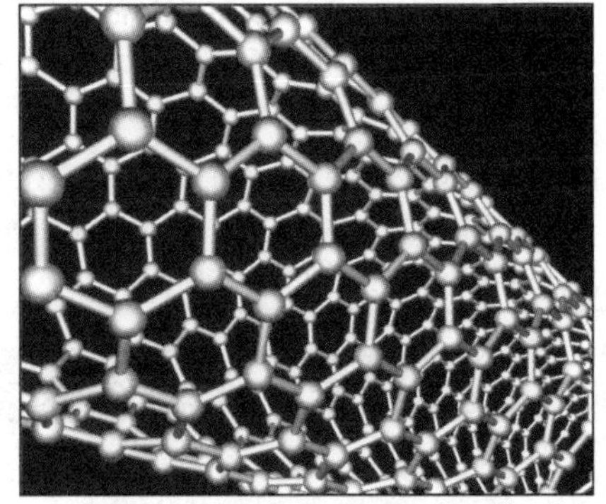
Scientists have used test tubes like this nanoparticle to catalyse new reactions.

- Nanoparticles are being made into tiny capsules that contain drugs. Others make good **biosensors** of particular toxins at very low levels. Nanoparticles could be designed to **process information** in computers millions of times faster than current components. Scientists expect to make new **materials** from nanoparticles that are harder, lighter and stronger than any known materials. Nanoparticle **catalysts** are expected to speed up manufacturing processes.

B–A*

- There is concern that nanoparticles can pass undetected into the body through the lungs, the digestive system and the skin, and that their effects are unknown.

Questions

Grades D-C

1 How do the halogens form ions?

Grades B-A*

2 Predict what would happen if bromine was bubbled through a solution of potassium chloride. Explain your prediction.

Grades D-C

3 How might nanoparticles be used to speed up chemical reactions?

Grades B-A*

4 Explain why there are valid concerns about the safety of nanoparticles.

Grades

Smart materials

Smart materials

How Science Works

You should be able to:
- relate the properties of substances to their uses
- suggest the type of structure of a substance, given its properties
- evaluate the developments and applications of new materials, e.g. nanomaterials and smart materials.

D–C

- A **smart material** has a particular **property**: it **responds** to an environmental change by changing in some way that is **reversible**.
- **Photochromic** materials used in sunglasses darken on exposure to strong light.
- **Thermochromic** materials respond to changes in **temperature** by changing **colour**, as in paper thermometers used on children's foreheads.
- **Electroluminescent** materials emit light of different colours when an alternating current passes through them, as used in some advertising signs.
- Smart materials include alloys that have shape memory (see page 33).

Compounds

Compounds and ratios of elements in them

D–C

- A **mixture** contains different substances that are not chemically combined and can be in any proportions. The substances can be separated.

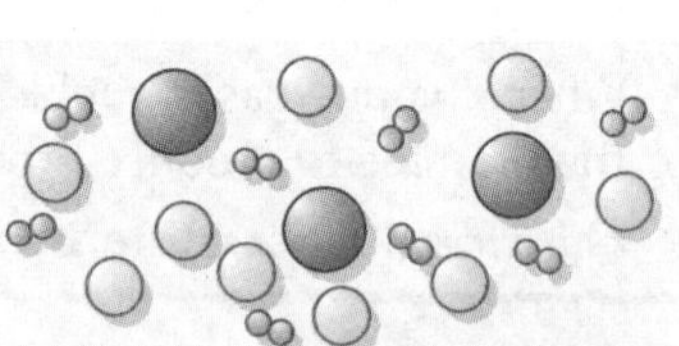

Substances in this mixture are not chemically combined or in fixed ratios.

- A **compound** contains atoms of more than one **element**, **chemically combined** and joined by **chemical bonds**. Atoms in a compound are in a fixed **ratio**.

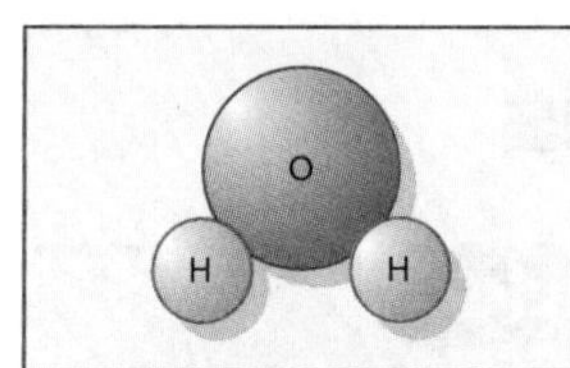

Hydrogen and oxygen in water are in the ratio 2 : 1.

- The **covalently bonded** molecule water has the formula H_2O. Its hydrogen and oxygen atoms are in the ratio 2 : 1.
- The giant ionic structure of salt, sodium chloride, has the formula NaCl. Its sodium ions (Na^+) and chloride ions (Cl^-) are in the ratio 1 : 1.

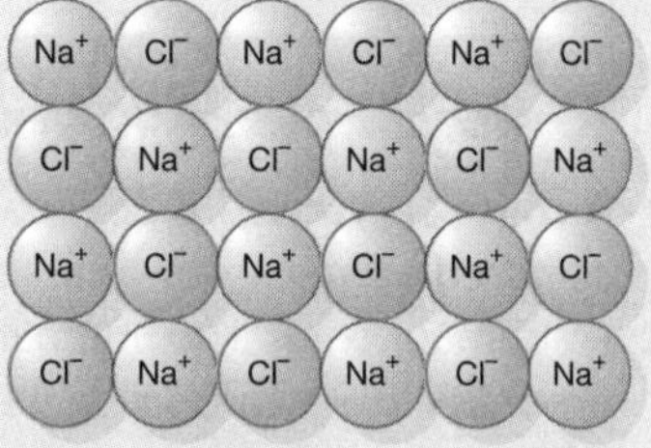

Salt contains sodium and chloride ions.

Top Tip!

A compound consists of atoms of two or more elements that are chemically combined.

Writing balanced equations

B–A*

- The **formula equation** that represents a chemical reaction must be **balanced**. The number of atoms in the products must equal the number of atoms in the reactants. If there are charges involved, these must also balance.

How Science Works

You should be able to: write balanced chemical equations for reactions.

- This **word equation** represents the reaction to form water:
  hydrogen + oxygen → water
- A formula equation for the reaction written as $H_2 + O_2 \rightarrow H_2O$ **IS WRONG**. There are 4 reactant atoms and 3 product atoms (there is 1 oxygen atom less in the product).
- To have 2 oxygen atoms on both sides, the hydrogen atoms must be doubled on both sides:
  $2H_2 + O_2 \rightarrow 2H_2O$

Questions

Grades D-C

1 Describe how a photochromic material behaves when exposed to light.

Grades D-C

2 Explain the difference between a 'compound' and a 'mixture'.

Grades D-C

3 The ionic compound magnesium chloride has the formula $MgCl_2$. What does this tell you about the compound?

Grades B-A*

4 Balance this equation: $N_2 + H_2 \rightarrow NH_3$

Percentage composition

C2A DISCOVER BUCKMINSTERFULLERENE!

Relative formula mass and percentage of elements in compounds

- The **relative formula mass** (**M_r**) of a molecule or compound is the sum of the **relative atomic masses** (**A_r**) of all the atoms in the **numbers** shown by the formula.
- The formula used to calculate the **percentage mass** of an **element** in a **compound** is:

D–C

$$\textbf{\% mass of an element in a compound} = \frac{\textbf{relative atomic mass of the element} \times \textbf{number of atoms of element in the formula}}{\textbf{relative formula mass of compound}} \times \mathbf{100}$$

- The formula of the **compound** potassium nitrate is KNO_3.

Relative atomic masses: K = 39, N = 14, O = 16

Relative formula mass of $KNO_3 = 39 + 14 + (3 \times 16) = 101$

% mass of potassium in potassium nitrate = $\frac{39 \times 1}{101} \times 100 = 38.6\%$

How Science Works

You should be able to: calculate chemical quantities involving formula mass (M_r) and percentages of elements in compounds.

Moles

Top Tip!

The percentage composition is calculated by mass.

Moles and masses

- There is the same number of particles in 1 **mole** of any substance:

a mole of a substance is the relative formula mass in grams of that substance

D–C

- Formula of the compound water = H_2O

Relative formula mass of $H_2O = (2 \times 1) + (1 \times 16) = 18$. So mass of 1 mole of water molecules is 18 g

- Water is formed when hydrogen is burnt in air (oxygen), see page 96:

The balanced equation for the reaction is: $2H_2 + O_2 \rightarrow 2H_2O$

This shows that 2 moles of hydrogen molecules react with 1 mole of oxygen molecules.

- Magnesium oxide is formed when magnesium metal burns in air (oxygen):

The balanced equation for the reaction is: $2Mg + O_2 \rightarrow 2MgO$

This shows that 2 moles of magnesium atoms react with 1 mole of oxygen molecules.

- Balanced symbol equations and moles are used to calculate masses of reactants and products.

$$\textbf{number of moles of a substance} = \frac{\textbf{mass of substance}}{\textbf{mass of 1 mole of substance}}$$

B–A*

- What is the mass of magnesium oxide formed when 12 g of magnesium are burned in oxygen? A_r values: magnesium = 24, oxygen = 16. We use the formula to calculate the number of moles:

Number of moles of magnesium atoms in 12 g = 12 ÷ 24 = 0.5 moles

Relative formula mass of magnesium oxide, MgO = 24 + 16 = 40

Balanced equation: $2Mg + O_2 \rightarrow 2MgO$

This shows that for every mole of magnesium atoms, the *same* number of moles of magnesium oxide molecules is formed, in this case 0.5 moles.

mass of magnesium oxide = number of moles of magnesium oxide × relative formula mass of magnesium oxide

Mass of magnesium oxide = 0.5 × 40 = 20 g

Questions

Grades D-C

1 Calculate the per cent by mass of carbon in methane, CH_4.

Grades D-C

2 What is the mass, in grams, of 1 mole of methane?

Grades B-A*

3 This is an unbalanced equation for the complete combustion of methane: $CH_4 + O_2 \rightarrow CO_2 + H_2O$. Balance the equation.

Grades B-A*

4 If you had 16 g of methane, how many grams of carbon dioxide would you get if it all reacted with oxygen?

Yield of product

Theoretical yield and actual yield

- In a **chemical reaction**, no atoms are gained or lost. Yet the **reactants** do not always give as much **product** as expected from the formula equation. The diagram shows some reasons.

why the actual amount of product may be less than the theoretical amount
- the reaction is reversible and does not go to completion
- some of the product may be lost when it is separated
- some of the reactants may react in an unexpected way

D–C

- The maximum amount of product that could form in a reaction is the **theoretical yield**. The amount that actually forms is the **actual yield**.
- We say a reaction has a high **atom economy** if the actual yield is high.
- It is important for **sustainable development** and for economic reasons to use reactions with a high atom economy.

How Science Works

You should be able to:
- calculate chemical quantities involving empirical formulae, reacting masses and percentage yield
- calculate the atom economy for industrial processes
- evaluate sustainable development issues related to this economy.

Percentage yield, ratios and empirical formulae

B–A*

- The cost of making a chemical depends on the **percentage yield**:

$$\textbf{percentage yield} = \frac{\textbf{actual yield}}{\textbf{theoretical yield}} \times \textbf{100}$$

- A reaction has an actual yield of 3.2 g and the theoretical yield is 4.0 g:

$$\text{percentage yield} = \frac{3.2}{4.0} \times 100 = 80\%$$

- **Ratios** of elements in compounds are explained on page 96. The simplest ratio of atoms in a substance is called its **empirical formula**.
- The compound that living things use in respiration to give energy is **glucose**:

Its formula is $C_6H_{12}O_6$. The ratio of elements C : H : O = 6 : 12 : 6

The simplest ratio is 1 : 2 : 1. So the empirical formula for glucose is CH_2O.

Reversible reactions

You need to be able to explain what a reversible reaction is.

Examples of reversible reactions

D–C

- In **reversible reactions**, the **products** react to produce the original **reactants**. Where A and B are reactants and C and D are products: **A + B** $\rightleftharpoons$ **C + D** (see also page 99)
- Heating the white solid **ammonium chloride** gives the gases **ammonia** and **hydrogen chloride**. Ammonium chloride re-forms when the gases cool down:
ammonium chloride $\rightleftharpoons$ **ammonia + hydrogen chloride**
- White **anhydrous copper sulfate** reacts with water to give blue **hydrated copper sulfate** and re-forms on heating:
hydrated copper sulfate $\rightleftharpoons$ **anhydrous copper sulfate + water**
- In the reaction used to test whether water is present, blue **anhydrous cobalt chloride** reacts with water to form pink **hydrated cobalt chloride**. The anhydrous compound re-forms on heating:
hydrated cobalt chloride $\rightleftharpoons$ **anhydrous cobalt chloride + water**

Questions

Grades D-C

1 State **two** reasons why the quantity of a product from a reaction may not be as great as expected.

Grades B-A*

2 A calculation predicted that 250 kg of product would be formed in a reaction. However, only 210 kg were formed. Calculate the percentage yield.

Grades D-C

3 Nitrogen and hydrogen can react to produce ammonia. This is a reversible reaction. Write the word equation for this reaction.

Grades B-A*

4 Using the equation $N_2 + 3H_2 \rightleftharpoons 2NH_3$ we can predict that 28 g of nitrogen could yield 34 g of ammonia. In practice, much less ammonia is produced. Explain why.

Grades

C2A DISCOVER BUCKMINSTERFULLERENE!

Equilibrium 1

Reversible reactions, equilibrium and reaction conditions

D–C

- A general equation for **reversible reactions** is: **A + B $\rightleftharpoons$ C + D**
  - In the **forward** reaction, substances A and B react to form substances C and D.
  - In the **reverse** reaction, C and D react to form A and B.
- A reversible reaction does not go to completion (i.e. not all of A and B change to C and D). Eventually, the forward and reverse reactions reach the *same* **rate** – the proportions of A, B, C and D become steady. The reaction has reached **equilibrium**.
- If conditions *do not change* (e.g. temperature or pressure or amount of A and B), we say that the reaction is happening in a **closed system**.
- If conditions *change*, the proportions of A, B, C and D change. For example:

B–A*

  - if A and B are gases (large volume) and C and D are liquids (small volume), applying **pressure** will speed up the forward reaction rate (the equilibrium moves right)
  - if the forward reaction gives out heat (is exothermic), raising the **temperature** still further will speed up the reverse reaction rate (the equilibrium moves left)
  - if C and D are **removed** as soon as they are formed (or if more A and B are added), the forward reaction rate will speed up.
- Adding a **catalyst** increases both reaction rates – the equilibrium is reached faster but its position is not affected by the catalyst – amounts of A, B, C and D are unaffected.

Haber process

Conditions used in the Haber process

D–C

- Ammonia, NH_3, is used to make fertiliser. It is produced in the **Haber process**, a **continuous** industrial process (see diagram). Nitrogen (from air) combines with hydrogen (from natural gas):

  $N_2(g) + 3H_2(g) \rightleftharpoons 2NH_3(g)$

  The reverse reaction also takes place.
- To produce a reasonable yield of ammonia quickly, the **conditions** used in the Haber process are:
  - a high pressure of approximately 200 atmospheres to drive the forward reaction (four molecules of reactant gases give two molecules of ammonia gas)
  - a compromise temperature of about 450 °C to reach equilibrium quickly (the reaction is exothermic, but at cool temperatures the reaction rate would be far too slow)
  - an iron catalyst to speed up the reaction rate
  - cooling the reaction mixture to liquefy and remove the ammonia
  - recycling the unreacted nitrogen and hydrogen.

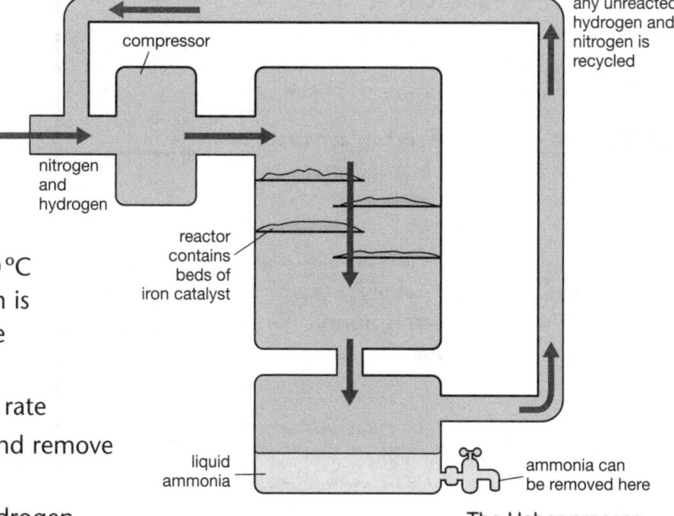

The Haber process.

Questions

Grades D-C

1 What effect does a catalyst have on the amount of product formed in a reversible reaction?

Grades B-A*

2 Explain what is meant by the term ‘equilibrium’ in a reversible reaction.

Grades D-C

3 Name the catalyst that is used in the Haber process.

Grades B-A*

4 Ammonia manufacturers keep the temperature for the Haber process at about 450 °C, not hotter. Explain why they do this.

C2a summary

Sub-atomic particles

An **atom** contains **sub-atomic particles**:
- a central **nucleus** of **protons** and **neutrons**
- **electrons**, which orbit the nucleus.

The **mass number** of an element = number of protons + number of neutrons.

The **atomic number** = number of protons = number of electrons (i.e. no net charge).

The **relative atomic mass**, A_r, of an element is its mass compared with the mass of a carbon-12 atom, where A_r for carbon = 12.

Isotopes are atoms of an element with different numbers of **neutrons**. Isotopes have different mass numbers.

In **nuclear** reactions, *nuclei* of atoms either split or join together and energy is released:
- in nuclear **fission**, the nucleus of one atom of a large element splits to give two nuclei of smaller elements
- in nuclear **fusion**, the nuclei of two small atoms join together to form a new element.

Electrons are arranged around the nucleus in **shells**, which represent **energy levels**: shell 1 contains 2 electrons; shell 2 contains 8 electrons.

Electron notation shows the **arrangement** of electrons in shells, e.g. Mg: 2, 8, 2.

Structure, properties and uses

Metal atoms *lose* electrons to form **positive** ions. **Non-metal** atoms *gain* electrons to form **negative** ions.

Ionic bonding is the attraction between **oppositely charged ions**.

Ionic compounds: are giant ionic **lattices**; are solids at room temperature (high m.p.); conduct electricity when molten or dissolved.

An **alkali metal** (Group 1 of the periodic table) loses one electron to form a positive ion when it reacts with a non-metal (e.g. a halogen) to form a **salt** that is an ionic compound.

A **nanoparticle** is a tiny manufactured **structure** with special properties due to the arrangement of atoms. Nanoparticles have a high **surface to volume ratio**.

They are used as **biosensors** and may process information, act as catalysts and provide new, very hard, light materials.

Delocalised electrons in **metals** and **graphite** enable them to **conduct** heat and electricity.

Non-metal atoms can **share** pairs of electrons to form **covalent bonds** between atoms with **stable outermost shells**.

A **halogen** (Group 7 of the periodic table) exists as a **molecule** of two covalently bonded atoms.

Simple molecular elements (e.g. oxygen) and **compounds** (e.g. water) with covalent bonds:
- do not conduct electricity
- have low melting and boiling points
- are often gases at room temperature
- when solid, are brittle or waxy
- have weak **intermolecular** forces
- are often insoluble in water.

Giant covalent structures (e.g. diamond, silicon dioxide) are macromolecules that: have very high melting points; are very hard; do not conduct electricity.

Smart materials have particular properties:
- **photochromic** materials react to light
- **thermochromic** materials respond to changes in temperature
- **electroluminescent** materials emit light when a.c. passes through them.

Composition, yield and equilibrium

The **relative formula mass** of a compound or molecule = sum of **relative atomic masses** of all atoms in its **formula**.

A **mole** of a substance is the **relative formula mass** in **grams** of that substance.

Atoms in a **compound** are in fixed **ratios**. The simplest ratio of atoms in a substance is called its **empirical formula**. The **formula equation** for a chemical reaction must be **balanced**.

$$\text{\% yield} = \frac{\text{actual yield}}{\text{theoretical yield}} \times 100$$

High **atom economy** is achieved if a high proportion of the reactants end up as useful products. This is important for **sustainable development** and to limit costs.

Reversible reactions can proceed in both directions. They are affected by: **pressure** and **temperature**, removal of products and presence of a **catalyst**.

Reversible reactions carried out in a **'closed system'** will eventually reach **equilibrium**.

The **Haber process** is a reversible reaction:

nitrogen + hydrogen ⇌ ammonia

$$\text{\% mass of an element in a compound} = \frac{\text{relative atomic mass of the element} \times \text{no. of atoms of element in the formula}}{\text{relative formula mass of compound}} \times 100$$

Rates of reactions

Measuring the rates of reactions

- The **rate of a chemical reaction** can be measured as the amount of:
  - **reactant** used up in a set period of time: **reaction rate = amount of reactant used up ÷ time**
  - **product** formed in a set period of time: **reaction rate = amount of product formed ÷ time**
- Example: **magnesium(s) + hydrochloric acid(aq) → magnesium chloride(aq) + hydrogen(g)**

D–C

- At timed intervals, measure **one** of the following: **volume** of gas formed (cm^3); **number** of bubbles formed; reduction in **mass** of magnesium; time for all the magnesium to **dissolve**.
- To increase the rate of this reaction, we can: increase the **temperature** or the **concentration** of the acid; use **powdered** metal.
- Collision theory (page 102) explains how conditions affect reaction rate.

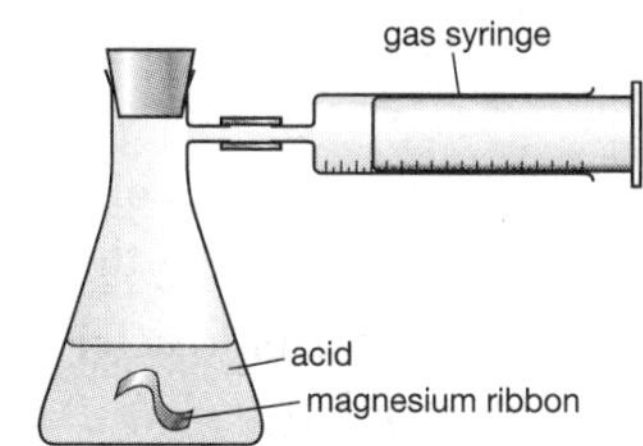

Measuring the volume of gas formed using a gas syringe.

Reaction conditions to record

Top Tip!

When talking about acids you should use the word *concentration* to describe how much acid is in the solution.

B–A*

- To be useful, results must allow someone else to repeat the experiment and obtain similar results. For the reaction rate of magnesium and hydrochloric acid, these **starting** conditions should be recorded:
  - the **mass** of the magnesium, namely the 'amount' of reactant in the first equation above; and whether the magnesium is in ribbon or powder form
  - the **concentration** and **volume** of the hydrochloric acid used.
- **Concentrations** of solutions are given in **moles** per cubic decimetre, mol/dm^3. (Page 97 gives more on moles.)
- Other conditions can include the room temperature and whether a catalyst is used.

Following the rate of reaction

Measuring the loss of mass of reactant where a gas is formed.

Methods of measuring reaction rate

D–C

- If a **product** of a reaction is a **gas**:
  (**1**) measure its **volume** at timed intervals: collect the gas in an upturned burette, measuring cylinder or syringe; or count gas bubbles fed into a test tube of water; or
  (**2**) measure the **mass** of gas lost from the reacting mixture at timed intervals.
- If two reacting solutions form a solid that clouds the mixture, measure the time for a cross beneath a flask to disappear (page 102), or measure the cloudiness at intervals using a sensor and **datalogger**.

B–A*

- To measure gas volume, the gas must be **insoluble** in the reaction mixture, and form slowly enough for its volume to be measured reliably.
- To measure loss of mass of a solid, the reaction should not be so vigorous as to splash or boil the liquid, and should not take too long.

Questions

Grades D-C

1 Describe how you could use a gas syringe to find the rate of reaction between a piece of magnesium and some hydrochloric acid.

Grades B-A*

2 The relative formula mass of HCl is 36.5. How many grams of HCl are there in 1 mole? How many grams of HCl are there in 1 dm^3 of hydrochloric acid with a concentration of 2 moles per dm^3?

Grades D-C

3 Suggest **one** way, apart from your answer to question 1, in which you could measure the rate of reaction between magnesium and hydrochloric acid.

Grades B-A*

4 Explain why measuring the volume of gas formed is not a good way of measuring rate of reaction if the gas is **soluble** in the reaction mixture.

Grades

Collision theory

When explaining why a factor such as higher temperature increases the rate of a reaction, do not say there are 'more collisions' – say that there are '*more frequent* collisions'.

Increasing the rate of collisions

- For a chemical reaction to occur, particles must **collide** with **sufficient energy**. The minimum energy particles need to react is called **activation energy**. (Each reaction has its own activation energy value.)
- **Collisions** that result in a reaction are more frequent when there is an increase in:
  - **temperature** of reactants: the **kinetic energy** of particles increases, they collide more frequently and with greater energy (see also 'Heating things up' below) D–C
  - **concentration** of solutions: in a solution of two reactants, doubling the concentration of *one* reactant doubles the **frequency** of **collisions** with the other reactant (see also page 103)
  - **pressure** on reacting gases: gas particles are squeezed closer together, increasing the frequency of collisions
  - **surface area** of reacting solids: collisions occur at surfaces, so particles collide more frequently when solids are crushed or powdered (see also page 103).
- A **catalyst** lowers the activation energy required for particles to react (see page 104).

More about gases, temperature and pressure

- Gases **expand** when their temperature increases or when pressure on them is reduced: both events cause the molecules to move further apart. Gases **contract** when their temperature decreases or when pressure is increased: both events cause the molecules to move closer together. B–A*
- At the *same* temperature and pressure, the same number of molecules of any gas will take up the same space. So, at 25 °C and atmospheric pressure, 1 dm^3 of oxygen has the *same number of molecules* as 1 dm^3 of hydrogen, or of any other gas.

Heating things up

Temperature and reaction rate

- **Solid** sulfur forms when sodium thiosulfate and hydrochloric acid react:

  $$Na_2S_2O_3(aq) + 2HCl(aq) \rightarrow S(s) + SO_2(g) + 2NaCl(aq) + H_2O(l)$$

  Sulfur clouds the water, gradually obscuring a cross drawn on paper under the flask. D–C
- The reaction is repeated at temperatures 10 °C apart. The reaction time (time taken to obscure the cross) is recorded.
- The graph shows that:
  - for each 10 °C increase, the reaction time is halved
  - the temperature change 20–30 °C has more effect than 30–40 °C.

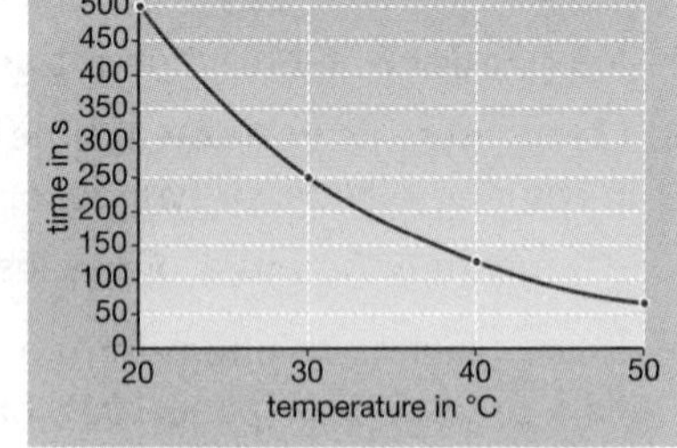

Graph for the reaction of sodium thiosulfate and hydrochloric acid showing the time for a cross to disappear at temperatures at 10 °C intervals.

- Since a temperature rise of 10 °C halves the reaction time, the frequency of effective collisions must double with each 10 °C rise. B–A*

Questions

Grades D-C

1 If you double the concentration of hydrochloric acid in which magnesium is reacting, by how much would this increase the frequency of collisions between particles of reactants?

Grades B-A*

2 What can you say about the number of molecules in 1 dm^3 of chlorine gas and in 1 dm^3 of nitrogen gas?

Grades D-C

3 Use the graph to calculate how much faster the reaction was at 30 °C than at 20 °C.

Grades B-A*

4 Explain why raising temperature by 10 °C doubles reaction rate.

Grind it up, speed it up

How Science Works

You should be able to: interpret graphs showing the amount of product formed (or reactant used up) with time, in terms of the rate of the reaction.

Surface area and reaction rate

D–C

- Carbon dioxide **gas** is produced when calcium carbonate (marble) and hydrochloric acid react:
  $CaCO_3(s) + 2HCl(aq) \rightarrow CO_2(g) + CaCl_2(aq) + H_2O(l)$
- Each flask (see diagram) contains the same mass (in **excess**) of marble chips, but of two different sizes, and the same volume of acid of the same concentration.
- As carbon dioxide bubbles off, the fall in mass is measured at 30 s intervals.
- The **graph** shows that:
  - the reaction rate is highest at the start
  - the reaction finishes for the smaller chips first.
- The reading at 1 min gives the **initial reaction rate** as *mass* of gas per minute:
  $$\textbf{rate of reaction} = \frac{\textbf{amount of product formed}}{\textbf{time}}$$
- The rate of reaction is greater for the smaller chips than the larger chips.

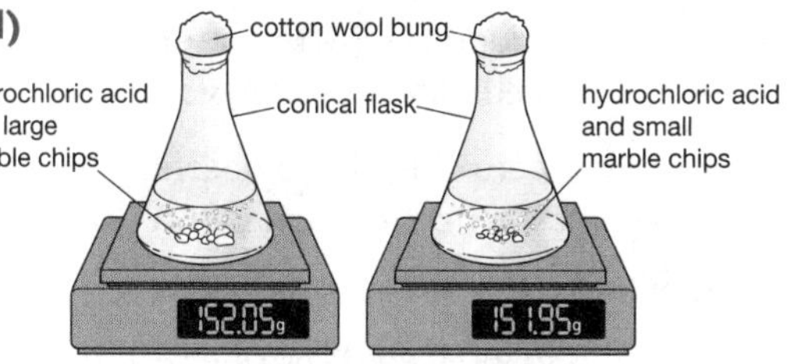

Marble chips reacting with acid.

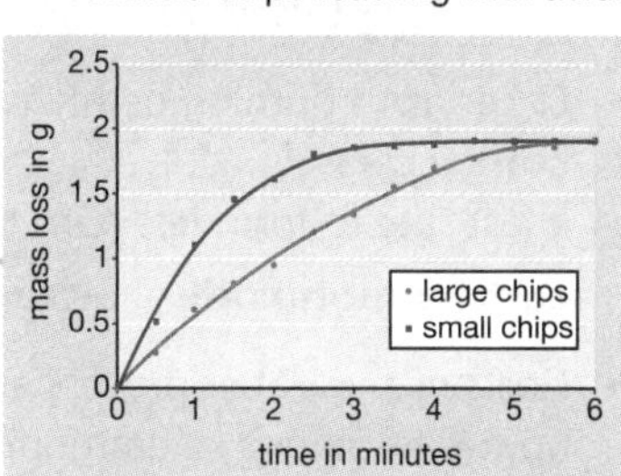

Graph to show the mass of carbon dioxide gas lost when acid reacts with marble chips of two different sizes.

Concentrate now

Top Tip!

1 mol/dm^3 hydrochloric acid has twice as many particles in the same volume of acid as 0.5 mol/dm^3 acid. Do not just say that a more concentrated solution has more particles in it. You must say there are *more particles in the same volume.*

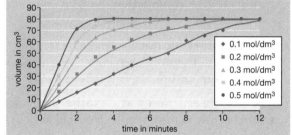

Concentration and reaction rate

D–C

- Hydrogen **gas** is produced when magnesium reacts with sulfuric acid:
  $Mg(s) + H_2SO_4(aq) \rightarrow H_2 + MgSO_4(aq)$
- In five experiments, the same **mass** of magnesium reacts with the same volume of hydrochloric acid (in **excess**) at five different concentrations.
- Gas volume is recorded at 1-min intervals for each of the five concentrations of acid. All other conditions are kept constant.
- The **graph** shows that:
  - the reaction rate is highest at the start
  - the reaction ceases for the highest acid concentration first.
- The rate of reaction is greatest for the highest acid concentration.

Graph to show the volume of hydrogen formed with five different concentrations of sulfuric acid.

B–A*

- Equal volumes of solutions of the same molar concentration (measured in mol/dm^3) contain the same number of moles of solute, that is, the same number of particles.
- There are twice as many acid particles in 0.4 mol/dm^3 sulfuric acid to collide with magnesium particles as in the 0.2 mol/dm^3 acid, so the reaction rate is double for the higher concentration acid.

Questions

Grades D-C

1 Explain why measuring the mass of gas lost after 5 min would not give you any useful information about the rate of the reaction between hydrochloric acid and marble chips.

Grades B-A*

2 Explain why the lines for both sizes of chips level off at the same height on the graph.

Grades D-C

3 Explain why using 2 mol/dm^3 acid rather than 1 mol/dm^3 of acid would double the rate of reaction.

Grades B-A*

4 What can you say about the number of sodium ions in 500 cm^3 of a 1 mol/dm^3 solution of sodium chloride, compared with the number of potassium ions in 250 cm^3 of a 2 mol/dm^3 solution or potassium chloride?

Catalysts

Catalysts and reaction rate

Top Tip!

All the catalysts in this section contain transition metals. Transition metals and their compounds make good catalysts.

- **Catalysts** increase reaction rates, but are not themselves used up in reactions.
- Catalysts include **transition metals** and their compounds. Different reactions need different catalysts.
- Hydrogen peroxide (H_2O_2) decomposes slowly to water and oxygen, but very fast when manganese(IV) oxide is added:

  hydrogen peroxide → water + oxygen
- Weighing manganese(IV) oxide before and after the reaction proves that it is not used up.
- Gases from burning petrol contribute to **acid rain**. **Catalytic converters** contain catalysts that make these gases react at a fast rate to form less harmful gases:

  carbon monoxide + nitrogen dioxide → carbon dioxide + nitrogen
- Costs of the Haber process are reduced by using an iron catalyst to speed up the formation of ammonia gas from nitrogen and hydrogen. Catalysts reduce costs in many industrial processes.

D–C

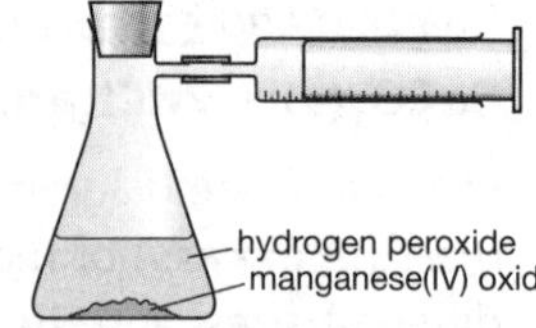

To follow the decomposition of hydrogen peroxide, the volume of oxygen that forms is measured.

How Science Works

You should be able to: explain and evaluate the development, advantages and disadvantages of using catalysts in industrial processes.

- Molecules are **adsorbed** onto the surface of the catalyst that weakens their bonds and so lowers their **activation energy** (page 102).
- After the reaction, the products leave the catalyst surface: they are **desorbed**.

B–A*

Activation energy for a forward reaction, with and without a catalyst.

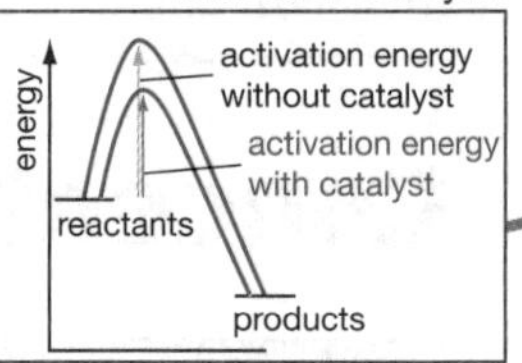

Energy changes

Exothermic and endothermic reactions

- In an **exothermic** reaction, **energy** is transferred *to* the surroundings. Energy is given out, often as heat, as in all **combustion** reactions and in many **oxidations** and **neutralisations** (**acid + alkali → salt + water**).
- In an **endothermic** reaction, energy is transferred *from* the surroundings. Energy is taken in, often as heat, as in **thermal decompositions** (e.g. **calcium carbonate → calcium oxide + carbon dioxide**).
- A **reversible reaction** that is **exothermic** in one direction is **endothermic** in the opposite direction. In both directions the same amount of energy is transferred.
- When water is added to **anhydrous copper sulfate** (see page 98), the white powder turns blue as it becomes **hydrated copper sulfate**. The reaction (a test for water) is exothermic.

  anhydrous copper sulfate + water $\underset{\text{endothermic}}{\overset{\text{exothermic}}{\rightleftharpoons}}$ **hydrated copper sulfate**
- When the hydrated copper sulfate is heated, the crystals *take in* an equal amount of heat that drives off the water. This reaction is endothermic.

Top Tip!

Exit means the way **OUT** and **Exo**thermic reactions give **OUT** energy. Endothermic is the opposite.

D–C

Questions

Grades D-C

1 Name the group of metals that make good catalysts.

Grades B-A*

2 What is activation energy, and how do catalysts affect it?

Grades D-C

3 What can you say about the amount of energy taken in or given out in a reversible reaction?

Grades D-C

4 Imagine you have some anhydrous copper sulfate in a test tube. You add some water to it. What will you feel if you hold the tube? Why?

Grades

C2B DISCOVER ELECTROLYSIS!

Equilibrium 2

Reaching a balance point

D–C

- The thermal decomposition of calcium carbonate is a **reversible reaction** (see page 98):

calcium carbonate $\underset{\text{exothermic}}{\overset{\text{endothermic}}{\rightleftharpoons}}$ **calcium oxide + carbon dioxide**

- When a reversible reaction occurs in a closed system (see page 99), **equilibrium** is reached when the reactions occur at exactly the same rate in each direction. The relative amounts of all the reacting substances at equilibrium depend on the **reaction conditions**.

B–A*

- If the **temperature** is *raised*, the **yield** from the endothermic reaction increases (the forward reaction above) and the yield from the exothermic reaction decreases (the reverse reaction above).
- If the temperature is *lowered*, the yield from the endothermic reaction decreases (the forward reaction above) and the yield from the exothermic reaction increases (the reverse reaction above).

Industrial processes

How Science Works

You should be able to:

- describe the effects of changing the conditions of temperature and pressure on a given reaction or process
- evaluate the conditions used in industrial processes in terms of energy requirements.

The Haber process

- In **gaseous** reactions, an increase in **pressure** will favour the reaction that produces the least number of molecules, as shown by the symbol equation for that reaction.
- **Ammonia**, required to manufacture **nitrate** fertilisers, is produced in the Haber process (see also page 99) according to the reaction:

$$N_2(g) + 3H_2(g) \underset{\text{endothermic}}{\overset{\text{exothermic}}{\rightleftharpoons}} 2NH_3(g)$$

number of molecules: 1 nitrogen 3 hydrogen 2 ammonia

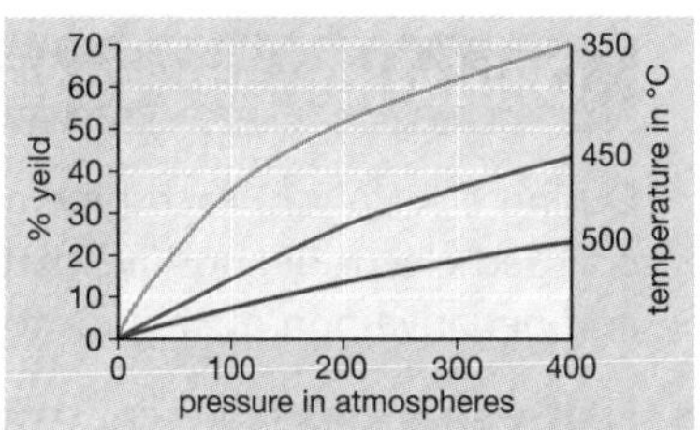

Graph showing the effect of conditions on yield in the Haber process.

B–A*

- For a good **yield**:
  - the reaction is **exothermic**, so low temperature increases yield; however, this lengthens the time to reach equilibrium
  - high pressure drives the reaction forward, but is expensive to apply
  - removing the product (ammonia) and replenishing the reactants (nitrogen and hydrogen) both drive the reaction forward (see diagram on page 99).
- For a good **rate** but lower yield, higher temperatures are used to produce ammonia faster.
- The factors of temperature and pressure, together with reaction rates, are important when determining the optimum conditions in industrial processes. In the Haber process, **optimum conditions** are: temperature 450 °C; pressure 200 atmospheres; use of an iron catalyst.
- **Yield** is 30%. This means that 30% of the starting materials are converted to product. This is acceptable because the **reaction rate** (see page 101) is reasonable and costs are relatively low.
- Reducing energy costs and energy wastage contribute to **sustainable development** as well as serving economic purposes. Using non-vigorous conditions where possible reduces the energy released into the environment.

Questions

Grades D-C

1 Look at the equation for the thermal decomposition of calcium carbonate. Why is this called 'thermal decomposition'?

Grades B-A*

2 What could you do to increase the yield from the endothermic direction of a reversible reaction?

Grades B-A*

3 Use the graph to find the conditions of temperature and pressure at which the greatest percentage yield of ammonia is achieved.

Grades B-A*

4 Look at the equation for the Haber process. Use the information in it to explain why a high pressure increases the yield of ammonia.

Free ions

Top Tip! A metal and a non-metal react to form an ionic solid of positive and negative ions held tightly together in a giant lattice.

Electrolysis of ionic compounds

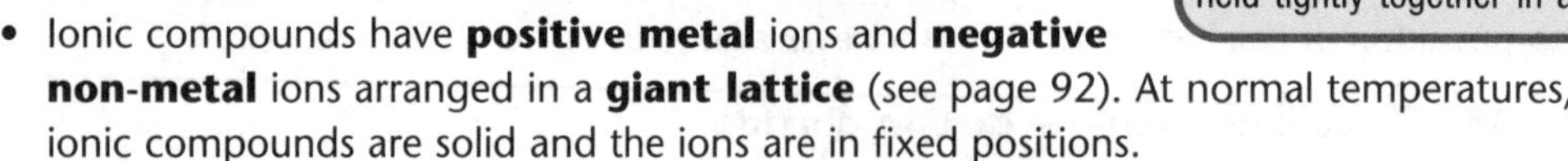

D–C

- Ionic compounds have **positive metal** ions and **negative non-metal** ions arranged in a **giant lattice** (see page 92). At normal temperatures, ionic compounds are solid and the ions are in fixed positions.
- When **molten**, the ions are free to move, and **electrolysis** (see page 31) splits the compound into its **elements**. Electrolysis of molten lead bromide produces molten lead and bromine gas:

$$PbBr_2(s) \xrightarrow{\text{heat}} PbBr_2(l) \xrightarrow{\text{electrolysis}} Pb(l) + Br_2(g)$$

- Ionic substances in **aqueous solutions** can be electrolysed because the ions are free to move in water.

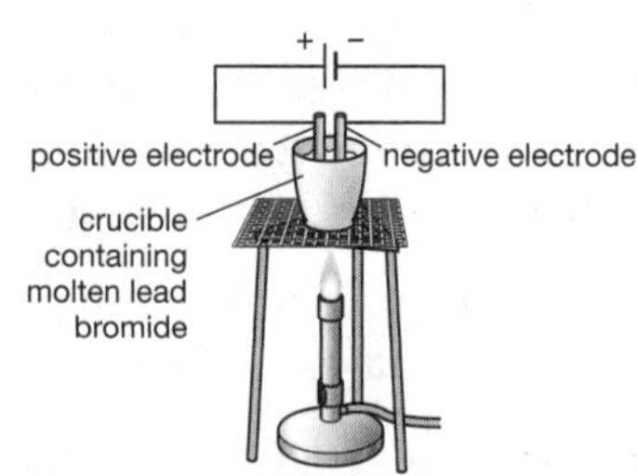

Electrolysis of molten lead to bromide.

B–A*

- In an aqueous solution of sodium chloride, Na^+ and Cl^- ions move towards oppositely charged electrodes. Chlorine gas forms, but not sodium metal.
- Electrolysis also splits water molecules to H^+ and OH^- ions. The H^+ ions join Na^+ ions at the negative electrode, but only hydrogen gas is formed there, because sodium is more **reactive** than hydrogen (see the reactivity series, page 107).
- Sodium and hydroxide ions, Na^+ and OH^-, stay in solution. Sodium hydroxide, chlorine and hydrogen are all important industrially.

Top Tip! Use the periodic table to help with questions about charges on ions.

Electrolysis equations

Oxidation and reduction in electrolysis

D–C

- During electrolysis, each type of ion moves towards the *oppositely* charged electrode: positive metal ions to the **negative** electrode, and negative non-metal ions to the **positive** electrode.

How Science Works

You should be able to: complete and balance supplied half equations for the reactions occurring at the electrodes during electrolysis.

B–A*

- **Half-equations** represent **oxidations** and **reductions** that happen at electrodes (see also page 31). For **molten** sodium chloride:
  - each positive **sodium ion** moves to the negative electrode where the ion *gains* an electron – it is **reduced** – to form an uncharged atom of sodium **metal**: $Na^+(aq) + e^- \rightarrow Na(s)$
  - each negative **chloride ion** moves to the positive electrode where the ion *loses* an electron – it is **oxidised** – to form an uncharged chlorine atom. This shares electrons with another chlorine atom to form a molecule of chlorine **gas**: $2Cl^-(aq) \rightarrow Cl_2(g) + 2e^-$

Each sodium ion gains an electron and each chloride ion loses an electron.

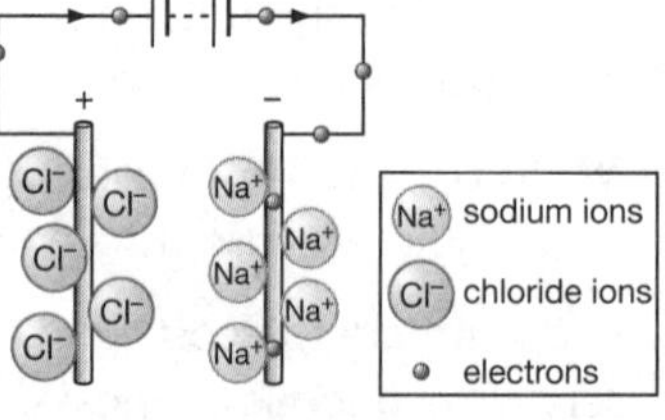

When molten sodium chloride is electrolysed, sodium ions (Na^+) and chloride ions (Cl^-) move to the electrodes.

Questions

(Grades D-C)

1 Explain why lead ions go to the negative electrode when lead bromide is electrolysed.

(Grades B-A*)

2 Explain why hydrogen is formed at the negative electrode when a sodium chloride solution is electrolysed.

(Grades D-C)

3 Describe what happens to chloride ions at the positive electrode, when sodium chloride solution is electrolysed. Are they reduced or oxidised?

(Grades B-A*)

4 Complete this half equation to show what happens to lead ions when molten lead bromide is electrolysed. Are the lead ions reduced or oxidised?

$Pb^{2+}(aq)$ + ... → ....

Uses for electrolysis

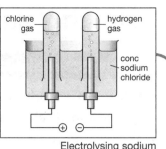

Sodium chloride and using electrolysis

Electrolysing sodium chloride (salt) solution.

- If there is a mixture of ions, the products formed in electrolysis depend on the **reactivity** of the elements involved. The **reactivity series** (below) arranges metals and hydrogen in order of their reactivity.

D–C

- Electrolysing an aqueous **solution** of sodium chloride produces **chlorine** gas at the positive electrode. **Hydrogen** gas (not sodium) is formed at the negative electrode.
- Na^+ and OH^- ions remain in solution as ions of **sodium hydroxide**. This is used to make soap, detergents and paper. **Hydrogen** is a fuel, and is used to make ammonia and to hydrogenate vegetable oils (page 42). **Chlorine** is used to make bleach, chlorinated solvents and PVC.

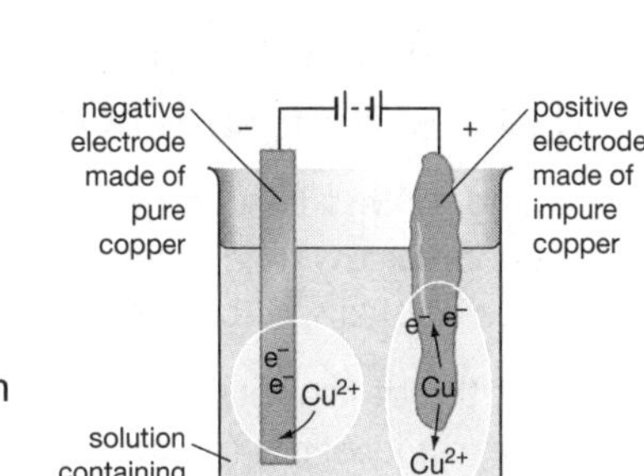

Copper transfers from the positive to the negative electrode.

B–A*

- Sodium is more reactive than hydrogen. When a solution of sodium chloride is electrolysed, Sodium ions stay in solution, while hydrogen ions are more readily reduced:

  $2H^+(aq) + 2e^- \rightarrow H_2(g)$
- To purify copper metal, impure copper is made the positive electrode in a solution containing copper ions, Cu^{2+}. Copper from the positive electrode forms ions:

  $Cu(s) \rightarrow Cu^{2+}(aq) + 2e^-$

  The copper ions move through the solution to the pure copper negative electrode where they gain two electrons and become copper atoms building up on the electrode:

  $Cu^{2+}(aq) + 2e^- \rightarrow Cu(s)$

How Science Works

You should be able to: predict the products of electrolysing solutions of ions.

Acids and metals

Which metals react with acids to make salts?

Top Tip!

If you are asked whether a metal reacts with acids, look at the reactivity series.

- Any **metal** *higher* than hydrogen in the **reactivity series** can react with an acid, **displacing** the less reactive hydrogen in the acid. A soluble **salt** and hydrogen gas are formed.
- Potassium and sodium are too reactive to be safe when reacting with sulfuric acid.
- Copper and silver (lower than hydrogen) will not react with dilute sulfuric acid.

D–C

- A **salt** forms when a metal or an **ammonium** ion replaces hydrogen in an acid. Salts from acids: sulfuric acid → sulfates; nitric acid → nitrates; hydrochloric acid → chlorides.
- To make the salt zinc sulfate (and hydrogen), the **reagents** are zinc metal and sulfuric acid:

  $Zn(s) + H_2SO_4 \rightarrow ZnSO_4 + H_2(g)$

  Zinc added in **excess** uses up all the acid so that the solution contains only zinc sulfate. Any remaining zinc is filtered, and the solution is evaporated, leaving zinc sulfate crystals.

Reactivity series

most reactive
K
Na
Mg
Al
Zn
Fe
Pb
H
Cu
Ag
least reactive

Top Tip!

Learn the formulae of these acids:
sulfuric acid = H_2SO_4
nitric acid = HNO_3
hydrochloric acid = HCl

Questions

Grades D-C

1 Give **two** uses of the gas formed at the positive electrode when a solution of sodium chloride is electrolysed.

Grades B-A*

2 Explain why, when purifying copper using electrolysis, the impure copper is made the *positive* electrode.

Grades D-C

3 Explain why you need to filter the mixture when the reaction finishes, if you want to get pure zinc chloride.

Grades B-A*

4 Write a balanced equation for the reaction between magnesium and sulfuric acid.

Making salts from bases

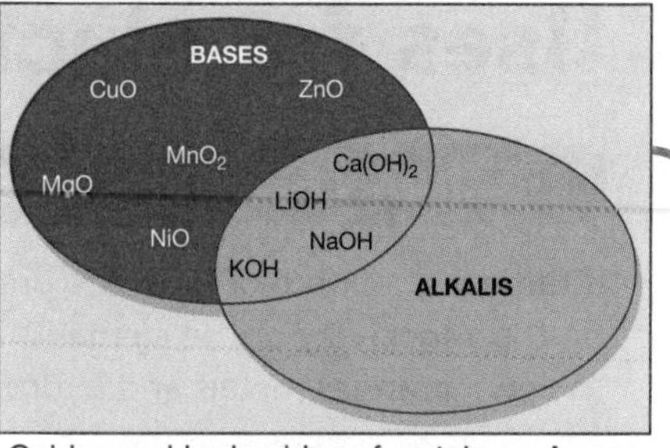

Oxides and hydroxides of metals are **bases**, but only soluble bases are **alkalis**.

Bases, alkalis and reaction with acids

D–C

- **Bases** react with acids to form **salts**:

  base + acid → salt + water

- The insoluble base **zinc oxide** reacts with warm acid:

  zinc oxide + hydrochloric acid → zinc chloride + water

  $$ZnO(s) + 2HCl(aq) \rightarrow ZnCl_2(aq) + H_2O(l)$$

- To make metal salts, appropriate reagents are chosen. To make **magnesium chloride** salt:

  $$MgO(s) + 2HCl(aq) \rightarrow MgCl_2(aq) + H_2O(l)$$

- Adding **excess** base to the acid ensures that all the acid reacts, leaving only magnesium chloride in solution. Excess magnesium oxide is filtered off and the water evaporated. In general, salt solutions can be crystallised to produce the solid salt.

B–A*

- Using the basic metal **oxide** (or hydroxide) is a way to prepare the salt of a **metal** that is too low in the reactivity series (see page 107) to react with the acid.
- Using excess of an **insoluble** base to make a salt ensures that all the acid reacts and that, when the excess base is filtered off, only the required salt remains in solution.

Acids and alkalis

Using indicators; fertiliser from ammonia

D–C

- An **indicator** shows – by changing colour – how acid or alkaline a solution is. When added to an alkali, an indicator changes colour when just enough acid (and no more) is added to react with all the alkali. At this point of **neutralisation**, the solution contains a **salt** and water only.
- **Phenolphthalein** is a suitable indicator to follow the reaction of sodium hydroxide with hydrochloric acid which forms sodium chloride (common salt).
  - Exactly 25 cm^3 of sodium hydroxide of known molar concentration is placed in a conical flask, and two or three drops of phenolphthalein are added. The mixture is pink.
  - Hydrochloric acid of approximately the same molar concentration is added to the flask from a burette to record the volume used. At neutralisation, one additional drop of acid turns the indicator colourless.

  sodium hydroxide + hydrochloric acid → sodium chloride + water

pH	14	1	7

  At pH 14 phenolphthalein is pink. At pH 7 or below it is colourless.

B–A*

- **Ammonia** gas, NH_3, is a very soluble **base** that forms an alkaline solution in water.
- Nitrogen is required by plants, but ammonia solution is an unsuitable fertiliser as it would make the soil too alkaline for crops.
- Ammonia reacts with acids to form ammonium salts. Being neutral, **ammonium salts** make good fertilisers. Ammonia in solution reacts with sulfuric acid to give **ammonium sulfate**:

  $$2NH_3(aq) + H_2SO_4 \rightarrow (NH_4)_2SO_4$$

Questions

Grades D-C

1 Name an oxide and an acid that would react to produce zinc sulfate.

Grades B-A*

2 Explain why it is best to use an *insoluble* base when deciding to make a salt by reacting the base with an acid.

Grades D-C

3 Explain why, if you want to make a pure sample of sodium chloride, you would first react the base and acid using an indicator, and then do it again *without* the indicator.

Grades B-A*

4 Ammonium nitrate, NH_4NO_3, is widely used as a fertiliser. How could ammonium nitrate be made from ammonia? Write a balanced equation for the reaction.

Neutralisation

> **Top Tip!**
> Acids are solutions that contain $H^+(aq)$.
> Alkalis are solutions that contain $OH^-(aq)$.

Measuring acidity and alkalinity

D–C

- Hydrogen ions, **$H^+(aq)$**, make aqueous solutions acidic. All **acids** contain hydrogen ions.
  Hydroxide ions, **$OH^-(aq)$**, make aqueous solutions alkaline. All **alkalis** contain hydroxide ions.
  In a **neutralisation** reaction, the hydrogen ions and hydroxide ions react to form water.
- The **pH scale** is a measure of the **strength** of acids and alkalis. pH depends on the amount of H^+ or OH^- ions released in water. A pH of 1 indicates a strong acid and a pH of 14 indicates a strong alkali.

B–A*

- Strong acids (e.g. hydrochloric acid) and strong alkalis (e.g. sodium hydroxide) split completely into ions:
  $HCl(aq) \rightarrow H^+(aq) + Cl^-(aq)$; $NaOH(aq) \rightarrow Na^+(aq) + OH^-(aq)$
- Hydrochloric acid reacts with sodium hydroxide: **$HCl(aq) + NaOH(aq) \rightarrow NaCl(aq) + H_2O(aq)$**
- In solution, NaCl(aq) is really split into ions. Showing all the ions in the reaction:
  $H^+(aq) + Cl^-(aq) + Na^+(aq) + OH^-(aq) \rightarrow Na^+(aq) + Cl^-(aq) + H_2O(aq)$
- So the $Na^+(aq)$ and $Cl^-(aq)$ ions do not change; the only change is: **$H^+(aq) + OH^-(aq) \rightarrow H_2O(l)$**
- In all **neutralisation** reactions, **hydrogen** ions and **hydroxide** ions react to form **water**.
- Neutralisations always produce a **salt**:
  acid + metal → metal salt + hydrogen
  acid + ammonia → ammonium salt
  acid + soluble base → metal salt + water
  acid + insoluble base → metal salt + water

> **How Science Works**
> You should be able to:
> - suggest methods to make a named salt
> - explain and evaluate processes that use the principles described in this unit.

Precipitation

Forming a precipitate, and hard water

- **Insoluble salts** can be formed by mixing two **soluble** salts that contain the appropriate ions. The solid **precipitate** of the salt is filtered, washed with distilled water and dried.
- To form the insoluble yellow compound lead chromate, compounds containing lead ions and chromate ions are required, for example:
  lead nitrate + potassium chromate → lead chromate + potassium nitrate
  $Pb(NO_3)_2(aq) + K_2CrO_4(aq) \rightarrow PbCrO_4(s) + 2KNO_3(aq)$

D–C

- **Hard water** contains dissolved **calcium** ions. To remove them, adding soluble sodium carbonate forms insoluble calcium carbonate that precipitates out:
  soluble calcium ion + soluble carbonate ion → insoluble calcium carbonate
- Before treatment, the **water supply** contains nitrate ions from fertilisers and phosphate ions from washing powders. **Phosphate** ions are removed using calcium ions (or iron or aluminium ions):
  soluble calcium ion + soluble phosphate ion → insoluble calcium phosphate
- **Heavy metals** are removed as insoluble **carbonates**.
- Several **metals** can be identified from their insoluble **hydroxides** which have characteristic colours. To form the hydroxide precipitate, sodium hydroxide is added to water containing the metal ions.

Questions

Grades D-C

1 Write a word equation to show what happens when any acid reacts with any alkali.

Grades B-A*

2 Write a balanced equation to show what happens when any acid reacts with any alkali.

Grades D-C

3 Explain why adding calcium ions to water can remove phosphate ions from it.

Grades B-A*

4 Silver nitrate is soluble, but silver chloride is insoluble. Suggest how this can be used to test for the presence of a chloride in a solution.

C2b summary

Rates of reaction

Collision theory states that **particles** must collide with sufficient energy in order to react.

The **minimum energy** required for a reaction to take place (a **successful collision**) is the **activation energy**.

The **rate of a reaction** can be measured as:
- the amount of **reactant used up** in a set time
- or the amount of **product formed** in a set time.

Collisions that result in a reaction are **more frequent** with an **increase** in:
- **temperature**
- **concentration** of a solution
- **pressure** in gas(es)
- **surface area** of a solid.

Using a **catalyst** lowers the activation energy required for particles to react.

Types of reaction

In an **exothermic** reaction, **energy** is transferred *to* the surroundings. **Exothermic** reactions give OUT energy.

In an **endothermic** reaction, energy is transferred *from* the surroundings. **Endothermic** reactions take IN energy.

If a **reversible reaction** is exothermic in one direction, it is endothermic in the opposite direction, e.g.

anhydrous copper sulfate + water $\underset{\text{endothermic}}{\overset{\text{exothermic}}{\rightleftharpoons}}$ **hydrated copper sulfate**

In reversible reactions, **equilibrium** is reached at a point when the rate of the reverse reaction **balances** the rate of the forward reaction.

The relative amounts of all the reacting substances at equilibrium depend on the **reaction conditions**.

Optimum conditions for the **Haber process** are:
- temperature 450 °C
- pressure 200 atmospheres
- iron catalyst.

Ions

When **molten** or **dissolved** in water, the **ions** in **ionic compounds** are **free** to move.

In **electrolysis**, an **electric current** passed through an ionic compound in which ions are free to move breaks it down into its **elements**.

During **electrolysis**:
- **positive** metal ions **gain electrons** to form atoms
- **negative** non-metal ions **lose electrons** to form atoms and molecules.

Electrolysis of **sodium chloride** (common salt) solution makes hydrogen, chlorine and sodium hydroxide:
- each positive **sodium ion** is **reduced** to form uncharged sodium **metal**:
  $Na^+(aq) + e^- \rightarrow Na(s)$
- each negative **chloride ion** is **oxidised** to form an uncharged chlorine atom that shares electrons with another chlorine atom to form chlorine **gas**:
  $2Cl^-(aq) \rightarrow Cl_2(g) + 2e^-$

OIL RIG:
Oxidation **I**s **L**oss
Reduction **I**s **G**ain – of electrons.

Making salts

Metal oxides and hydroxides are **bases** and react with **acids** to form **salts**.

Soluble bases are **alkalis**.

In **neutralisation** reactions, hydrogen ions and hydroxide ions react to form water:

$H^+(aq) + OH^-(aq) \rightarrow H_2O(l)$

Neutralisations always produce a **salt**:
- **acid + metal → metal salt + hydrogen**
- **acid + ammonia → ammonium salt**
- **acid + soluble base → metal salt + water**
- **acid + insoluble base → metal salt + water**

Some **metals** can be reacted with **acids** to make **salts**. The **reactivity series** of metals shows whether a metal will react with an acid.

Salts from acids: sulfuric acid → sulfates; nitric acid → nitrates; hydrochloric acid → chlorides.

Insoluble salts can be made as **precipitates** when two solutions are mixed together.

Ion impurities can be removed from the **water supply** by making them into **insoluble salts**, e.g. phosphate ions from washing powders can be removed using calcium ions to form insoluble calcium phosphate.

Grades

See how it moves!

Distance-time graphs

D–C

- A **distance-time graph** shows the distance an object moves in a period of time. **Gradient** (slope: distance ÷ time) = **speed**.
- The graph shows Emma's journey from home to school:
  - A-B: walks to bus stop; *slow*
  - B-C: waits for bus; stationary
  - C-D: on the bus; *fast*
  - D-E: still on the bus; slower
  - E-F: off at school bus stop, waits for friend; stationary.

Emma's journey to school.

B–A*

- For a journey in which speed varies:

$$\textbf{average speed} = \frac{\textbf{total distance travelled}}{\textbf{total time taken}}$$

- Where speed changes gradually, the graph is curved. To find the gradient at any point, draw a tangent at that point (see graph).

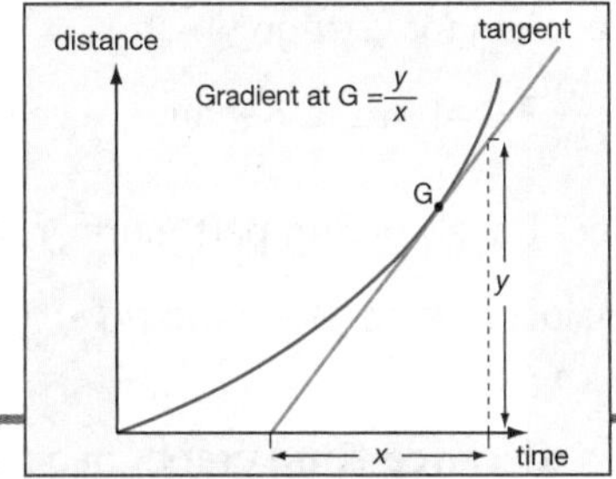

Distance/time graph for an object that is speeding up.

Top Tip!

A distance-time graph that slopes downwards shows an object coming back towards its starting point, not one that is slowing down.

How Science Works

You should be able to:

- construct distance-time graphs for a body moving in a straight line when the body is stationary or moving with a constant speed
- calculate the speed of a body from the slope of a distance-time graph.

Speed isn't everything

Velocity, acceleration and direction

D–C

- **The velocity, *v*, of an object is its speed in a given direction.**
- The diagram shows the velocities of three objects in an *upward* direction.

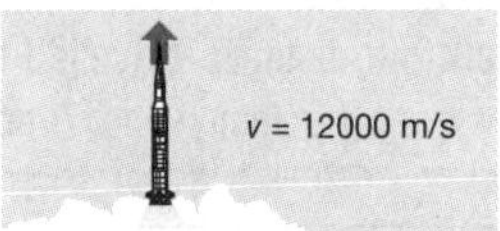

The **positive velocity** of the rocket tells us that it is moving at 12 000 m/s upwards, away from the ground.

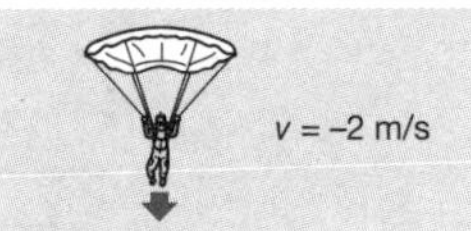

The parachute has a **negative velocity** because it is travelling *towards* the ground and velocity is measured *away* from the ground.

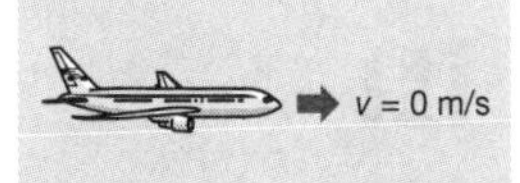

The aeroplane might have a very high speed, but its velocity *away from the ground* is 0 m/s. It is not moving towards or away from the ground.

Top Tip!

If a question doesn't tell you which way velocity is being measured, you can choose which way is positive, but keep it the same for the whole question.

- Any **change** in velocity (speeding up or slowing down) is called **acceleration**. The value is negative for slowing down. Like velocity, acceleration must have a **direction**.

$$\textbf{acceleration (in m/s}^2\textbf{)} = \frac{\textbf{change in velocity (in m/s)}}{\textbf{time taken for change (in s)}}$$

B–A*

- A student rotates a bung on a string at a constant speed. Since the *direction* of the bung constantly changes, so does its *velocity*: we say it is *accelerating*.
- While the bung is rotating, a force is constantly pulling it towards the centre of the circle, so the *direction* of the acceleration is towards the centre.
- The same applies to the planets round the Sun.

Circular motion.

Questions

Grades D-C

1 Use the average speed equation to calculate Emma's average speed for her journey to school.

Grades B-A*

2 Use the graph to calculate Emma's speed between C and D.

Grades D-C

3 A stationary car starts to move, and reaches a speed of 20 m/s in 20 s. Use the acceleration equation to calculate its acceleration in m/s^2.

Grades B-A*

4 A spacecraft is orbiting the Earth at a steady speed. Is its velocity changing? Is it accelerating? Explain your answers.

Velocity-time graphs

Top Tip! When you describe the line on a velocity-time graph showing constant velocity, say that the line is 'horizontal', not just 'straight'.

Grades

Velocity-time graphs and acceleration

- A **velocity-time** graph shows how the velocity of an object changes with time. The **gradient** of a velocity-time graph represents **acceleration** (see page 111).
- The **area** under a velocity-time graph shows the **distance** travelled.

The graph shows the changing velocity of a cyclist:

- slope A: positive acceleration
- horizontal B: constant velocity
- slope C: negative acceleration
- horizontal D: constant velocity.

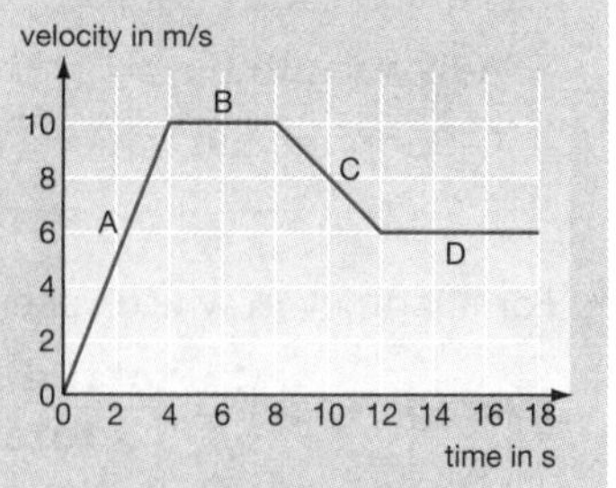

Velocity-time graph for a cyclist's journey.

D–C

- **i** What is the **acceleration** during part-journey C?

Use acceleration (in m/s²) = $\frac{\textbf{change} \text{ in velocity (in m/s)}}{\text{time taken for change (in s)}}$

acceleration = $\frac{6 - 10 \text{ (m/s)}}{12 - 8 \text{ (s)}} = \frac{-4 \text{ m/s}}{4 \text{ s}}$ = –1 m/s, a negative acceleration

- **ii** What is the **distance** the object travels during part-journey B?

Distance = area under graph = velocity (in m/s) × time (in s)
distance = 10 (m/s) × (8 – 4) (s) = 40 m

Top Tip! The slope (gradient) of a velocity-time graph shows the acceleration.

- To calculate **speed** from a *curved* **distance-time** graph, page 111, draw a *tangent* to the curve (hypotenuse of a right-angled triangle). Speed is given by the *gradient* (slope): height ÷ base = $y \div x$
- To calculate **acceleration** from a *curved* **velocity-time** graph, draw a tangent to the curve, and take *gradient* values as above.
- To calculate **distance** travelled from a *sloping* **velocity-time** graph, work out the *area* under the triangle from: $\frac{1}{2}$base × height

B–A*

How Science Works

You should be able to:

- construct velocity-time graphs for a body moving with a constant velocity or a constant acceleration
- calculate the acceleration of a body from the slope of a velocity-time graph
- calculate the distance travelled by a body from a velocity-time graph.

Let's force it!

Adding forces

The resultant force acting on the car moves it forwards.

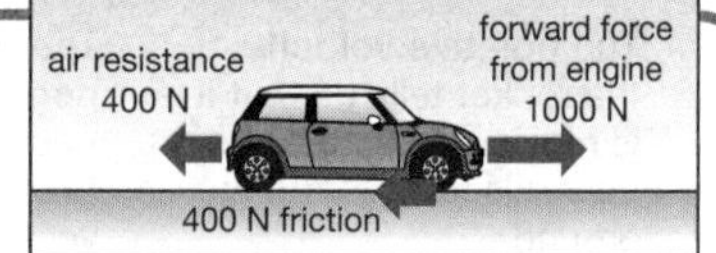

- All the different forces on an object act together as a single, **resultant force**.

The resultant force has the same effect as all the forces acting on the object.

D–C

- *Add* forces acting in the *same* direction, *subtract* forces acting in the *opposite* direction. For the car:

resultant force = forward force – air resistance – friction = 1000 N – 400 N – 400 N = 200 N

- If the resultant force is zero, **stationary** objects remain stationary, and moving objects maintain the same **speed** in the same **direction**. If the resultant force is *not* zero, stationary and moving objects **accelerate** in the direction of the resultant force.
- If forces do not act in the same or opposite directions, a **triangle of forces** gives the **resultant force**. The length of each side is drawn proportional to the **size** of each force.

B–A*

The third side of the triangle gives the resultant force.

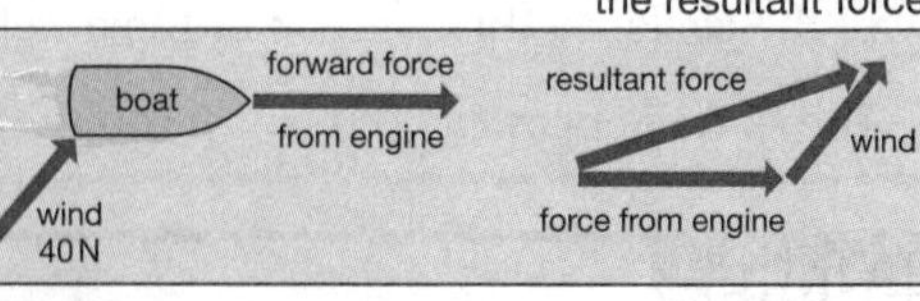

Questions

Grades D-C

1 Sketch a velocity-time graph to show a car moving at a steady speed, slowing down and stopping at traffic lights, then accelerating back up to the same steady speed.

Grades B-A*

2 Use the velocity-time graph for the cyclist's journey to calculate the distance travelled between 0 and 4 s.

Grades D-C

3 Tom is trying to push a box along the ground. He pushes with a force of 10 N. Friction between the box and the ground exerts a force of 10 N in the other direction. What happens to the box?

Grades B-A*

4 In the 'triangle of forces' diagram measure the 'resultant force' arrow and work out its value in newtons.

Grades

Force and acceleration

Unbalanced forces and acceleration

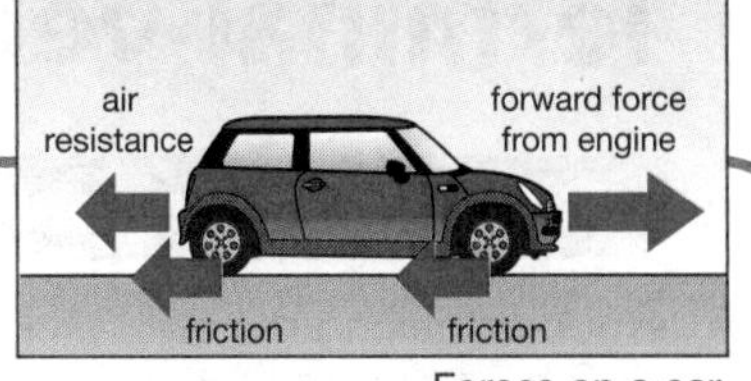

Forces on a car.

- A car travels at a **constant** speed when the forwards **driving force** exactly **balances** the backwards **frictional forces** of air and road.
- With **unbalanced** forces, the car experiences a **resultant** force. It **accelerates** in the *direction* of the resultant force:
  - pressing the accelerator increases the driving force, increasing speed
  - braking increases the frictional force, reducing speed (negative acceleration).
- For the same mass, a bigger force increases acceleration. For the same force, a bigger mass reduces acceleration.

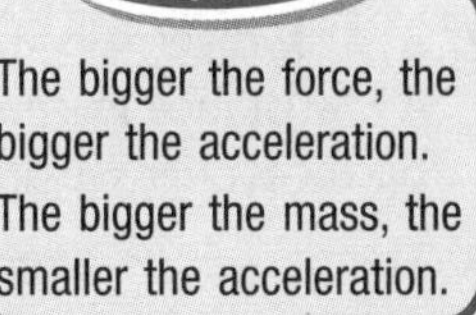

The bigger the force, the bigger the acceleration. The bigger the mass, the smaller the acceleration.

D–C

- Different **forces** and different **masses** are used with the apparatus shown. In each trial, the **force** is made constant: the **force meter** measures **force** (newtons) and the **ticker-timer** measures **velocity** (metres/second).

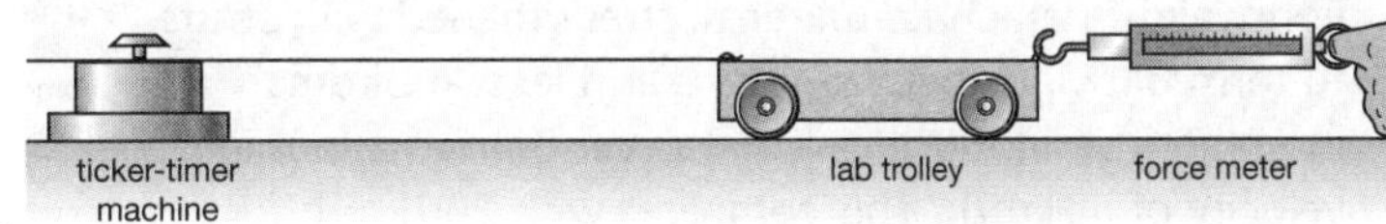

Apparatus to investigate how acceleration changes when force changes.

- Velocity-time graphs drawn from the results show that **acceleration** is:
  - given by the **slope** (velocity ÷ time)
  - **directly proportional** to force, $a \propto F$
  - **inversely proportional** to mass, $a \propto \frac{1}{m}$
- **resultant force (N) = mass (kg) × acceleration (m/s^2): $F = ma$**

Balanced forces

Balanced forces and stationary and moving objects

- **For every force there is an equal and opposite force.**
- **Balanced** forces have the *same* value but act in *opposite* directions.

D–C

- **Stationary** objects with balanced forces include: the tug-of-war teams in the diagram; something floating – the water's **upthrust** balances the downward force of the object's weight.

The two teams are stationary, pulling with equal and opposite forces.

- A parachutist falling at **constant velocity** is a **moving** object experiencing balanced forces.

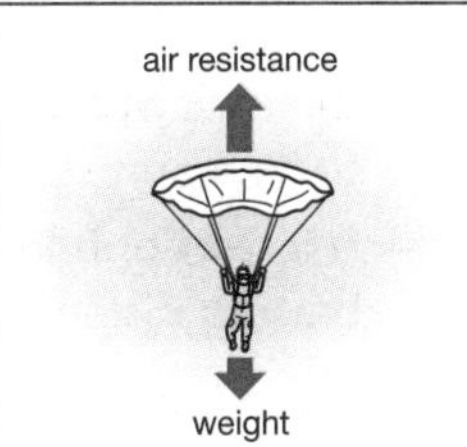

A parachutist falling at constant velocity: his weight acting downwards balances the **air resistance** acting upwards.

Acceleration in a lift

B–A*

- When a lift is stationary, the upward force of the floor on a passenger's feet is equal and opposite to the downward force due to the person's weight.
- When the lift accelerates upwards, so does the person. The upward force of the floor on the person increases. This value is equal and opposite to the person's increased downward force, which makes the person feel heavier than usual.

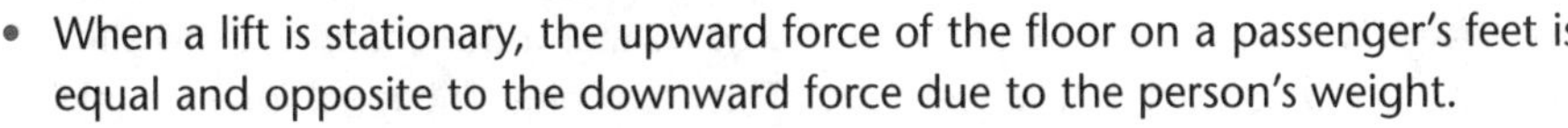

Questions

Grades D-C

1 Complete this sentence: Acceleration is .... proportional to force.

Grades B-A*

2 An object with a mass of 10 kg is accelerating at 10 m/s^2. Calculate the force that is acting on it.

Grades D-C

3 A resultant force of 2 N is acting on an object of 1 kg. Calculate the acceleration of the object.

Grades B-A*

4 What can you say about the forces of gravity and air resistance on a sky-diver who is accelerating downwards?

Terminal velocity

Reaching terminal velocity

Grades D–C

- An object falling through air eventually reaches a top speed, its **terminal velocity**. At this speed, the force of **weight** downwards balances the force of **air resistance** upwards: the **resultant** force is zero.
- As a skydiver falls through the air, she accelerates. Air resistance increases from zero until it equals (and balances) her weight. She reaches terminal velocity.
- The parachute opens. Air resistance increases, exerting an upward force greater than the skydiver's weight. She slows down until weight = air resistance at a lower terminal velocity than before.

Grades B–A*

- In a modification of freefall, birdmen wear suits of lightweight parachute material, strengthened with struts to form batwings between arms and legs. A birdman jumping from 4500 m can glide over 8 km at a terminal velocity of 40 km/h. This compares with 200 km/h for a skydiver and 20 km/h for a parachutist.

Top Tip!

When a parachute opens it *reduces* the resultant downward force – but the resultant force is still downwards until terminal velocity is reached.

Velocity-time graph for a skydiver.

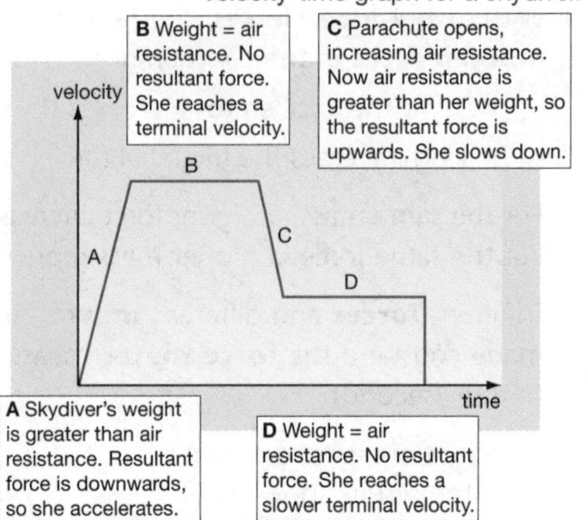
velocity
time
A
B
C
D
B Weight = air resistance. No resultant force. She reaches a terminal velocity.
C Parachute opens, increasing air resistance. Now air resistance is greater than her weight, so the resultant force is upwards. She slows down.
A Skydiver's weight is greater than air resistance. Resultant force is downwards, so she accelerates.
D Weight = air resistance. No resultant force. She reaches a slower terminal velocity.

Stop!

Stopping distance

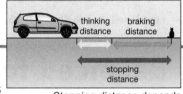

thinking distance
braking distance
stopping distance

Stopping distance depends on the speed of the car.

Grades D–C

- When a driver has to stop suddenly, the **stopping distance** depends on the distance the car travels during the driver's reaction time (**thinking distance**), and then while the brakes are applied (**braking distance**):
  stopping distance = thinking distance + braking distance
- A car's **speed** and the driver's **reaction time** both affect the thinking distance:
  thinking distance = speed × reaction time
- Drinking alcohol, taking other drugs or being tired prolongs the reaction time.
- **Braking distance** depends on:
  - **speed squared**; doubling the speed increases the braking distance four times
  - the car's **mass**; the car's **kinetic energy** is proportional to its mass
  - **friction** of brakes and tyres; greater friction shortens the braking distance, while icy or wet roads lengthen it.

How Science Works

You should be able to:
- draw and interpret velocity-time graphs for bodies that reach terminal velocity, including a consideration of the forces acting on the body
- calculate the weight of a body using: weight (N) = mass (kg) × gravitational field strength (N/kg)

Grades B–A*

- The tread on tyres ensures that water on the road squirts out sideways along the channels, and allows the rubber of the tyres to make good contact with a wet road. This maximises frictional forces, which enable the driver to make an emergency stop.
- Without channels for water to escape by, a braking car with bald tyres is liable to skid on a film of water, greatly increasing its stopping distance.

Questions

Grades D-C

1 Explain how a parachute reduces the terminal velocity of a person falling through the air.

Grades B-A*

2 Suggest why a 'birdman' reaches a higher terminal velocity than a parachutist.

Grades D-C

3 Thinking distance is an important component of the time taken for a driver to bring a car to a stop in an emergency. State **two** factors that might *increase* the thinking distance.

Grades B-A*

4 Why is a car with bald tyres liable to skid on wet roads?

Grades

Moving through fluids

The cyclist must apply greater force than on the road because the drag force of water is greater than air resistance.

Forces on objects moving in fluids

D–C

- Liquids and gases are **fluids** (they flow). They apply frictional forces on objects moving through them – **resistance** in air and **drag** in water. Frictional forces increase as: the **speed** of a moving object increases; the fluid gets **denser**.
- Birds and fish have a **streamlined** shape to reduce friction with air or water.
- An object falling through a liquid accelerates due to the force of **gravity**, until it reaches a **terminal velocity**. Then, the resultant force is zero and:

upthrust (liquid's resistance force) = object's weight due to gravity

Density and viscosity of liquids

B–A*

- A liquid exerts a **drag** force: an object moving through the liquid has to push the liquid particles aside. Drag is partly to do with density (mass per cubic metre): to move the same distance, an object must displace a greater **mass** of a denser fluid.
- Drag also depends on a liquid's **viscosity** (stickiness). **Oil** is less dense than water, but its viscosity is greater because its particles are more tightly bonded and harder to push apart.

Energy to move

How Science Works

You should be able to: discuss the transformation of kinetic energy to other forms of energy in particular situations.

Kinetic energy

D–C

- **Kinetic energy** is the energy of **movement**. Other forms of energy (e.g. chemical energy in food and petrol) are **transformed** into kinetic energy.
- Kinetic energy is **transformed** when: a hammer hits a horseshoe and **does work** flattening it; a gong is hit and produces **sound**; rubbing your hands produces **heat** (wasted energy); see diagram also.
- **Energy transfer diagrams** show how energy is **transferred** and **transformed**. Some energy is usually transformed into heat (wasted energy).

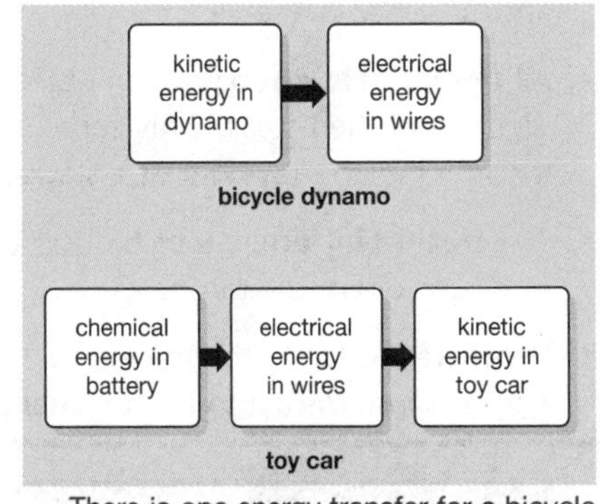

There is one energy transfer for a bicycle dynamo. There are two energy transfers for the toy car. In each stage, energy is also transformed.

Flywheels store kinetic energy

B–A*

- A flywheel is very heavy. A large input of energy makes it spin very fast; it then stores that kinetic energy with very little energy loss. (It is the mechanical equivalent of a battery.)
- If the electrical supply fails, the energy in a flywheel can immediately be used to generate electricity. This is vital to supply electricity during power failures, for example, at telecommunications centres and airports: Mexico International Airport has replaced slower-acting generators with flywheels.

Questions

Grades D-C

1 Why do fish need a more streamlined shape than birds?

Grades B-A*

2 Explain why it would be harder to swim through treacle than through water.

Grades B-A*

3 A child plays on a swing. At which point on the swing's movement does it have: **a** most potential energy; **b** most kinetic energy?

Grades B-A*

4 How can flywheels help to maintain electricity supplies?

Working hard

Work, energy and force

- When a **force** moves an object through a **distance**:

  work done (J) = energy transferred (J)

- In the drawings: the man does work against **air resistance**; the gardener does work against **friction**; the dog owner does work against **gravity**.

Each person is doing work by moving or lifting something.

D–C

- Pushing a trolley a **distance** of 20 m requires more **work** than pushing it 10 m.

  work done = force × distance moved in the direction of the force

  (joules, J) (newtons, N) (metres, m)

- In a train, Ruth uses a force of 150 N to lift a bag 1.5 m onto the luggage rack. How much work does she do?

  work done = 150 × 1.5 = 225 J

Both men raise their blocks the same amount. But the man on the right also does work against the friction between the ramp and the block.

How much energy?

Transferring and transforming energy

- An object's **kinetic energy** increases when its **mass** increases and/or when its **speed** increases. Kinetic energy decreases when mass and/or speed decrease.
- All the **kinetic energy** of an object is *transferred* if it collides with another object and then *stops*. Only some kinetic energy has been transferred if it continues to move but at a slower speed.

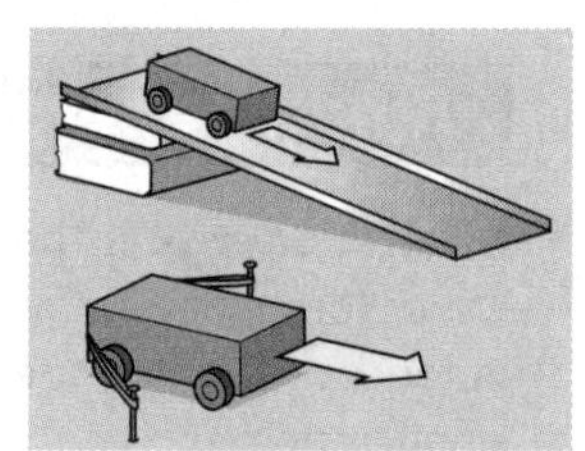
Two energy transformations.

D–C

- The **potential energy** of the top trolley is *transformed* into kinetic energy as it runs down the ramp.
- **Work done** to stretch the elastic band behind the trolley is stored as **elastic potential energy**. This is *transformed* into kinetic energy when the trolley moves forwards.

Calculating kinetic energy

- The kinetic energy of a body can be calculated from:

  kinetic energy = $\frac{1}{2}$ × **mass** × **speed²**

  joules, J kilograms, kg (metres/second)², (m/s)²

  This is written as: **kinetic energy** = $\frac{1}{2}mv^2$

B–A*

- Gary bowls a 1.5 kg bowl at a speed of 8 m/s. What is its kinetic energy?

  kinetic energy of bowl = $\frac{1}{2}$ × 1.5 × (8 × 8) = 48 J

Questions

Grades D-C

1 Jake lifts two parcels, each weighing 15 N, onto a desk 1 m above the ground. How much work does he do?

Grades B-A*

2 Michael is pushing a toy car along the ground. The car has a mass of 0.5 kg and he pushes it for 2 m. Michael says the work he has done on the car must be 0.5 × 2 J. Why is he wrong?

Grades D-C

3 Kay uses an elastic band as a catapult to fire a paper pellet across the room. Where did the kinetic energy in the pellet come from?

Grades B-A*

4 Calculate the kinetic energy of an aeroplane with a mass of 120 000 kg that is travelling at 100 m/s.

Momentum

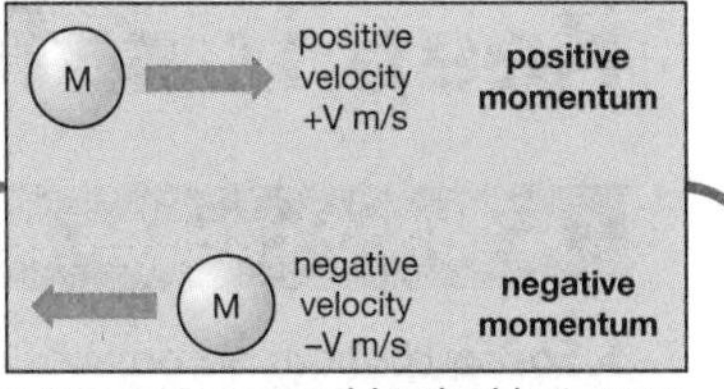

The two masses are identical but move in opposite directions. Therefore the momentum of one is positive and the momentum of the other is negative.

Momentum and collisions

- Objects have **momentum** when they move. Momentum depends on an object's **mass** and **velocity**:

  momentum = mass × velocity
  (kg m/s) (kg) (m/s)

- Momentum is a **vector** quantity since it has **direction** as well as **magnitude** (size). Depending on direction, its value is positive or negative (see diagram above right).
- Momentum is *not* the same as **kinetic energy**. Kinetic energy is a **scalar** quantity – it has magnitude but not direction. Its value is always positive.

Top Tip!

Kinetic energy is a scalar quantity; momentum is a vector quantity.

D–C

- Trolleys of known **mass** are used to investigate collisions and **momentum**. Ticker tape (page 113) gives their **velocities** just before and after collision.

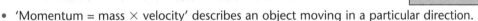

v (v = 0) before; v_1 after

An **inelastic** collision in which two objects collide and stick together. (They do not bounce apart after collision, as happens in **elastic** collisions.)

- When *no external forces* act on an object, the **Law of Conservation of Momentum** applies:

  total momentum before collision = total momentum after collision Momentum is conserved.

- A 1 kg mass moving at 10 m/s has an inelastic collision with a stationary 4 kg mass. What is the velocity after collision?

before		after	rearranged
(1 × 10) + (4 × 0)	=	(1 × v) + (4 × v)	10 + 0 = 5v
			v = 10/5 = 2 m/s

How Science Works

You should be able to: use the conservation of momentum to calculate the mass, velocity or momentum of a body involved in a collision or an explosion.

Off with a bang!

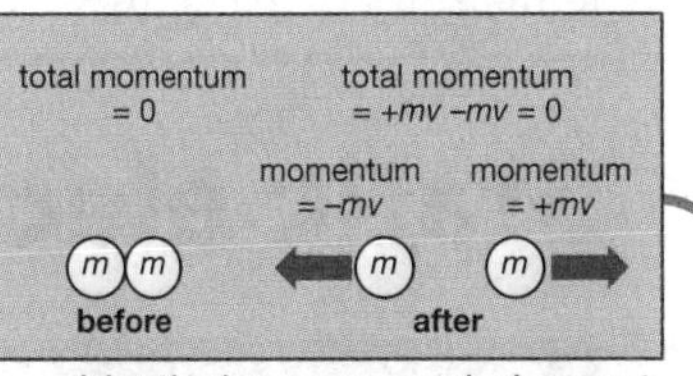

Identical masses explode apart. Positive momentum = negative momentum, so net momentum is zero.

Momentum and explosions

D–C

- Before an object **explodes**, nothing moves – momentum is zero. When it explodes, pieces fly in all directions. The net momentum of all the pieces will be zero. **Momentum is conserved.**

A lighter gun would recoil more.

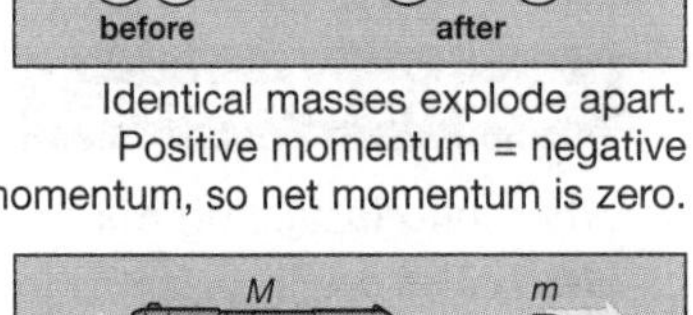

Momentum of spinning objects

B–A*

- 'Momentum = mass × velocity' describes an object moving in a particular direction.
- A **spinning** object, moving on the spot, has a different kind of momentum. Imagine a skater spinning with her arms outstretched and then pulling them in to her body. As she brings some of her mass closer to the centre, she spins faster.
- Momentum must be conserved. Since the skater's velocity increased and mass stayed the same, we conclude that 'spinning' momentum must also involve the *distance* of mass from the centre of spin.

Questions

Grades D-C

1 How does momentum differ from kinetic energy?

Grades D-C

2 A 2 kg mass moving at 2 m/s hits a stationary 4 kg mass. What is the momentum of each mass before the collision?

Grades B-A*

3 After they collide inelastically, the two masses continue in the same direction, travelling at the same speed as each other. What is this speed?

Grades D-C

4 Two inflated balloons, one small and one large, are resting on a table. What is the momentum of the balloons? One of the balloon bursts and shoots off the table. What can you say about the momentum of the moving balloon and the air that bursts out of it?

Keep it safe

The force exerted by the brakes changes the momentum of the car more slowly than the force exerted by the tree.

Grades

A force causes a change in momentum

- A **force** *changes* an object's **momentum** by changing its **velocity**:

  change in momentum = mass × change in velocity
  (kg m/s) (kg) (m/s)

  A force can also change an object's momentum by changing its **direction**, because momentum is a vector quantity.

- A change in momentum depends on the size of the **force** and the **time** over which it acts. The force is given by:

  force = change in momentum ÷ time
  (kg m/s^2) (kg m/s) (s)

- The diagrams show that a *small* braking force changes momentum *slowly*, and a *large* force from the tree changes momentum *quickly*.
- In a head-on collision (lower diagram), the car bonnet (**crumple zone**) collapses and slows the *rate of change* of the car's momentum, absorbing some of the car's **kinetic energy**.

How Science Works

You should be able to: use the ideas of momentum to explain safety features.

D–C

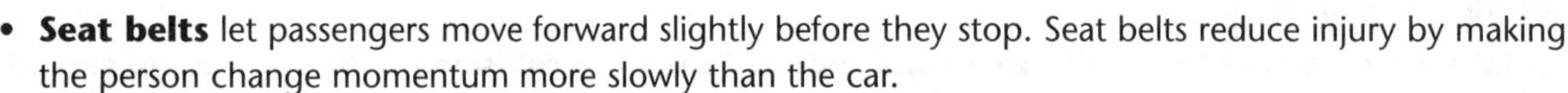

- **Seat belts** let passengers move forward slightly before they stop. Seat belts reduce injury by making the person change momentum more slowly than the car.
- A small car of mass 1200 kg travels at 15 m/s. The driver applies the brakes and the car takes 3 s to slow down to 10 m/s. What is the change in momentum, and what force is exerted by the brakes?

  change in momentum (kg m/s) = mass (kg) × change in velocity (m/s)
  change in momentum = 1200 × (15 – 10) = 6000 kg m/s
  force (kg m/s^2) = change in momentum (kg m/s) ÷ time (s) = 6000 ÷ 3 = 2000 kg m/s^2
  Remember that 1 kg m/s^2 is the same as 1 newton, so force = 2000 N

B–A*

Static electricity

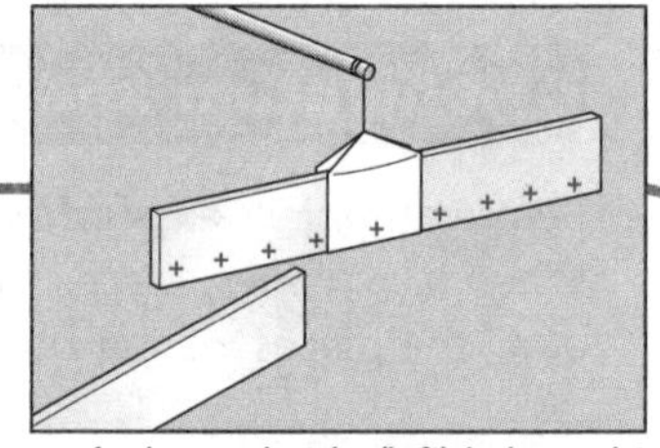

A charged strip (left) is brought close to a positively charged cellulose acetate strip.

Static electrical charge

- When two insulating materials are rubbed together, negatively charged **electrons** are transferred from one to the other, making both materials electrically charged. Their charges are *equal* and *opposite* **static** electrical charges.
- Like charges **repel**: both positive or both negative. Unlike charges **attract**: positive + negative.
- An **electrical insulator** is a material that holds static electrical charge (it is not a conductor).
- The suspended strip of cellulose acetate in the diagram above carries a positive charge. If the strip approaching it:
  - is also charged cellulose acetate, it will be positive and be repelled – *like* charges repel.
  - is charged polythene, it will be negative and be attracted – *unlike* charges attract.

D–C

Questions

Grades D-C

1 Use the idea of momentum to explain how wearing a seat belt helps to avoid head injuries in a car crash.

Grades B-A*

2 A vehicle of mass 2000 kg is travelling at 18 m/s. When the driver brakes, the car slows down to 10 m/s in a time of 5 s. Calculate the change in momentum. Then use your answers, and the information about time taken, to calculate the force applied by the brakes.

Grades D-C

3 A camper's shirt sleeve rubs against the fabric of her tent. The tent fabric becomes negatively charged. What has moved, and in which direction, to cause the tent fabric to become negatively charged?

Grades B-A*

4 Explain why static electricity normally only builds up on insulators, such as plastic.

Charge

Investigating charge

- Charges exert a **force** on each other: like charges **repel** and unlike charges **attract**.
- **Insulators** (e.g. plastics, glass) hold electrical charge.
- **Electrical conductors** (e.g. metals, water), allow charge (electrons) to flow as an **electrical current** in a closed circuit. They hold charge if insulated or if not part of a closed circuit.

D–C

- An **electroscope** shows if an object has an electrical charge. The cap and rod are metal and leaf is gold; they are on an insulating base.
- The electroscope also shows whether a charge is positive or negative.

a The gold leaf electroscope is **uncharged**. It has equal numbers of **positive** and **negative** charges all over it.
b The negative electrical charge on the strip repels the negative electrical charge on the electroscope. Charges move as far away as they can.
c The rod and the gold leaf both get a negative electrical charge. They move away from each other. The gold leaf rises.
d The more electrical charge the strip has, the more the gold leaf rises.

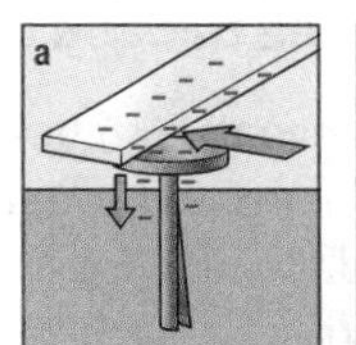

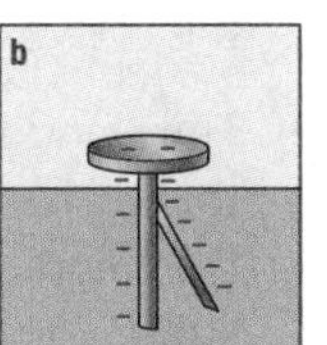

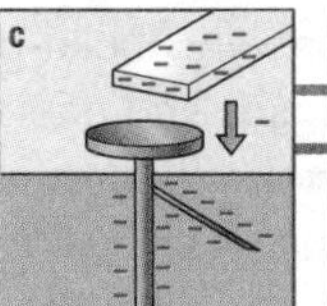

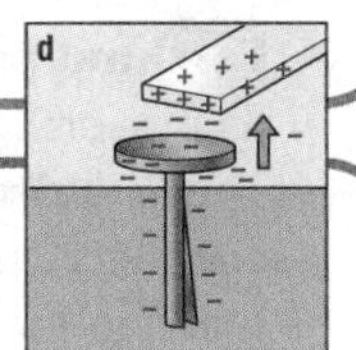

a Wipe a polythene strip known to be negatively charged across the cap. The negative charges move as far apart as possible down the electroscope.
b Remove the polythene strip. The electroscope now has a negative electrical charge. The gold is repelled and rises.
c Another negatively charged object held near the electroscope repels more negative charges down the rod. The gold leaf rises more.
d Holding a positively charged object near the electroscope attracts negative charges up to the cap. The gold leaf loses charge and falls.

Electrostatic induction

B–A*

- A balloon (of insulating material) gains a negative electrical charge when rubbed with wool. It will then 'stick' to the ceiling.
- The plaster of a ceiling is also an insulator. The charge on the balloon induces negative charge to move to the ceiling's upper surface, leaving its lower surface positively charged. This process is called **electrostatic induction**.

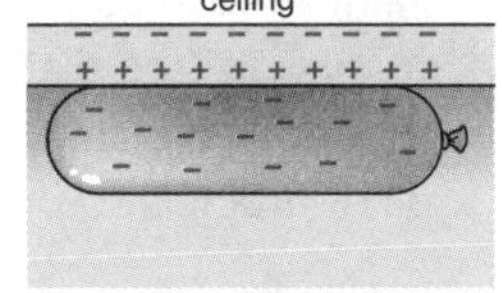

The charged balloon induces an opposite charge in the adjacent surface of the ceiling.

Van de Graaff generator

Redistribution of charge

D–C

- In a **Van de Graaff generator**, two materials rub together and build up equal and opposite static electrical charges. The dome gains a large but safe **positive charge**.
- A girl stands on an **insulating** mat and touches the dome: the charge is equally redistributed over her and the dome. Her charged hairs repel each other, move apart and stand on end. If she touches an **earthed conductor**, the charge will flow to earth, discharging her hair.

The charge causes the hairs to repel each other.

Questions

Grades D-C

1 Look at the diagrams of the electroscope. Explain why the gold leaf goes up after the polythene strip is wiped across the cap.

Grades B-A*

2 Suzanne strokes her cat on a hot, dry day. When she touches the tip of the cat's ear, she and the cat feel a small electric shock. Can you suggest what is happening?

Grades D-C

3 Why do the hairs on the girl's head repel each other, when she touches the dome of the Van de Graaff generator?

Grades D-C

4 What happens if you connect a charged object to earth with a conductor?

Sparks will fly!

Grades

Electrostatic charge, dangers and uses

- When the girl on the **insulating** mat (see page 119) touches the Van de Graaff generator dome, the charge becomes evenly distributed between dome and girl.
- When a conducting object which is **earthed** (connected to the earth) approaches the dome of a charged Van de Graaff generator, a **spark** (electrical charge) jumps across, and charge flows to earth as an electrical current.
- This happens when the charge on any insulated object is great enough. The greater the charge, the greater is the charge difference (**potential difference** or voltage) between the object and earth.
- When a charged object is connected to **earth** by a **conductor**, the object is **discharged** as charge flows though the conductor.
- The cloud's negative static charge **induces** a positive charge on the earthed lightning conductor. This charge streams up towards the cloud, discharging it and reducing the chance of a lightning strike.
- **Earthing** prevents build-up of charge in an operating theatre, where flammable fluids and gases are used. Plastics can become charged. They are **discharged** safely by keeping air humid (water conducts electricity) and installing a floor of material that conducts (earths) the charge.
- Paint droplets gain positive charge as they stream through the spray gun (see diagram on right). The car bodywork is earthed, so a negative charge is **induced** near the spray. It attracts the paint droplets so they are deposited evenly over the bodywork.

D–C

- Power stations use **smoke precipitators** (right) to clean flue gases and reduce pollution. A **high voltage** is applied to a grid of wires in the chimney, and they emit electrons into the space between them. **Smoke** particles in the flue gases gain negative charge and are repelled by the wires. Electrostatic forces attract them to earthed metal plates lining the chimney. The particles stick to the plates and are collected.
- The drum inside a **photocopier** is positively charged. When you press Copy, light reflected from white areas of a sheet being photocopied illuminates the drum. Charge from those areas flows away, leaving positive charge on the drum only where there is colour on the sheet. These areas of the drum therefore pick up **negatively charged** black toner, which is then transferred to the copy.

B–A*

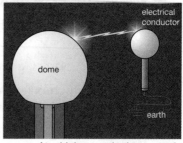

electrical conductor
dome
earth

At a high enough charge on the dome, a spark jumps across to the electrical conductor.

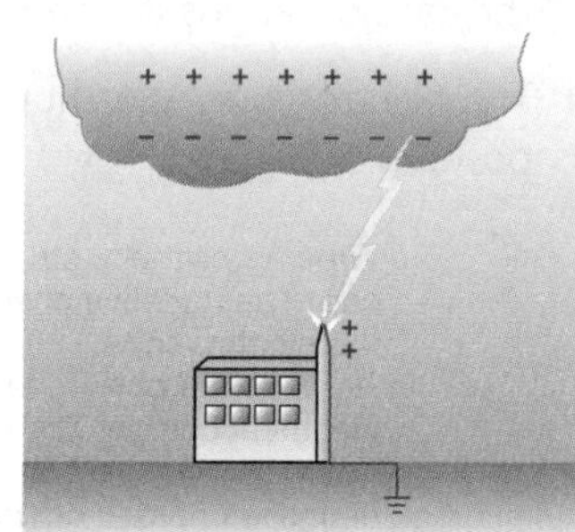

The earthed conductor discharges the cloud.

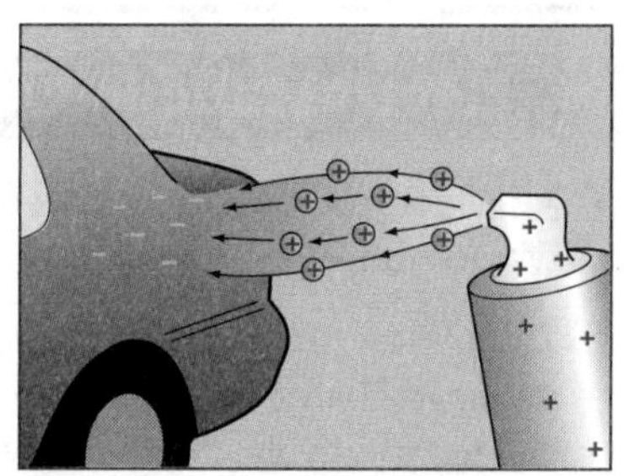

Positively charged paint droplets are attracted to the car's bodywork.

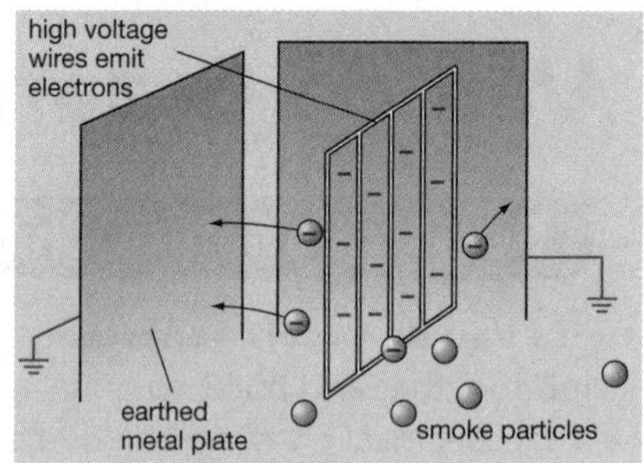

A smoke precipitator uses static charge to remove ash and dust from flue gas in power stations.

How Science Works

You should be able to:

- explain why static electricity is dangerous in some situations and how precautions can be taken to ensure that the electrostatic charge is discharged safely
- explain how static electricity could be useful.

Questions

Grades D-C

1 High-pressure water hoses that are used to clean tanks containing flammable oil are always made of metal and not plastic. Can you suggest why?

Grades D-C

2 Explain why the floor in an operating theatre is made of a conducting material, not an insulator.

Grades D-C

3 The toner in a photocopier has a negative charge. Why do particles of toner stick to some parts of the drum but not others?

Grades B-A*

4 How do the smoke particles in a smoke precipitator acquire a charge, and why do they move away from the wires?

P2a summary

Motion and forces

Distance-time graphs help us to 'picture' how an object moves. The **slope** of the graph shows the **speed**. If speed changes **gradually**, the slope is **curved**.

average speed = $\frac{\text{total distance travelled}}{\text{total time travelled}}$

The **velocity** of an object tells us its **speed** in a given **direction**. **Acceleration** is the **rate of change** of velocity.

acceleration (m/s^2) = $\frac{\text{change in velocity (m/s)}}{\text{time taken for change (s)}}$

Velocity-time graphs show how the velocity changes:

- the **slope** shows the **acceleration**
- the **area** under the graph shows the **distance travelled**.

Forces can add up or cancel out to give a **resultant force**.

The resultant force has the same effect as all the forces acting on the object.

When the resultant force is *not* zero, an object accelerates. Acceleration is:

- directly proportional to **force**, $a \propto F$
- inversely proportional to **mass**, $a \propto \frac{1}{m}$

resultant force (N) = mass (kg) × acceleration (m/s^2)

$F = ma$

For every force there is an equal and opposite force.

Balanced forces have the same value but act in opposite directions, e.g. a person floating in water has **stationary** balanced forces.

An object moving at **constant velocity** or **terminal velocity**, e.g. a parachutist, has balanced forces.

Using a **force** to do **work** on an object gives it **energy**:

work done = energy transferred

The amount of work done depends on **force** and **distance**:

work done (J) = force (N) × distance moved in the direction of the force (m)

stopping distance = thinking distance + braking distance

thinking distance = speed × reaction time

Stopping distance increases with:

- increasing **speed**
- a greater **mass**
- reduced **friction**
- increased **reaction time**.

Frictional forces arise between a fluid and a moving object. **Air resistance** in gases and **drag** in liquids increase as: the object's **speed** increases; the fluid gets **denser**.

An object **falling** through a **fluid**, **accelerates** until it reaches **terminal velocity**, when the **resultant force** is **zero**:

upthrust (fluid's resistant force) = object's weight due to gravity

Every **moving** object has **kinetic energy** (a **scalar** quantity) that can be transformed into other forms.

An object's kinetic energy increases when its:

- **mass** increases
- **speed** increases.

kinetic energy = $\frac{1}{2}mv^2$

Every **moving** object has **momentum** that depends on its **mass** and its **velocity**:

momentum (kg m/s) = mass (kg) × velocity (m/s)

Momentum (a **vector** quantity) has **size** and **direction** and is **conserved** in collisions and explosions.

Forces change an object's momentum by changing its **velocity** or **direction**. The bigger the force on a moving object, the larger its **rate of change** of momentum.

change in momentum (kg m/s) = mass (kg) × change in velocity (m/s)

force (N) = $\frac{\text{change in momentum (kg m/s)}}{\text{time taken for the change (s)}}$

Static electricity

Rubbing **electrical insulators** together can build up **static electricity**, because **electrons** transfer from one material to another.

If an object **gains** electrons, it has a **negative** charge. If it **loses** electrons, it has a **positive** charge. Similar (like) charges **repel**. Opposite charges **attract**.

A **gold leaf electroscope** shows the **size** and **type** of charge on an object.

Static electricity can be dangerous if **sparks** cause flammable materials to **ignite** or **explode**. **Earthing** prevents a build-up of static charge.

Paint sprayers, smoke precipitators and photocopiers are all useful **applications** of static electricity.

Grades

Circuit diagrams

Standard symbols and circuit diagrams

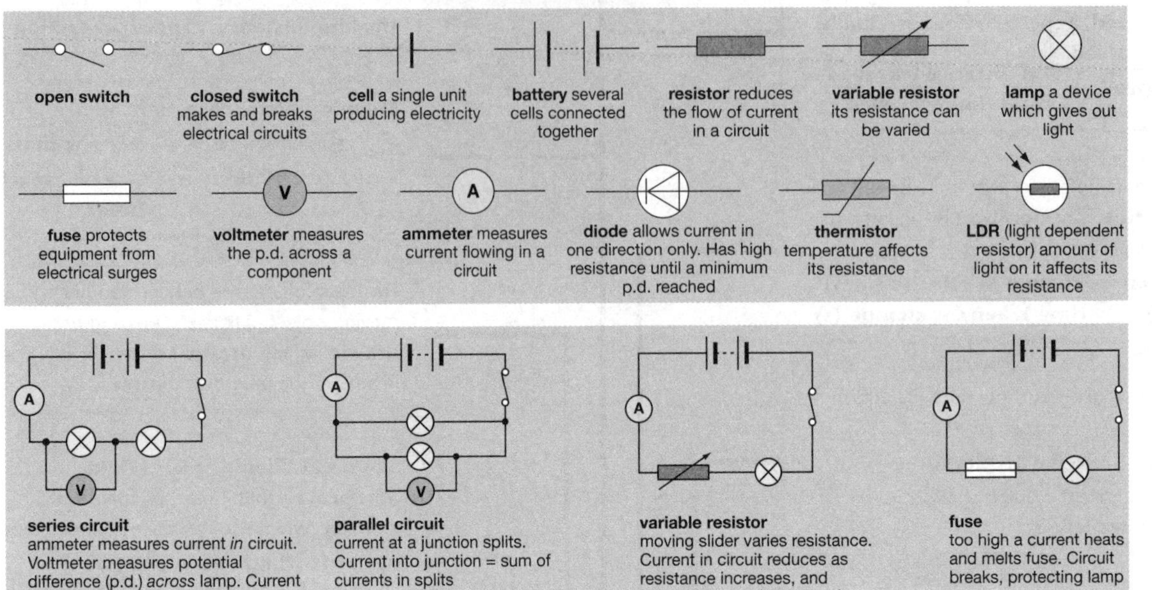

open switch
closed switch makes and breaks electrical circuits
cell a single unit producing electricity
battery several cells connected together
resistor reduces the flow of current in a circuit
variable resistor its resistance can be varied
lamp a device which gives out light
fuse protects equipment from electrical surges
voltmeter measures the p.d. across a component
ammeter measures current flowing in a circuit
diode allows current in one direction only. Has high resistance until a minimum p.d. reached
thermistor temperature affects its resistance
LDR (light dependent resistor) amount of light on it affects its resistance
V
A
series circuit
ammeter measures current *in* circuit. Voltmeter measures potential difference (p.d.) *across* lamp. Current is *same* at any point in circuit
parallel circuit
current at a junction splits. Current into junction = sum of currents in splits
variable resistor
moving slider varies resistance. Current in circuit reduces as resistance increases, and increases as resistance reduces
fuse
too high a current heats and melts fuse. Circuit breaks, protecting lamp

D–C

- **Conservation of energy** (see page 50) applies to electrical circuits: the energy the cell *delivers* = the energy the components *use*. **Current** is the energy carrier.
- In the **series** circuit above, current is the same at all points. (See also page 124.)
- In the **parallel** circuit, the *sum* of currents through the lamps = current from the cell. (See also page 125.)

B–A*

How Science Works

You should be able to: interpret and draw circuit diagrams using standard symbols.

Resistance 1

Electrons as energy carriers

- The outermost electrons of atoms in a conductor are **free** electrons – free from their fixed parent atoms, which are left as positive ions. **Electrical current** is the flow of these electrons.
- The free electrons slow down when they **collide** with the fixed positive ions and their bound electrons. The more collisions there are in a conducting material, the greater is its **resistance** and the smaller is the current.

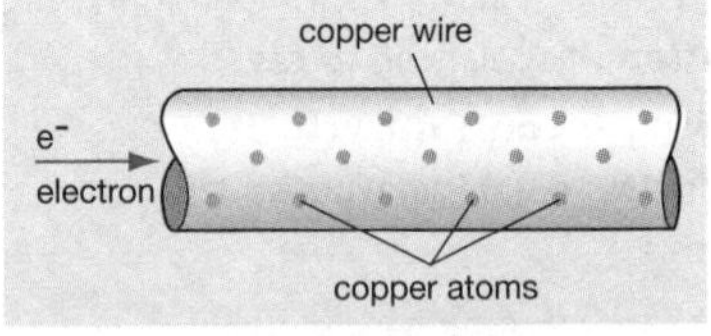

copper wire
e^-
electron
copper atoms

Free electrons travel between the fixed positive ions of a copper wire.

D–C

- As they have free electrons, metals (particularly copper) are good **conductors** with low resistances.
- **Insulators** have no free electrons, so are very poor conductors with high resistances.
- A **potential difference** (**voltage**) exerts a **force** that pushes the free electrons along.

Top Tip!

Potential difference is the amount of 'push' a battery has to move current around a circuit.

Unit of potential difference = volt (V)

Symbol for potential difference = *V*

Questions

Grades D-C

1 Draw a series circuit containing two ammeters, one lamp and a voltmeter measuring the potential difference across the lamp.

Grades B-A*

2 What can you say about the current in the different branches of a parallel circuit?

Grades D-C

3 Why are metals good conductors?

Grades D-C

4 What is the name for the force that pushes electrons along in a circuit?

Resistance 2

Factors affecting resistance

D–C

- Electrical **resistance** is due to electron collisions (see page 122). The number of collisions: *increases* as the **length** of a wire increases; *decreases* as the wire's **thickness** increases (more electrons are free to flow). Resistance is also different for different materials.
- Increasing **temperature** *increases* a wire's **resistance**: as **temperature** rises, fixed positive ions gain **kinetic energy** and **vibrate** more vigorously. They repel and deflect free electrons more frequently and with greater energy.

Ohm's Law

Ohm's Law: resistance, potential difference and current

D–C

- **Georg Ohm** investigated how **potential difference (p.d.)** relates to **current**, using the circuit shown. As he adjusted the variable resistor, he recorded potential difference across the **resistor** and the current through it.

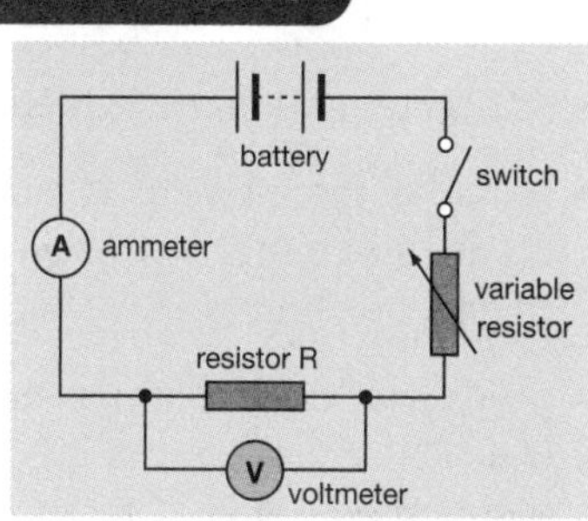

Georg Ohm's electrical circuit.

- A graph of p.d. against current (mid-right) shows that current is *directly proportional* to potential difference across the resistor. This is Ohm's **straight-line graph**. It illustrates Ohm's Law, and gives the equations:

potential difference (volt, V) = current (ampere, A) × resistance (ohm, Ω)
resistance = p.d. ÷ current

- As voltage increases, so does the current, while the resistance of resistor R remains constant. So, to find the **resistance** of any component, put it in place of resistor R, measure the current through it and the p.d. across it, and calculate from: resistance = p.d. ÷ current.

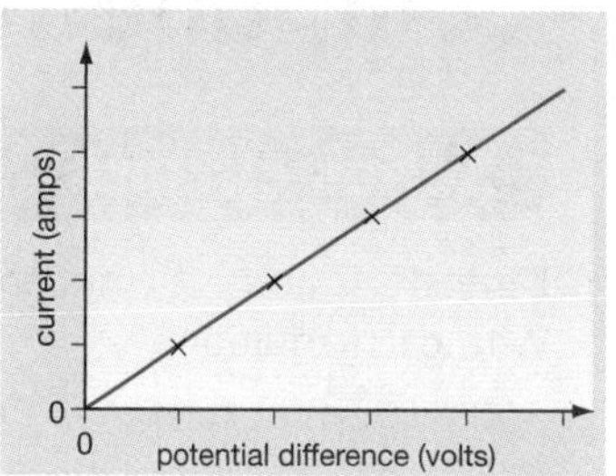

Ohm's graph for potential difference plotted against current.

- For a *fixed* p.d. (voltage), the greater the resistance, the smaller the current.
- A **filament lamp** behaves differently: with increasing p.d., the current levels off (see lower right) because the resistance increases as the filament gets hotter.

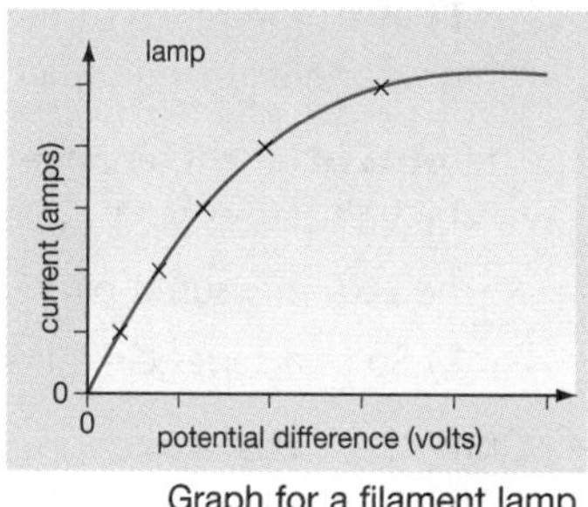

Graph for a filament lamp.

B–A*

- See the graphs below. In the first two, the graph line in the bottom left quarter is for current in the *reverse* direction. As the **diode** graph shows, no current flows in one direction (left of *y*-axis); current starts to flow (in the other direction) only when a certain p.d. is reached.

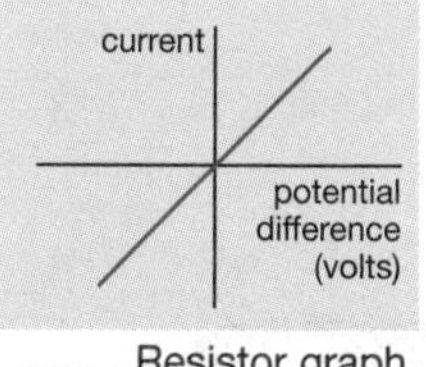

Resistor graph.

Filament lamp graph.

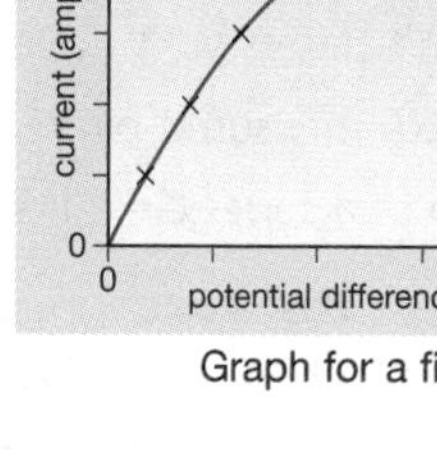

Diode graph.

Questions

Grades D-C

1 When a current flows through the metal wire in a filament lamp, it gets hot. How does this affect its resistance?

Grades D-C

2 Write down the equation that links current, potential difference and resistance (Ohm's Law). Use it to calculate the resistance of a circuit component in which a current of 3 A is flowing, and across which there is a potential difference of 6 V.

Grades D-C

3 Why does a filament lamp not obey Ohm's Law?

Grades B-A*

4 Explain the shape of the curve for a filament lamp above.

More components

Grades

Thermistors and LDRs

- Thermistors and LDRs contain **semiconductor** materials whose resistance *decreases* as they receive **energy**. It frees more electrons to flow and so **current** *increases*.

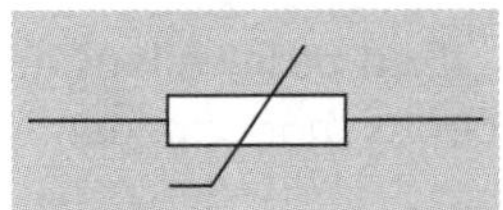
Symbol for a thermistor.

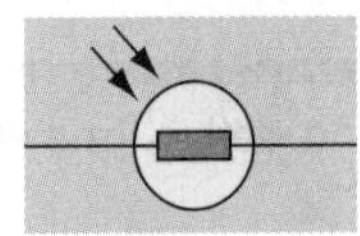
Symbol for an LDR.

How Science Works

You should be able to: apply the principles of basic electrical circuits to practical situations.

D–C

- When the **temperature** of a **thermistor** is increased, its resistance *decreases*.
- Electronic thermometers and fire alarms use thermistors to detect temperature changes.
- When **light** falls on a **light dependent resistor** (**LDR**), its resistance *decreases*.
- LDRs are used in smoke detectors, automatic light controls and burglar alarms.

B–A*

- A digital thermometer is much safer than a glass thermometer. It contains a thermistor. As the thermistor warms up, current flow increases. This is transformed into a digital readout.
- Automatic light control systems in offices ensure that employees work in good light. An LDR circuit registers loss of natural light as a decrease in current. This turns the lighting on. When natural light is restored, the LDR triggers the 'off' switch.

Components in series

Series circuits, resistance and batteries

D–C

- Lamps 1, 2 and 3 are added to a **series circuit** one at a time. With each addition:
  - the lamps get dimmer. Each lamp has a resistance. Adding lamps in a series circuit **increases resistance**, so the **current** decreases – this is measured by ammeters (A_1–A_4).
  - (**1**) In a series circuit, the value of the **current** through all components is the *same*.

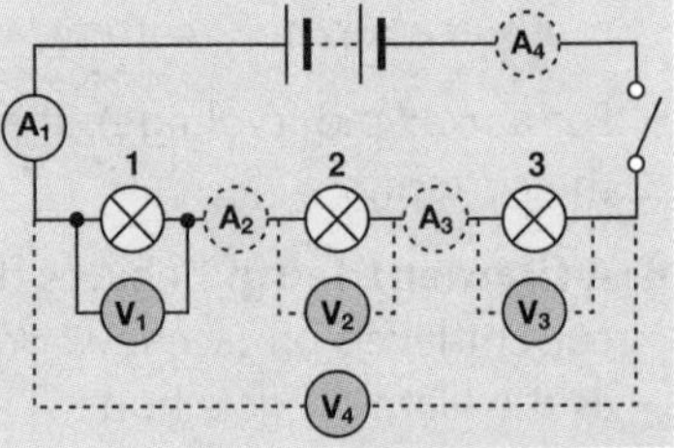

Lamps connected in series: measuring current and p.d.

- The **potential difference** (p.d.) across each lamp is measured by the voltmeters V_1, V_2 and V_3, and across all three lamps by V_4.
  - The p.d. measured by V_4 = *sum* of p.d.s. ($V_1 + V_2 + V_3$) = p.d. (voltage) of the battery.
  - (**2**) So in a series circuit, **total p.d.** is *shared* between all components.
- Ohm's Law, **resistance = p.d. ÷ current**, gives the resistance of each lamp from p.d. and current measurements.
  - (**3**) In a series circuit, **total resistance** = sum of resistances of components.
- A **battery** is several **cells** connected in series:
  - with all cells connected plus to minus, p.d. of battery = sum of p.d.s of cells.

Top Tip!

Remember: In a series circuit, total p.d. is the *sum* of the p.d.s of each cell. This p.d. is *shared* between the components.

Questions

Grades D-C

1 Give **one** use of an LDR.

Grades B-A*

2 With reference to resistance, explain this use of an LDR.

Grades D-C

3 Explain why the brightness of a lamp in a series circuit decreases if more lamps are added in the circuit.

Grades D-C

4 A cell has a p.d. of 1.5 V. What is the p.d. across four of these cells connected in series?

Components in parallel

Current and potential difference in parallel circuits

D–C

- In a **parallel circuit** (see diagram), the current takes different routes at **junctions**, and then joins up again.
- Lamps 1, 2 and 3 are added to a **parallel circuit**:
  - at each addition, the **brightness** of the lamps remains the same
  - with all three lamps added, ammeters A_2, A_3 and A_4 show that the current is the same through each lamp: $\mathbf{A_2 = A_3 = A_4}$
- Ammeters A_1 and A_5 measure the same **total current**. It equals the *sum* of the currents through the separate lamps: $\mathbf{A_1 = A_5 = A_2 + A_3 + A_4}$
  - (**1**) In a parallel circuit, **total current** is the sum of the currents through the separate components: current is *shared* between all components.
- Voltmeters (see diagram) show that the **potential difference** (p.d.) is the same across the battery and each lamp: $\mathbf{V_1 = V_2 = V_3 = V_4}$
  - (**2**) In a parallel circuit, the **potential difference** across all components (including the battery) is the *same*.

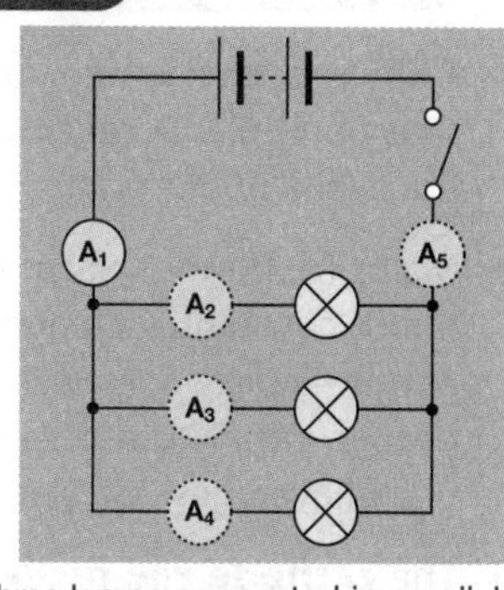

Three lamps connected in parallel.

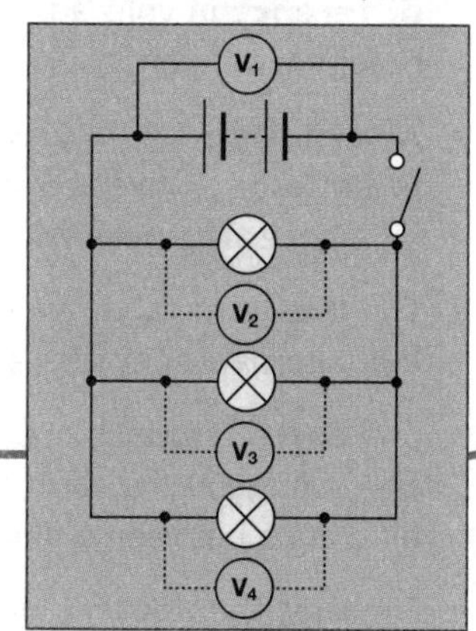

Measuring potential difference.

The three-pin plug

Structure and wiring a three-pin plug

D–C

- A **cable** and a **three-pin plug** connect most electrical appliances to the mains. The cable carries the three wires, live, neutral and earth, or 'twin and earth'. **Insulating** material encloses the cable and all of the plug except the pins.
- In a **three-pin plug**, the wires are colour coded:
  - **yellow** and **green** striped wire is **earth**, connected to the top pin
  - **blue** wire is **neutral**, connected to the left pin
  - **brown** wire is **live**, connected through the fuse to the right pin
  - **fuse** (see page 126) joins the brown wire and the right pin.
- When the **fuse** blows, it isolates the live wire and no current flows.

B–A*

- Appliances with a plastic outer casing and no touchable metal parts use **twin-core cable**, with live and neutral wires only. They are **double insulated** appliances: they have a fuse and the casing is an insulator.

  They carry the following symbol: 
- Hairdryers, electric drills and lawn mowers are double insulated appliances.

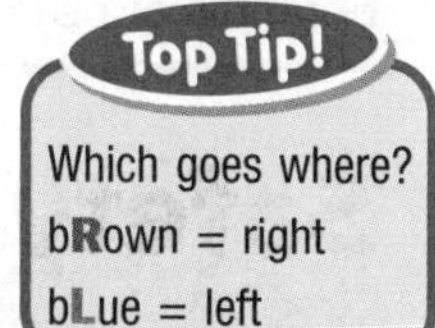

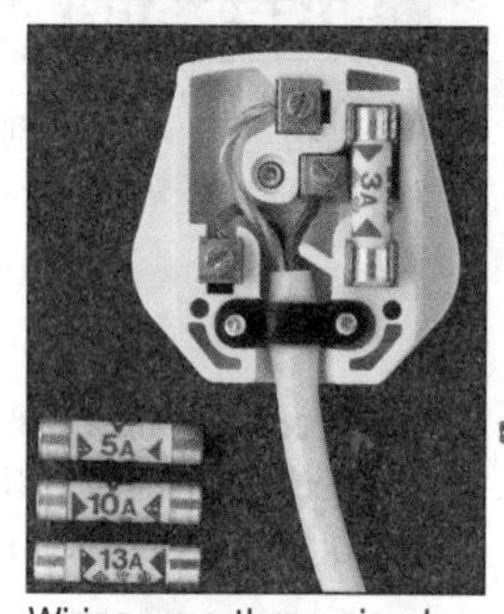

Wiring up a three-pin plug.

How Science Works

You should be able to: recognise errors in the wiring of a three-pin plug.

Questions

Grades D-C

1 In the first circuit diagram, the current flowing through ammeter A_1 is 12 A. What is the current measured by ammeter A_5?

Grades D-C

2 In the first circuit diagram, the cells provide a potential difference of 4.5 V. The three lamps are identical. What is the p.d. across each lamp?

Grades D-C

3 Why does a plug contain a fuse?

Grades B-A*

4 Explain why you can use a twin-core cable, without an earth cable, for appliances that have an outer plastic casing.

P2B DISCOVER NUCLEAR FUSION!

Grades

Domestic electricity

Direct current and alternating current

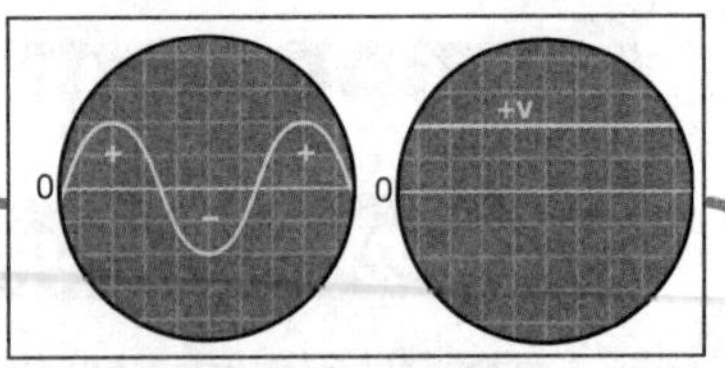

Oscilloscope trace for voltage of alternating current (left) and direct current (right) against time.

- **Cells** and **batteries** supply **direct current**, **d.c.**, in one direction only. **Direct** current gives a straight-line, positive trace (see upper right).
- Mains (domestic) electricity supplies **alternating current**, **a.c.** It constantly changes direction, oscillating from plus to minus – positive to negative – (see wave traces) at 50 times per second, 50 Hz – its **frequency**. The potential difference of a.c. averages 230 V, often written as (220–250) V.
- The **cathode ray oscilloscope** (**CRO**) displays **potential difference** in volts against time. The *x*-axis is at zero potential difference.
- **Alternating** current has an S-shaped trace. The maximum height of the wave above the mid line gives the **peak** potential difference. The peak value for mains electricity is about 250 V.

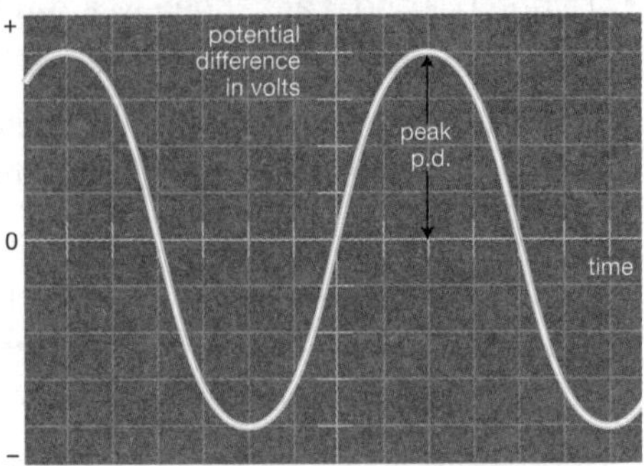

An oscilloscope trace for alternating current.

D–C

- The **live** terminal of the mains supply has a p.d. (voltage) that alternates between +230 V and –230 V.
- The **earth** terminal is connected to the ground. The voltage at the earth terminal is zero (no potential difference between a plug's 'earth' pin and the ground).
- Similarly, the **neutral** terminal (which is connected to the ground at the substation, see page 55) has a potential of nearly zero volts.

How Science Works

You should be able to:

- recognise dangerous practice in the use of mains electricity
- compare potential differences of d.c. supplies and the peak potential differences of a.c. supplies from diagrams of oscilloscope traces
- determine the period and hence the frequency of a supply from diagrams of oscilloscope traces.

B-A*

Safety at home

Fuses and earth wires

- An electrical fault may cause a surge in the current that could damage an appliance and injure or kill someone. Fuses and circuit breakers are designed to break the circuit instantly and cut off the current.
- The **fuse** in a plug joins the live wire to its pin. A current that exceeds the fuse rating will heat and melt the fuse wire.
- In **circuit breakers**, an electromagnet or electronic mechanism immediately breaks the circuit when there is a hazardous increase in current. Circuit breakers are used in fuse boxes.
- A **residual current device** (**RCD**) is a circuit breaker that detects any fault as a difference between the current in the live and neutral wires. It then breaks the circuit in less than 0.05 seconds. An RCD is used in appliances such as lawn mowers and hedge trimmers.
- Appliances with metal casings have an **earth wire**. If the live wire touches the metal casing, a large current flows from the live wire through the earth wire. The surge in current melts the fuse and breaks the circuit.

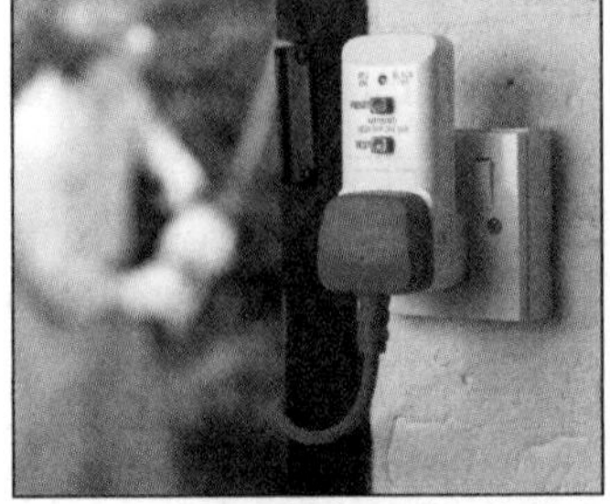
Residual current device.

D–C

Top Tip!

The fuse and earth wire together protect the appliance and the user.

Questions

Grades D-C

1 What does the trace of direct current look like on an oscilloscope screen?

Grades B-A*

2 On the oscilloscope trace for an alternating current, each small square represents 58 V. What is the peak voltage of this a.c. supply?

Grades D-C

3 Suggest why having an RCD in a circuit is even safer than a fuse.

Grades D-C

4 The earth wire route for an electric current has a very low resistance. How does this help to reduce the risk of you getting a shock from an electrical appliance?

Grades

P2B DISCOVER NUCLEAR FUSION!

Which fuse?

How Science Works

You should be able to: calculate the current through an appliance from its power and the potential difference of the supply, and from this determine the size of the fuse needed.

Calculating the correct fuse rating

- The **rate** at which energy is transformed is measured in joules per second, also known as watts: 1 J/s = 1 watt. The rate of energy transfer is called **power**:

  (**1**) **power (watts, W) = $\frac{\textbf{energy transformed (joules, J)}}{\textbf{time taken (seconds, s)}}$**

  – The **power rating** (W) is the amount of energy an electrical appliance uses per second (J/s). A 100 W light bulb transforms electrical energy to light and heat at a rate of 100 J/s.

D–C

- The power rating determines the **fuse** (in amperes, A) that an appliance requires:

  (**2**) **power = p.d. × current**, so **current = power ÷ p.d.** (mains p.d. = 230 V)

- A blender has a power rating of 1000 W. What fuse does it require?

  First work out the **current** the blender requires:

  current = power ÷ p.d = 1000 W ÷ 230 V = 4.3 A

  A 3 A fuse would melt when the blender is switched on, and a 13 A fuse would not melt if a fault occurred. The correct fuse is 5 A.

B–A*

- **Current** is the **flow of electrical charge**. The amount of charge carried by a current of 1 A flowing for 1 s = 1 **coulomb**. So current = charge ÷ time, and:

  charge (coulomb, C) = **current** (ampere, A) × **time** (seconds, s)

- The **energy transformed** in a circuit depends on the p.d. (amount of force, see page 122) and the charge being forced along:

  energy transformed (joule, J) = **potential difference** (volt, V) × **charge** (coulomb, C)

- When an electrical charge flows through a resistor (e.g. a lamp filament), some of the **electrical energy** is transformed into **heat energy**.

Radioactivity

particle	relative mass	charge
proton	1	+1
neutron	1	0
electron	0	–1

Atoms and atomic particles

D–C

- The table shows characteristics of the particles in an atom (see page 26).
- An atom has no net charge: **number of protons = number of electrons**
- Atoms may lose or gain electrons to become **ions** (charged particles).
- All atoms of a particular element have the *same* number of protons. Atoms of different elements have *different* numbers of protons.
- The number of: protons in an atom is its **atomic number**; protons + neutrons is its **mass number**.
- The number of neutrons in the atoms of an element may vary, hence the mass number varies. The different forms of an element are **isotopes**.
- The nuclei of some isotopes are unstable. They emit energy as particles or rays called **radiation** (see page 128).

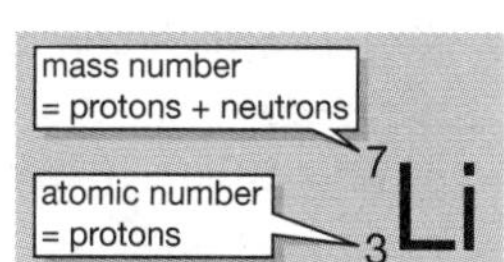

The atomic number shows which **element** an atom is, and the mass number shows which **isotope** it is.

Questions

Grades D-C

1 A hairdryer has a power rating of 1500 W. If the mains supply is 230 V, what current flows through the hairdryer?

Grades B-A*

2 A charge of 2 C passes though a point in a wire in 4 s. What is the current?

Grades B-A*

3 The potential difference across the wire in Question 2 is 1.5 V. Calculate the amount of energy transformed in this 4 s.

Grades D-C

4 This is the full symbol for boron: $^{11}_{5}B$. How many protons does it have in its nucleus? How many neutrons does it have in its nucleus? What is its mass number?

Alpha, beta and gamma rays 2

Grades

Types of radiation

D–C

- In **radioisotopes**, the balance of protons and neutrons is **unstable**. To reach stability, the nucleus emits **radiation** which carries energy.
- **Alpha (α) radiation** consists of 2 protons and 2 neutrons – a helium nucleus.
- **Beta (β) radiation** is a very fast, high energy electron.
- **Gamma (γ) radiation** is an electromagnetic wave.
- For details on alpha, beta and gamma radiations (see page 63).

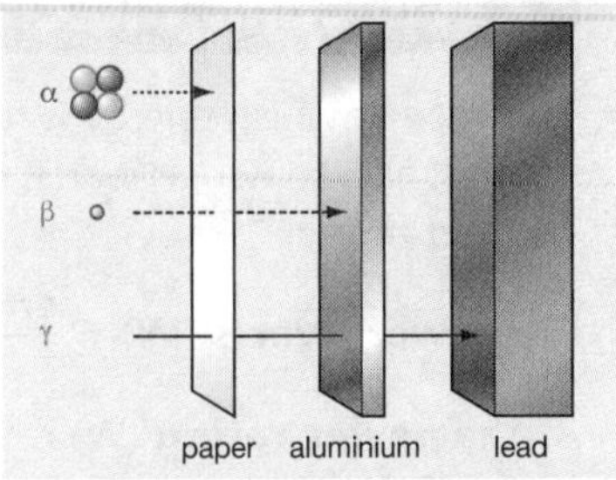

The ability of different radiations to penetrate different materials depends on their energy and mass.

B–A*

- When either an alpha or a beta particle leaves an atom, its nucleus becomes that of a new element.
- The radioisotope $^{228}_{88}$radium decays by emitting an **alpha** particle, $^{4}_{2}$helium (or $^{4}_{2}\alpha$). The nucleus of the radium atom therefore loses 2 protons and 2 neutrons and becomes an isotope of radon, $^{224}_{86}$Ra:

$$^{228}_{88}\mathrm{Ra} \rightarrow {}^{224}_{86}\mathrm{Rn} + {}^{4}_{2}\alpha$$

- The radioisotope $^{218}_{84}$polonium decays by emitting a **beta** particle, $^{0}_{-1}$e (or $^{0}_{-1}\beta$) *from its nucleus*, and becomes an isotope of astatine, $^{218}_{85}$At. This happens when a neutron (no charge) changes to a proton (+1 charge). So the mass number doesn't change, but the atomic number increases by 1:

$$^{218}_{84}\mathrm{Po} \rightarrow {}^{218}_{85}\mathrm{At} + {}^{0}_{-1}\beta$$

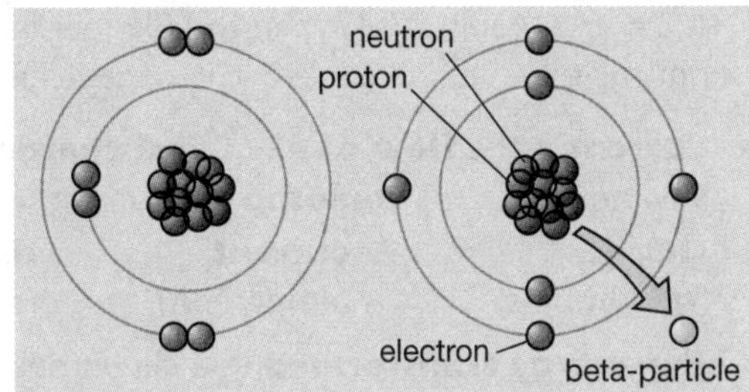

Left: the stable isotope carbon-12. Right: the radioisotope carbon-14 emitting a beta particle. This is the radioactive decay detected in radiocarbon dating (see page 64).

Background radiation 2

Sources of background radiation

D–C

- There is always a low level of natural **background radiation** in the environment. Living things have evolved to survive it.
- Its sources include radioactive materials in rocks and soil, and **cosmic rays** from the Sun. The Earth's magnetic field and the atmosphere shield us from most harmful cosmic rays.
- **Granite** rock contains radioactive materials that decay to give radon gas, itself radioactive and emitting harmful radiation (see page 63).
- Some background radiation is due to the use or manufacture of radioactive materials. Those who work with radioactive materials risk higher radiation levels and so monitor their exposure.

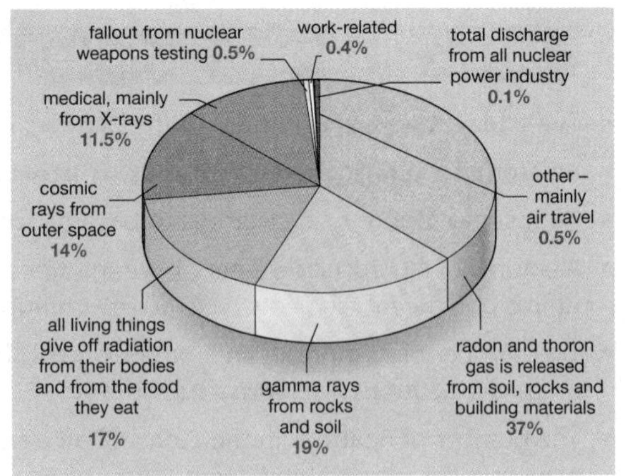

Sources of background radiation in the UK.

Questions

(Grades D-C)

1 State the charges on: **a** an alpha particle; **b** a beta particle; **c** gamma radiation.

(Grades B-A*)

2 The isotope of polonium $^{214}_{84}$Po decays by emitting an alpha particle. What is the new element that is formed?

(Grades B-A*)

3 Write an equation for the decay of $^{214}_{84}$Po.

(Grades D-C)

4 What is the major source of background radiation in the UK? How can this explain the high levels of background radiation in places such as Cornwall, where there is a lot of granite?

Inside the atom

Grades

Investigating the structure of the atom

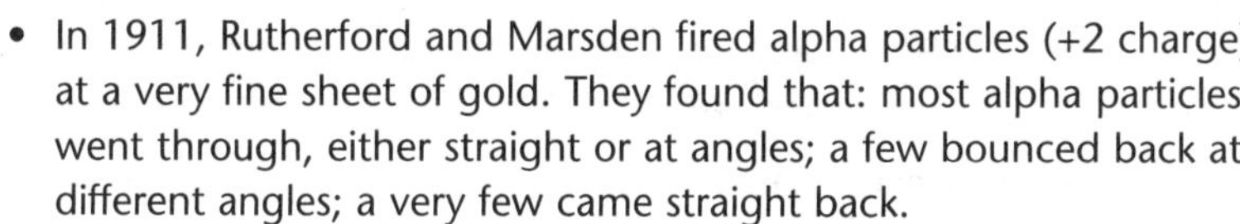

B–A*

- Scientists knew that atoms were neutral. In 1897, J.J. Thompson discovered negatively charged electrons, and suggested that this charge must be balanced by a positive charge. He proposed that an atom was a mass of positive matter studded with electrons – the 'plum pudding' model.

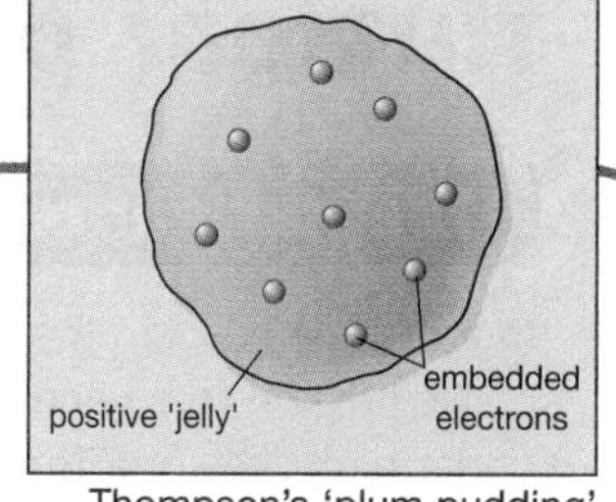

Thompson's 'plum pudding' model of the atom.

D–C

- In 1911, Rutherford and Marsden fired alpha particles (+2 charge) at a very fine sheet of gold. They found that: most alpha particles went through, either straight or at angles; a few bounced back at different angles; a very few came straight back.
- Their conclusions were that: most of an atom is empty space; it has a very small, dense **nucleus**; electrons are not part of the nucleus, but orbit round it.

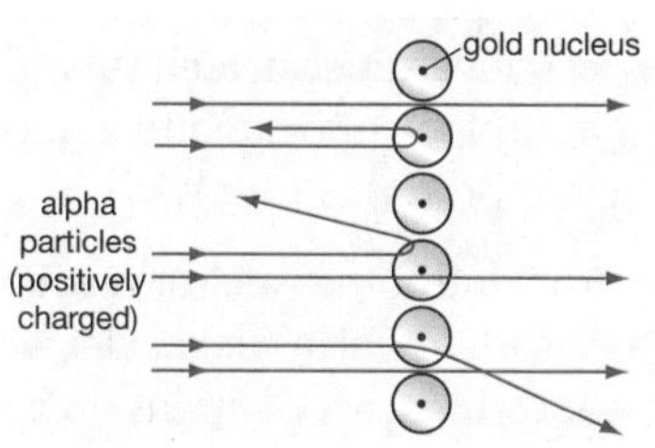

The scientists' model to explain what happened in their investigation.

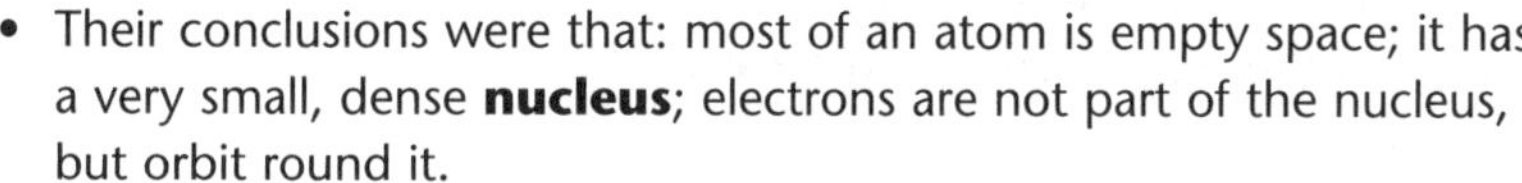

- Like charges repel so, when protons were discovered, scientists wondered why their positive charge did not push them apart.
- Protons do exert a strong **electrical force** of repulsion on each other.

B–A*

- However, a force 100 times stronger pulls them together. This is the **strong nuclear force** which attracts protons and neutrons and keeps the nucleus together.

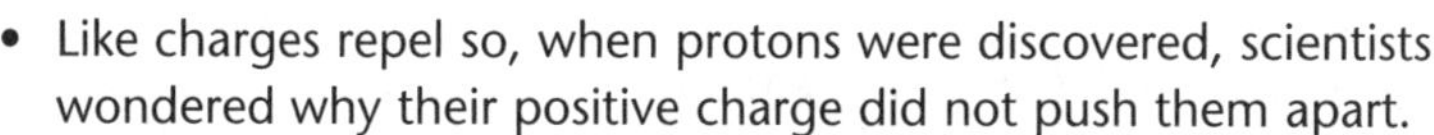

Rutherford and Marsden's scattering apparatus.

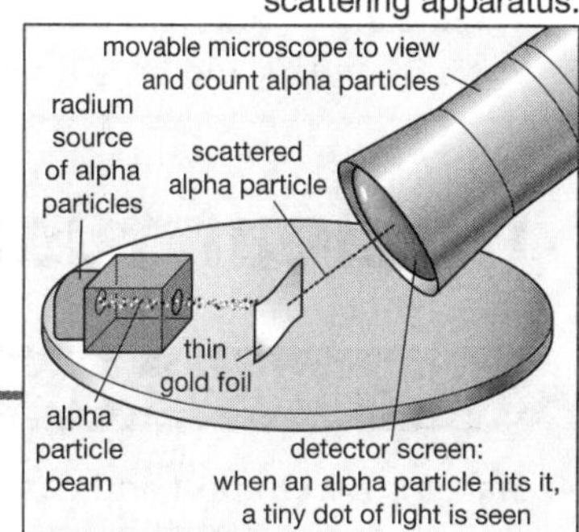

How Science Works

You should be able to:

- explain how the Rutherford and Marsden scattering experiment led to the 'plum pudding' model of the atom being replaced by the nuclear model.
- sketch a labelled diagram to illustrate how a chain reaction may occur.

Nuclear fission

Splitting the atom

D–C

- When an atomic nucleus splits in **nuclear fission**, some of its mass is converted to energy. Uranium-235, uranium-238 and plutonium-239 have large **unstable** nuclei and are used as fuel to produce **nuclear energy** in nuclear power stations.
- Neutrons are fired at material containing unstable atoms. The nucleus of an unstable atom absorbs a neutron, splits into two smaller nuclei and releases nuclear energy. It also produces two or three more neutrons that set up a **chain reaction**.

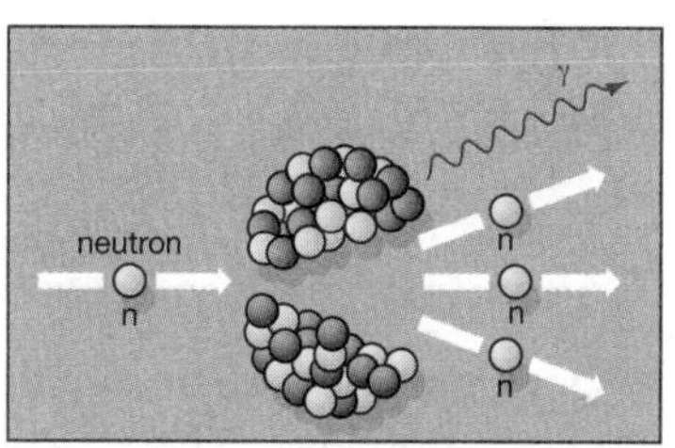

A single atom splitting.

- To produce large amounts of energy continuously, the uranium-235 has to be pure and at least tennis ball-sized, so that neutrons stay inside the mass and set up the chain reaction. The minimum amount is called the **critical mass**.

B–A*

- When an atom of uranium-235 absorbs a neutron, it undergoes nuclear fission to form barium-143 and krypton-90, and to produce three more neutrons: $^{235}_{92}U + ^{1}_{0}n \rightarrow ^{143}_{56}Ba + ^{90}_{36}Kr + ^{1}_{0}n + ^{1}_{0}n + ^{1}_{0}n$ + **ENERGY**

The neutrons continue the chain reaction.

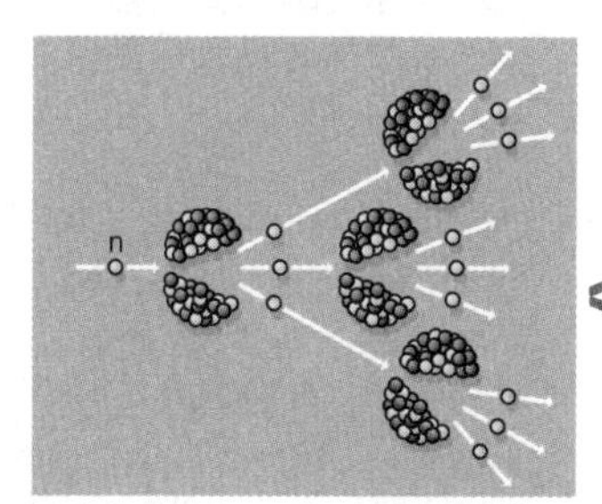

A chain reaction.

Questions

Grades D-C

1 How did the results obtained by Rutherford and Marsden suggest that atoms had a very small nucleus surrounded by empty space?

Grades B-A*

2 Explain why scientists were initially puzzled when they discovered that protons in the nucleus of an atom had a positive charge.

Grades D-C

3 What must a uranium-235 atom absorb, to make it undergo fission?

Grades B-A*

4 A uranium-235 atom can split in several ways. Complete this equation showing one way in which it can split apart.

$^{235}_{92}U + ^{1}_{0}n \rightarrow ^{?}_{56}Ba + ^{90}_{36}Kr + ^{1}_{0}n + ^{1}_{0}n$

Nuclear power station

Grades

The nuclear reactor

- In a nuclear reactor, the energy from nuclear fission (see page 129) heats water under high pressure. The pressurised hot water passes through a **heat exchanger** where it boils a separate water supply to produce super-heated steam. This turns a turbine connected to a generator that produces **electricity**.

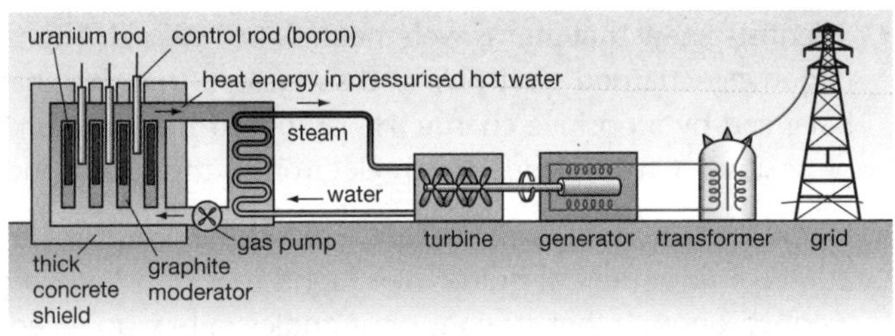

In a nuclear power station, energy from nuclear fission generates steam.

D–C

- To control the nuclear reaction, rods of graphite (a **moderator** material), slow down neutrons to a speed at which fuel nuclei will absorb them. Also, movable **control rods** of boron or cadmium absorb excess neutrons. In addition, water acts as a coolant.
- The spent nuclear fuel contains:
  - some uranium radioisotopes that can be 'enriched' (concentrated) and re-used as nuclear fuel
  - radioactive plutonium-240 that is not a fissile fuel and must be disposed of safely. One method is to set it in glass and bury it deep underground.

B–A*

Nuclear fusion

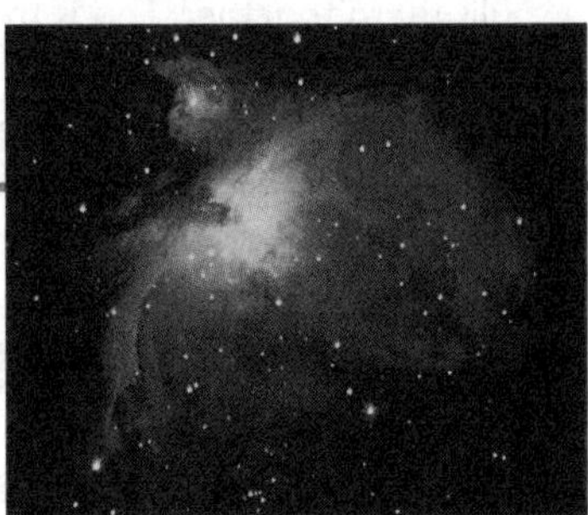

A nebula in which stars are born.

The birth of a star

- Stars form from dust and gas clouds that collapse under gravity. (See also page 66.) The mass of material heats up and, at 15 million Kelvin, two hydrogen nuclei react to form one helium nucleus in **nuclear fusion**:

$$\textbf{2 hydrogen nuclei} \xrightarrow{\textbf{nuclear fusion}} \textbf{1 helium nucleus + ENERGY}$$

D–C

- Similar reactions produce heat in stars and supernovae (see page 67).
- To start the fusion of hydrogen nuclei (protons), they need enough **kinetic energy** to overcome the force of repulsion between their like charges.
- Once started, the reaction itself produces enough energy to continue. As a result, the **reaction cycle** continues, generating even more energy.
- **Nuclear fusion** is a very clean and efficient way to produce energy from hydrogen, and scientists are trying to achieve it on Earth.
- Because hydrogen nuclei are charged particles, scientists use giant electromagnets to confine the superheated nuclei in a space away from container walls that would otherwise vaporise.
- However, they have so far reached the huge temperatures only for a fraction of a second. Also, it is not yet known how they will control the reaction.

B–A*

Top Tip!

In nuclear fission, the nucleus of a very large atom splits. In nuclear fusion, the nuclei of two small atoms join together. Both processes produce huge amounts of energy.

Questions

Grades D-C

1 What is the role of the control rods in a nuclear reactor?

Grades B-A*

2 Explain why it is difficult to deal with spent nuclear fuel.

Grades D-C

3 Explain the difference between 'nuclear fission' and 'nuclear fusion'.

Grades B-A*

4 Suggest why nuclear fusion is said to be a 'very clean' way to produce energy.

P2b summary

Circuits and resistance

Electrical current is the flow of **electrons**. The amount of charge carried by a current of 1 A flowing for 1 s = 1 **coulomb**.

charge (C) = current (A) × time (s)

Electrical energy is transformed to heat energy when charge flows through a **resistor**.

energy transformed (J) = potential difference (V) × charge (C)

A **potential difference** (**voltage**) exerts a **force** that pushes free electrons along.

Electrical **resistance** in wires is due to **free electrons** colliding with fixed, positive ions and their bound electrons.

Resistance in wires depends on their length, thickness, temperature and type of material.

Metals have free electrons and are **good conductors** with **low resistances**.

Insulators do not have free electrons and so are **poor conductors** with very **high resistances**.

Ohm's Law states that:

$$\textbf{resistance } (\Omega) = \frac{\textbf{potential difference (V)}}{\textbf{current (A)}}$$

The resistance of:

- a **light-dependent resistor** (LDR) decreases as light intensity increases
- a **thermistor** decreases as the temperature increases
- a **filament lamp** increases as the temperature of the filament increases.

Components in an electrical **circuit** can be connected in **series** and in **parallel**.

In a **series circuit**: current through all components is the *same* at all points; total p.d. is shared between all components; total resistance = sum of resistances of components.

In a **parallel circuit**: total current is the *sum* of currents through the separate components; the p.d. across all components is the *same*.

Domestic electricity

UK domestic electricity is an **alternating current** of **frequency** 50 Hz, average **p.d.** 230 V. An **oscilloscope** shows the p.d. and frequency of a.c. and d.c.

Fuses, **circuit breakers** and **earth wires** protect appliances from damage and people from injury or death. **Three-pin plugs** must be correctly wired and hold the correct fuse.

The **current** through an appliance and the **size of the fuse** required can be calculated from:

power (W) = current (A) × p.d. (V)

The **rate** at which energy is **transformed** is called **power**:

$$\textbf{power (W)} = \frac{\textbf{energy transformed (J)}}{\textbf{time taken (s)}}$$

Radioactive substances

Different forms of an **element** with the same number of **protons** but a different number of **neutrons** in their atoms are called **isotopes**.

Isotopes have the same **atomic number** but different **mass numbers**.

Isotopes with **unstable nuclei** emit **energy** as particles or rays called **radiation**. An **alpha particle** is a helium ion. A **beta particle** is a high energy electron. A **gamma ray** is an electromagnetic wave with no charge or mass.

Background radiation comes mainly from rocks, soil, cosmic rays, living things and medical X-rays.

Using scattering experiments, Rutherford and Marsden revealed the **structure** of the **atom**: a small dense **nucleus** of **protons** and **neutrons** (held together by strong **nuclear forces**); fast-moving **electrons** orbiting round the nucleus; and lots of empty **space**.

Nuclear fusion and nuclear fission

Nuclear fission is the **splitting** of a **nucleus**.

Neutrons are fired at unstable isotopes and **energy** is released. This produces more neutrons and sets up a **chain reaction**.

In **nuclear fusion**, two smaller **nuclei join** to form a larger one, as happens in **stars**:

2 hydrogen nuclei → 1 helium nucleus + ENERGY

Uranium-235 and **plutonium-239** are **fissionable** elements and are used to produce **nuclear energy** in nuclear power stations. The minimum amount of pure uranium-235 needed to produce large amounts of energy is called the **critical mass**.

Checklists – Science Unit 1

B1a Human biology

I know:

how human bodies respond to changes inside them and to their environment

- [ ] nerves and hormones coordinate body activities and help control water, ion and blood sugar levels, and temperature
- [ ] receptors transform energy from stimuli into electrical impulses, enabling information to travel rapidly along the nerves
- [ ] reflex actions are automatic and very fast; they involve sensory, relay and motor neurones, with synapses between them, in a reflex arc
- [ ] the menstrual cycle is controlled by the hormones oestrogen (secreted by the ovaries) FSH and LH (secreted by the pituitary gland). The concentrations of these hormones change during the menstrual cycle

what we can do to keep our bodies healthy

- [ ] a healthy diet contains the right balance of the different foods you need and the right amount of energy
- [ ] obesity increases the risk of arthritis, diabetes, high blood pressure and heart disease
- [ ] too much cholesterol in LDL in the blood increases the risk of heart disease
- [ ] the liver makes cholesterol if intake is too low

how we use/abuse medical and recreational drugs

- [ ] drugs affect people's behaviour and can damage the brain, and some hard drugs (heroin) are addictive. Some drugs are legal (alcohol) and some are illegal (cocaine, heroin)
- [ ] the substances in tobacco smoke cause many diseases, e.g. cancer and bronchitis
- [ ] alcohol is a depressant and hinders the activity of parts of the brain
- [ ] medical drugs are beneficial, but must be thoroughly tested before use

what causes infectious diseases and how our bodies defend themselves against them

- [ ] pathogenic microorganisms cause infectious diseases
- [ ] some white blood cells (phagocytes) ingest pathogens and kill them; and others (lymphocytes) make antibodies
- [ ] antibiotics kill bacteria inside the body but do not kill viruses; however, some bacteria can develop resistance to antibiotics
- [ ] immunisations and vaccinations offer protection from various diseases

B1b Evolution and environment

I know:

what determines where particular species live

- [ ] animals and plants are adapted to live in different habitats, to compete for resources and to survive attack from predators
- [ ] they may have adapted (thorns, poisons, warning colours) to cope with specific features of their extreme environment

why individuals of the same species are different from each other; and what new methods there are for producing plants and animals with desirable characteristics

- [ ] DNA is the genetic material that controls inherited characteristics
- [ ] reproduction can be sexual (genetic variation in offspring) or asexual (identical offspring)
- [ ] cloning and genetic engineering can be used to produce plants and animals with desirable characteristics
- [ ] concerns about GM organisms include: their safety for use, genes jumping to other species, and long-term consequences of eating them
- [ ] cross-species embryo transplantation may help to preserve endangered wild species

why some species of plants and animals have died out and how new species of plants and animals develop

- [ ] theories of evolution have changed over time; Darwin proposed the theory of evolution by natural selection
- [ ] fossils tell us how present-day species have evolved and how they compare to prehistoric species
- [ ] life on Earth may have originated: as a result of early light/air/water conditions; from a meteorite; or in deep oceans
- [ ] environmental and human change can cause extinction of species

how humans affect the environment

- [ ] increases in human population use up more resources and produce more waste and pollution
- [ ] human action contributes to acid rain, air pollution, water pollution, over-use of land, and loss of diversity in rainforests
- [ ] lichens, invertebrates and fish can be used as indicators of pollution
- [ ] increasing the greenhouse effect can lead to climate change
- [ ] sustainable development, e.g. using renewable energy resources and recycling, can help to safeguard the environment for future generations

C1a Products from rocks

I know:

how rocks provide building materials

- [ ] limestone, calcium carbonate, can be used to make cement, mortar, concrete and glass which are used as building materials
- [ ] an element consists of one type of atom; two atoms of the same element can join together to form a molecule; a compound consists of atoms of two or more elements joined together
- [ ] atoms are held together in molecules and lattices by chemical bonds, which involves giving, taking or sharing electrons
- [ ] the electronic configuration of an atom shows the number of its electrons and their arrangement in shells; atoms with a full outermost shell of electrons are stable
- [ ] how to interpret chemical equations in symbol form and balance equations in terms of numbers of atoms

how rocks provide metals and how metals are used

- [ ] metals are extracted from their ores, often oxides, by reduction with carbon (iron), electrolysis (aluminium, copper) or other chemical reactions (titanium)
- [ ] pure metals are soft and easily shaped because the atoms form a regular arrangement – the layers of atoms can slide easily over each other
- [ ] metals are mixed together to make alloys (e.g. iron plus other metals or carbon make steel)
- [ ] aluminium is expensive to produce, is often too soft on its own, but forms strong alloys with other metals
- [ ] copper is hard, strong, a good conductor and can be used for wiring and plumbing
- [ ] aluminium and titanium are resistant to corrosion and have a low density

how we get fuels from crude oil

- [ ] crude oil is a mixture of hydrocarbon compounds that can be separated by fractional distillation; some fractions can be used as fuels
- [ ] most of the compounds in crude oil are saturated hydrocarbons called alkanes, which have the general formula C_nH_{2n+2}
- [ ] burning fossil fuels releases useful energy but also harmful substances, e.g. sulfur dioxide causes acid rain; carbon dioxide causes climate change; smoke particles cause global dimming

C1b Oils, Earth and atmosphere

I know:

how polymers and ethanol are made from oil

- ☐ crude oil is made from long-chain hydrocarbons that can be cracked by thermal decomposition to form shorter-chain alkanes and alkenes
- ☐ alkenes are unsaturated hydrocarbons, they contain double carbon-carbon bonds and have the general formula C_nH_{2n}
- ☐ alkenes can be made into polymers, which are long-chain molecules created when lots of small molecules called monomers are joined together in polymerisation
- ☐ ethene can be reacted with steam in the presence of a catalyst to produce ethanol
- ☐ polymers can be used to make useful substances, e.g. waterproof materials and plastics, but many are not biodegradable

how plant oils can be used

- ☐ vegetable oils can be hardened to make margarine in a process called hydrogenation
- ☐ biodiesel fuel can be produced from vegetable oils
- ☐ oils do not dissolve in water; they can be used to produce emulsions, e.g. in salad dressings
- ☐ processed foods may contain additives to improve appearance, taste and shelf-life
- ☐ E-numbers identify permitted additives and must be listed; some additives can be harmful and may be banned
- ☐ chemical analysis can be used to identify additives and food colouring in foods, e.g. by paper chromatography

what the changes are in the Earth and its atmosphere

- ☐ the Earth has three main layers: the crust, mantle and core
- ☐ the Earth's atmosphere has changed over millions of years; many of the gases that make up the atmosphere came from volcanoes
- ☐ for 200 million years, the proportions of different gases in the atmosphere have been much the same as they are today
- ☐ human activities have recently produced further changes, e.g. the levels of greenhouse gases are rising

P1a Energy and electricity

I know:

how heat (thermal energy) is transferred and what factors affect the rate at which heat is transferred

- ☐ heat energy can be transferred by conduction, convection and thermal radiation
- ☐ thermal conductors (e.g. metals) transfer heat energy easily; thermal insulators (e.g. plastic, glass) do not
- ☐ dark, dull surfaces emit and absorb thermal radiation better than shiny, light surfaces
- ☐ the bigger the temperature difference between an object and its surroundings, the faster the rate at which heat is transferred

what is meant by the efficient use of energy

- ☐ energy is never created nor destroyed; some energy is usually wasted as heat
- ☐ the greater the percentage of the energy that is usefully transformed in a device, the more efficient the device is
- ☐ how to calculate the efficiency of a device:

$$\textbf{efficiency} = \frac{\textbf{useful energy output}}{\textbf{total energy input}}$$

why electrical devices are so useful

- ☐ they transform electrical energy to whatever form of energy we need at the flick of a switch
- ☐ the National Grid transmits electricity around the country at high voltages and low current to keep energy losses low
- ☐ dynamos produce electricity when coils of wire rotate inside a magnetic field
- ☐ how to work out the power rating of an appliance (the rate at which it transforms electrical energy)
- ☐ how to calculate the amount of energy transferred from the mains: **energy transferred = power × time**
- ☐ how to calculate the cost of energy transferred from the mains: **total cost = number of kilowatt-hours × cost per kilowatt-hour**

how we should generate the electricity we need

- ☐ we need to use more renewable energy sources, including wind, hydroelectric, tidal, wave and geothermal power
- ☐ most types of electricity generation have some harmful effects on people or the environment; there are also limitations on where they can be used

P1b Radiation and the Universe

I know:

what the uses and hazards of the waves that form the electromagnetic spectrum are

- ☐ from longest to shortest wavelength: radio waves, microwaves, infrared, visible light, ultraviolet, X-rays, gamma rays
- ☐ electromagnetic radiation has many uses in communication, e.g. radio, TV, satellites, cable and mobile phone networks
- ☐ communication signals can be digital or analogue
- ☐ some forms of electromagnetic radiation can damage living cells: ionising radiation (ultraviolet, X-rays and gamma rays) can cause cancer
- ☐ electromagnetic waves obey the wave formula: **wave speed = frequency × wavelength**

what the uses and dangers of emissions from radioactive substances are

- ☐ the uses and hazards of radioactive substances (which emit alpha particles, beta particles and gamma rays) depend on the wavelength and frequency of the radiation they emit
- ☐ background radiation is all around us, e.g. granite rocks can emit gamma rays and form radioactive radon gas
- ☐ the relative ionising power, penetration through materials and range in air of alpha, beta and gamma radiations
- ☐ the activity (count) rate of a radioisotope is measured as its half-life

about the origins of the Universe and how it continues to change

- ☐ the Universe is still expanding; in the beginning, matter and space expanded violently and rapidly from a very small initial point, i.e. the Big Bang
- ☐ red shift indicates that galaxies are moving apart; the further away a galaxy, the faster it is moving away from us
- ☐ telescopes on Earth and in space give us information about the Solar System and the galaxies in the Universe

Checklists – Additional Science Unit 2

B2a Discover Martian living!

I know:

what animals and plants are built from

- [ ] animal cells and plant cells have a membrane, cytoplasm and a nucleus; plant cells also have a cell wall and may have a vacuole and chloroplasts
- [ ] in multicellular organisms, different cells are specialised for different functions
- [ ] the chemical reactions inside cells are controlled by enzymes

how dissolved substances get into and out of cells

- [ ] diffusion is the net movement of particles of gas, or substances dissolved in a solution, from a region of high concentration to a region of lower concentration
- [ ] oxygen required for respiration passes through cell membranes by diffusion
- [ ] osmosis is the diffusion of water molecules through a partially permeable membrane

how plants obtain the food they need to live and grow

- [ ] green plants use chlorophyll to trap light energy from the Sun to photosynthesise
- [ ] leaves are specially adapted for photosynthesis – they can be broad, flat, thin and have lots of stomata
- [ ] the rate of photosynthesis is affected by light intensity, carbon dioxide concentration and temperature
- [ ] mineral salts in the soil are used to make proteins or chlorophyll; lack of a mineral ion results in a deficiency symptom in a plant

what happens to energy and biomass at each stage in a food chain

- [ ] energy passes along food chains but some energy is lost at every stage
- [ ] the shorter the food chain, the less energy is lost
- [ ] the mass of biomass at each stage in a food chain is less than it was at the previous stage; this can be shown in a pyramid of biomass
- [ ] reducing energy loss increases the efficiency of food production
- [ ] decomposers and detritus feeders feed on dead organisms and their waste

B2b Discover DNA!

I know:

what enzymes are and what their functions are

- [ ] enzymes are proteins that act as biological catalysts, speeding up chemical reactions, e.g. in washing powders and in industry
- [ ] each enzyme works at an optimum temperature and pH
- [ ] high temperatures or extremes of pH denature enzymes by affecting the shape of their active sites
- [ ] they are involved in respiration, photosynthesis, protein synthesis, and digestion

how our bodies keep internal conditions constant

- [ ] blood sugar levels are controlled by the pancreas, which makes insulin to bring down blood sugar levels
- [ ] waste products (carbon dioxide, urea) must be removed from the body
- [ ] sweating cools the body and maintains a steady body temperature
- [ ] if core body temperature is too high, blood vessels supplying the skin capillaries dilate so more blood flows through and more heat is lost
- [ ] if core body temperature is too low, blood vessels supplying the skin capillaries constrict to reduce the flow of blood through capillaries; muscles may shiver

some human characteristics show a simple pattern of inheritance

- [ ] some inherited characteristics are controlled by a single gene
- [ ] different forms of a gene are called alleles; in homozygous individuals the alleles are the same, in heterozygous individuals they are different
- [ ] how to construct/interpret a genetic diagram; and how to predict/explain the outcome of crosses between individuals for each possible combination of dominant and recessive alleles of the same gene
- [ ] in mitosis each new cell has the same number of identical chromosomes as the original
- [ ] cells in reproductive organs (testes and ovaries in humans) divide to form gametes by the process of meiosis
- [ ] sex chromosomes determine the sex of the offspring (male XY, female XX)
- [ ] stem cells can specialise into many types of cells

C2a Discover Buckminsterfullerene!

I know:

how sub-atomic particles help us to understand the structure of substances

- [ ] an element's mass number is the number of protons plus the number of neutrons in an atom
- [ ] an element's atomic number is the number of protons in an atom
- [ ] electrons arranged in shells around the nucleus have different energy levels and this can be used to explain what happens when elements react and how atoms join together to form different types of substances
- [ ] how to write balanced chemical equations for reactions
- [ ] metals consist of giant structures of atoms arranged in a regular pattern, with delocalised electrons

how structures influence the properties and uses of substances

- [ ] ionic bonding is the attraction between oppositely charged ions
- [ ] ionic compounds are giant lattice structures with high melting points that conduct electricity when molten or dissolved
- [ ] non-metal atoms can share pairs of electrons to form covalent bonds
- [ ] giant covalent structures are macromolecules that are hard, have high melting points but do not conduct electricity
- [ ] simple molecular elements (oxygen) and compounds (water) have weak intermolecular forces
- [ ] delocalised electrons in metals and graphite enable them to conduct heat and electricity as they are free to move through the whole structure
- [ ] nanoparticles are very small structures with special properties because of their unique atom arrangement

how much we can make and how much we need to use

- [ ] the relative masses of atoms can be used to calculate how much to react and how much we can produce, because no atoms are gained or lost in chemical reactions
- [ ] the percentage of an element in a compound can be calculated from the relative masses of the element in the formula and the relative formula mass of the compound
- [ ] how to calculate chemical quantities involving empirical formulae, reacting masses and percentage yield and how to balance symbol equations
- [ ] high atom economy (atom utilisation) is important for sustainable development and economic reasons
- [ ] reversible reactions carried out in a 'closed' system will eventually reach equilibrium

C2b Discover electrolysis!

I know:

how we can control the rates of chemical reactions

- ☐ the rate of a reaction can be found by measuring the amount of a reactant used or the amount of product formed over time
- ☐ reactions can be speeded up by increasing the: temperature; concentration of a solution; pressure of a gas; surface area of a solid; and by using a catalyst
- ☐ particles must collide with sufficient energy in order to react; the minimum energy required is the activation energy
- ☐ concentrations of solutions are given in moles per cubic decimetre (mol/dm^3); equal volumes of solutions of the same molar concentration contain the same number of particles of solute
- ☐ equal volumes of gases contain the same number of molecules

whether chemical reactions always release energy

- ☐ chemical reactions involve energy transfers
- ☐ exothermic reactions give OUT energy; endothermic take IN energy
- ☐ in reversible reactions, equilibrium is reached at a point when the rate of the reverse reaction balances the rate of the forward reaction
- ☐ the relative amounts of all the reacting substances at equilibrium depend on the conditions of the reaction; this principle is used to determine the optimum conditions for the Haber process

how can we use ions in solutions

- ☐ ions in ionic compounds move freely when molten or dissolved in water
- ☐ passing an electric current through an ionic compound breaks it down into its elements: this is called electrolysis
- ☐ at the negative electrode, positively charged ions gain electrons (reduction) and at the positive electrode, negatively charged ions lose electrons (oxidation)
- ☐ electrolysis of sodium chloride solution makes hydrogen, chlorine and sodium hydroxide
- ☐ how to complete and balance supplied half equations for the reactions occurring at the electrodes during electrolysis
- ☐ metal oxides and hydroxides are bases and react with acids to form salts
- ☐ soluble salts can be made from reacting an acid with a metal or a base and insoluble salts can be made by mixing solutions of ions
- ☐ in neutralisation reactions, H^+ ions from acids react with OH^- ions to produce water

P2b Discover nuclear fusion!

I know:

what the current through an electrical current depends on

- ☐ the symbols for components shown in circuit diagrams
- ☐ resistance is increased in long, thin, heated wires
- ☐ the current through a component depends on its resistance; the greater the resistance the smaller the current for a given p.d. across the component: **potential difference = current × resistance**
- ☐ in a series circuit: the total resistance is the sum of the resistance of each component; the current is the same through each component; the total p.d. of the supply is shared between the components
- ☐ in a parallel circuit: the total current through the whole circuit is the sum of the currents through the separate components; the p.d. across each component is the same

what mains electricity is and how it can be used safely

- ☐ mains electricity (230 V a.c., frequency 50 Hz) is very dangerous
- ☐ fuses and earth wires protect appliances and people from harm
- ☐ three-pin plugs must be wired correctly and hold the correct fuse
- ☐ how to interpret diagrams of oscilloscope traces

P2a Discover forces!

I know:

how we can describe the way things move

- ☐ how to calculate the speed of a body from the slope of a distance-time graph
- ☐ how to calculate the acceleration of a body from the slope of a velocity-time graph
- ☐ how to calculate the distance travelled by a body from a velocity-time graph

how we make things speed up or slow down

- ☐ to change the speed of an object, an unbalanced force must act on it
- ☐ forces can add up or cancel out to give a resultant force; when the resultant force is not zero, an object accelerates: **resultant force = mass × acceleration: $F = ma$**
- ☐ an object falling through a fluid accelerates until it reaches a terminal velocity, when the resultant force is zero
- ☐ the stopping distance of a car is the thinking distance plus the braking distance; this increases as the speed increases

what happens to movement energy when things speed up or slow down

- ☐ when force causes an object to move, energy is transferred and work is done
- ☐ every moving object has kinetic energy that can be transformed into other forms
- ☐ the kinetic energy of a body depends on its mass and its speed: **kinetic energy = $\frac{1}{2}$ × mass × speed^2**

what momentum is

- ☐ every moving object has momentum that depends on its mass and its velocity: **momentum = mass × velocity**
- ☐ momentum has size and direction; it is conserved in collisions and explosions
- ☐ how to use the equation: $\textbf{force} = \frac{\textbf{change in momentum}}{\textbf{time taken for the change}}$

what static electricity is, how it can be used and the connection between static electricity and electric currents

- ☐ rubbing electrical insulators together builds up static electricity because electrons are transferred
- ☐ if an object gains electrons it has a negative charge; if it loses electrons it has a positive charge
- ☐ electrostatic charges are used in photocopiers, smoke precipitators and paint sprayers
- ☐ when electrical charges move we get an electrical current
- ☐ if potential difference becomes high enough, a spark may jump across the gap between a body and any earthed conductor which is brought near it

why we need to know the power of electrical appliances

- ☐ $\textbf{power} = \frac{\textbf{energy transformed}}{\textbf{time taken}}$
  power = current × potential difference
- ☐ **energy transformed = potential difference × charge**
  charge = current × time
- ☐ most appliances have their power and the p.d. of the supply they need printed on them so we can calculate the current and fuse required

what happens to radioactive substances when they decay

- ☐ isotopes (elements with the same number of protons but a different number of neutrons) with unstable nuclei emit energy as radiation
- ☐ how the Rutherford and Marsden scattering experiment revealed the structure of the atom
- ☐ background radiation comes from rocks, soil, cosmic rays, living things and medical X-rays

what nuclear fission and nuclear fusion are

- ☐ nuclear fission (in nuclear reactors) is the splitting of an atomic nucleus
- ☐ nuclear fusion (how stars release energy) is the joining of two smaller nuclei to form a larger one

Answers

Page 4 Unit B1a Human biology

1 Motor.
2 Receptors in the retina transform light energy to energy in an electrical impulse in a neurone.
3 Adrenaline: heart, breathing muscles, eyes, digestive system. Reproductive hormones *or* oestrogen: ovaries.
4 It makes the heart beat faster. The blood flows faster, taking oxygen and nutrients to the muscles more rapidly. The muscles can respire faster and so release more energy, so you can run or fight faster and harder.

Page 5

1 The neurone secretes a chemical, which diffuses across the gap and produces an impulse in the next neurone.
2 Advantage: allows different responses to the same stimulus. Disadvantage: slows down the impulse.
3 The water evaporates and takes heat from the skin.
4 The water content of the body must remain roughly constant. On a hot day, more water is lost in sweat, so you need to drink to replace this lost water.

Page 6

1 It makes eggs mature in the ovaries, and it makes the ovaries secrete oestrogen.
2 Oestrogen levels rise during the first half of the cycle, causing the lining of the uterus to build up. They fall during the last part of the cycle, and the uterus lining breaks down when oestrogen is at its lowest at the start of the next cycle.
3 Advantages: fewer unwanted pregnancies; fewer abortions; slower population growth. Disadvantages: could encourage more sexual partners; could increase spread of sexually transmitted diseases.
4 FSH makes a woman produce several eggs, so more than one could be fertilised.

Page 7

1 The greater the muscle to fat ratio, the faster the metabolic rate.
2 One gram of fat contains twice as much energy as one gram of carbohydrate. Therefore if you eat too much fat, you run a greater risk of taking in too much energy, some of which will be stored in the body as fat.
3 Bones rub together as the joint moves, which is very painful. The joint may become too stiff to move.
4 It cannot avoid Type 1 diabetes, which does not seem to be affected by lifestyle. Eating well, keeping your weight down and staying fit can reduce the risk of developing Type 2 diabetes.

Page 8

1 Lack of protein in the diet.
2 Eat food containing less energy than the energy that you use up each day. Eat less, exercise more.
3 HDLs.
4 Some people's genes cause the liver to make lots of cholesterol, regardless of the diet. Statins inhibit an enzyme that the liver uses to make cholesterol.

Page 9

1 They experience withdrawal symptoms.
2 Heroin, cocaine, cannabis, ecstasy (there are others).
3 No-one thought to test the drug on pregnant women.
4 We can make sure that a pregnant woman does not take the drug. It is very useful for others, especially people with leprosy.

Page 10

1 If they inject the drug and share needles, they may introduce the AIDS virus into their body.
2 They alter chemicals in the brain, which causes craving for the drug. As long as the chemicals remain altered, the person will keep feeling the need to take the drug.
3 The liver.
4 It slows down activity in the brain. (It also slows down the transmission of nerve impulses.)

Page 11

1 There seems to be a link between them. When one increases, so does the other. It does not, though, necessarily mean that one *causes* the other.
2 It affects the genes that normally control cell division. If the gene mutates, it may behave differently and allow the cell to divide uncontrollably.
3 Doctors were no longer carrying pathogens from one patient to another.
4 Doctors have treated ulcers for many years, assuming they were caused by stress or over-secretion of acid into the stomach. It is difficult to alter people's long-held beliefs, without strong proof.

Page 12

1 Lymphocytes.
2 We now know that these children lack the gene to produce an enzyme that white blood cells need in order to destroy bacteria. If they are given the gene, then the cells can make the enzyme and their immune system can work normally.
3 Viruses reproduce inside body cells, so it is difficult to destroy the viruses without destroying the cells as well.
4 By 16.4 prescriptions per 100 people. Doctors are asked not to prescribe antibiotics unless absolutely necessary, to try to reduce the risk of bacteria becoming resistant to them.

Page 13

1 It was a new disease, so no-one had immunity to it and there were no drugs to combat it. International travel meant it could spread quickly all over the world.
2 Increase = 1400 – 410 = 990
So percentage increase = (990 × 100) ÷ 410 = 241%
3 Many parents did not let their children have the jab, so they were not immune against measles, mumps or rubella. They could have got one of these diseases or helped to spread it to others.
4 As many children were now not vaccinated, the mumps virus could survive in these non-vaccinated children and spread easily around the population.

Page 15 Unit B1b Evolution and environment

1 Hot, dry desert; no insulation under skin so heat can be lost more easily; long legs hold it high above the hot ground; very little urine produced, helping to conserve water; can drink large volumes of water and store them in its stomach, so when water becomes available, it can make best use of it.
2 Fennec fox is smaller than Arctic fox, giving it a larger surface area to volume ratio which increases rate of heat loss; its longer ears also increase surface area to volume ratio; it has less fat beneath its skin so heat can be lost more easily.
3 It could have leaves with a very large surface area; it can grow so that its leaves do not overlap; it could grow taller than other plants around it.
4 Although the kingsnake is not poisonous, its colouring is very similar to that of the poisonous coral snake, so predators avoid it.

Page 16

1 Sexual involves gametes, asexual does not. Sexual involves fertilisation, asexual does not. Sexual produces variation in the offspring, asexual does not. (NB –sexual reproduction does not always involves two parents, as many plants have one flower that produces both male and female gametes and they fertilise themselves.)
2 Dolly showed that clones could be made of adult mammals, which opens up the possibility of cloning adult humans. The ethical question is whether or not we should allow people to do this.
3 An inherited characteristic is caused by genes, inherited from your parents, e.g. hair colour. A non-inherited characteristic is caused by your environment and lifestyle, e.g. a tattoo, obesity.
4 State whether you think this is good or not, and give a clear reason, e.g. it is not right because it might cause employers to reject employees just because they had a higher risk of getting high blood pressure. Some people might find it almost impossible to get work, through no fault of their own.

Page 17

1 The new plants are all: genetically identical to the parent; look the same; grow at the same rate. It is a quick, cheap way of getting many new plants.
2 The new tea bushes will all: be genetically identical; grow at the same speed and to the same size; have the same good characteristics as the parent plant chosen as the source of the cuttings.
3 A nucleus from the cell to be cloned is placed inside an egg cell from which the nucleus has been removed.
4 Rare mammals may have only one or a few offspring at a time. If early embryos are split, then two or more can develop from one fertilised egg. If only a few females of the rare mammal are available, some embryos can be transplanted into females of other species, increasing the number of offspring born.

Page 18

1 Bacteria.
2 They may worry that the new type of soya might harm their health or that growing GM soya could allow the 'new' genes to spread to other plants growing nearby, upsetting ecosystems.
3 They have become adapted for eating different foods.
4 Lamarck thought that characteristics acquired during an organism's lifetime could be passed on to its offspring. Darwin thought that natural variation amongst organisms, inherited from their parents, formed the basis of natural selection; he did not believe that acquired characteristics were passed from parent to offspring.

Page 19

1 The black individuals were less likely to be eaten by birds, because they were better camouflaged. So they were more likely to reproduce, passing on the gene for black coloration to their offspring.
2 If most of the sparrows that survived had medium-length legs, they may have passed this characteristic on to their offspring. Few sparrows with long or short legs will have reproduced, so these characteristics would have been less likely to be passed on. So it is likely that most sparrows in future generations would have had medium-length legs.
3 They show the structures of organisms that lived long ago, which can be related to the structures of organisms that live today. Sometimes we can find sequences of fossils of different ages and can see how the structures have changed over time.
4 In shallow seas; in space or another planet, brought to Earth on a meteorite; in the deep oceans, near deep-sea volcanic vents.

Page 20

1 The species that is the better competitor may prevent the original species from getting enough resources to survive.
2 There was not enough time for the Tasmanian wolf to adapt; humans wiped them out so quickly; no way the wolf could ever adapt to avoid being shot and killed by humans that were determined to wipe it out.
3 Rate of increase is getting less; graph is getting flatter.
4 Resources that are needed for the growing population in one area may be taken from the environment in another, sparsely-populated area.

Page 21

1 Getting as much production from crops or animals as possible in a small area, using lots of fertilisers and pesticides.
2 Advantage: less land needed for waste disposal. Disadvantage: dioxins may be produced if incinerator is not managed properly.
3 DDT does not break down. It became concentrated up the food chain, so predators ended up with such large amounts in their bodies that it killed them.
4 Run-off from farmland contains fertilisers which cause too much growth of water plants; these die and are fed on by bacteria, which use up oxygen in the water, killing fish and other aquatic organisms.

Page 22

1 It causes respiratory illness. (It can trigger asthma attacks or cause bronchitis.)
2 Pollution by a country does not directly affect that country, so there may not be much incentive to clean up the waste. Most people on the ground are not aware of the pollution, so they do not push for the pollution to be controlled. It is expensive to get equipment up into space that could be used to clean up the pollution.
3 It reacts with calcium carbonate, causing the stone to break down.
4 Catalytic converters are not 100% efficient, so not all gases are converted, and not all cars have them.

Page 23

1 Heavily polluted.
2 Raw sewage provided nutrients for bacteria, which respired and used up oxygen. The sewage also provided nutrients for plants, which grew rapidly and then died, providing even more food for the bacteria and reducing oxygen levels further.
3 Decay organisms feeding on the wood respire, breaking down carbon compounds in the wood and releasing carbon dioxide to the air.
4 They have a very high biodiversity, so there are many different species of plants there that might provide as yet unknown chemicals.

Page 24

1 Carbon dioxide, methane.
2 Methane is over 20 times as effective as a greenhouse gas as carbon dioxide.
3 Use less electricity; walk or cycle to school rather than going in a car; turn the heating down; improve home insulation (there are others).
4 energy loss per second = 3.5 × 500 × (19 – 12) = 12 250 J

Page 26 Unit C1a Products from rocks

1 Zn
2 Iodine has similar properties to fluorine, chlorine and bromine, so Mendeleev thought that it should go into the same group, even if this meant putting it after an element with a greater atomic mass.
3 12. A magnesium atom has 12 protons and 12 electrons.
4 Any from: beryllium, magnesium, strontium, barium, radium. All of them have two electrons in their outer electron shell.

Page 27

1 Two more.
2 N≡N
3 Advantages: jobs; roads; local income. Disadvantages: habitat destruction; noise; traffic. (There are others.)
4 It is non-renewable; it formed millions of years ago and will not be replaced as quickly as we use it up.

Page 28

1 $CaCO_3 \rightarrow CaO + CO_2$
2 Limewater is a solution of calcium hydroxide in water. When carbon dioxide is added, it reacts with the calcium hydroxide to form calcium carbonate. This is not soluble, so it makes the liquid cloudy.
3 Concrete is a mixture of cement, sand and rock chippings; it is stronger than cement.
4 Sodium and calcium ions disrupt the regular arrangement of particles in glass, making it tougher.

Page 29

1 Oxygen is removed from it; the oxygen is taken by carbon, which is oxidised.
2 Iron is less reactive than carbon, so carbon will take oxygen away from iron.
3 Steel is harder than iron; also, different kinds of steels can be made with different properties for different uses.
4 Steel is a mixture of different atoms, which cannot form regular layers and so cannot slide over each other as easily as in pure iron.

Page 30

1 It is very hard (and can be sharpened to a fine edge).
2 Probably medium carbon steel, as it is fairly easy to shape but is hard, strong and cannot be easily bent by the forces that will be applied to it in use.
3 It has delocalised electrons which can move freely.
4 Metals have delocalised electrons and very good conductors of heat. Heat energy from your hands is conducted away rapidly by the metal, making your hands feel cold.

Page 31
1 Aluminium is more reactive than carbon, so carbon is not able to take oxygen away from the aluminium.
2 Positively charged aluminium ions each gain three electrons from the cathode, becoming neutral aluminium atoms.
3 It destroys habitats; pollutes the air with carbon dioxide; may spread diseases between migrant workers (there are other possible suggestions).
4 Less aluminium ore would need to be taken from the ground, reducing damage to habitats such as rainforests, where many species of animals and plants could become extinct.

Page 32
1 Titanium dioxide does not conduct electricity (because it is a covalent compound).
2 The process has several stages; reactions have to be carried out at high temperatures, using lots of energy.
3 There is a limited supply of high-grade ores, so eventually we will need to get it from low-grade ores.
4 In pure copper, all the atoms are the same size and form layers that easily slide over one another. In bronze, copper atoms are mixed with tin atoms; they are different sizes so do not form neat layers that can easily slide, making it hard.

Page 33
1 Alloys that go back to their original shape after being deformed; nickel-titanium alloys.
2 Any from: unreactive with fluids found in the body; non-toxic; strong, will not break or bend easily.
3 It does not produce carbon dioxide (which contributes to climate change) or sulfur dioxide (which produces acid rain).
4 They carry very heavy batteries, are usually slow-moving, and batteries have to be recharged frequently.

Page 34
1 Those with small molecules.
2 There are more intermolecular forces holding the molecules together, so more heat energy is needed to allow them to escape from one another.
3 C_5H_{12}
4 A molecule with the same number of each kind of atom, but with different arrangements of atoms.

Page 35
1 It contains carbon dioxide, which is found in the air and dissolves in rain drops to form carbonic acid.
2 The acidity stops young fish developing, there is less competition for older fish, so they grow larger.
3 The gases are sprayed with a limestone slurry, which reacts with sulfur dioxide forming solid calcium sulfate.
4 Most people are far more aware of aeroplanes than of ships, regularly see planes flying overhead, and many travel in planes for holidays/business. Fewer people travel by ship or see ships regularly.

Page 37 Unit C1b Oils, Earth and atmosphere
1 Breaking a substance down by heating it.
2 $C_{10}H_{22} \rightarrow C_3H_6 + C_7H_{16}$
3 They contain a double carbon-carbon bond, this can break and form a single carbon-carbon bond plus a bond with another atom.
4 It passes out of one fruit and affects another; it is a gas.

Page 38
1 It has an –OH group.
2 Heat the mixture to the boiling point of ethanol (which is lower than that of water). Collect the ethanol vapour and cool it so that it liquefies.
3 They have carbon-carbon double bonds, which can open and link to adjacent carbon atoms.
4 It would not corrode; it might be cheaper; it might be lighter (so less strong supports would be needed).

Page 39
1 Viscosity.
2 For a body part, e.g. a replacement lens (there are other examples); the part can be bent to put into the body, but it goes back to its normal shape afterwards.
3 Mix them with starch or cellulose.
4 The polymer will gradually break down inside the body, after the bone is healed.

Page 40
1 Dissolve the crushed plant material in a hydrocarbon solvent. Distillation may also be used.
2 Omega-3 fatty acids lower levels of fats in blood, reduce the risk of blood clots, lower blood pressure and generally prevent heart disease
3 Non-renewable fossil fuels are running out. We need to find more sources of renewable fuels, such as biofuels.
4 More land will be needed to grow the crops to make them. The price of the crops will go up, making it more expensive for people to buy them as food.

Page 41
1 The droplets of dispersed liquid scatter light rays.
2 An emulsifier stops a mixture separating into layers.
3 Add it to bromine water. If the bromine loses its colour, the liquid contains unsaturated compounds.
4 Unsaturated oils from plants are best; they can lower blood cholesterol levels.

Page 42
1 Extra hydrogen atoms are added to the carbon-carbon double bonds, so the molecules become straight instead of bent. They can pack together more tightly, forming a solid.
2 These 'healthier' option margarines don't contain trans fats, which have molecules containing some carbon-carbon double bonds. Diets with a high content of trans fats have been linked with heart disease.
3 The E-number identifies the additive, which will have been tested to make sure it is safe.
4 There is no strong evidence that tartrazine has these effects. Only a few children are affected, and if foods containing tartrazine are clearly labelled, parents can avoid it. Manufacturers may have put pressure on regulators to allow them to keep adding tartrazine to food, because the colour encourages more people to buy it.

Page 43
1 Spots of the colour are placed on a pencil line on paper, which stands in a solvent. The solvent moves up the paper, taking the colours with it. Some colours travel faster than others. The height of the spot of an unknown colour can be matched with height of a spot of a reference colour.
2 About 0.35.
3 Crust, mantle and core.
4 The magnetic field repels the 'solar wind', which is a stream of charged particles that would harm life on Earth if they all reached the ground.

Page 44
1 Heat generated by radioactive processes deep inside the Earth. (There is also heat left over from when the Earth was first formed.)
2 Two tectonic plates are slowly moving apart; magma wells up and forms new rocks on the sea floor.
3 At a subduction zone, an oceanic plate is gradually moving underneath a continental plate. Friction stops the plates moving smoothly. The plates move in sudden jerks, causing earthquakes.
4 The trench forms where the oceanic plate pushes under the continental plate, dragging material down with it.

Page 45
1 They have full outer electron shells.
2 Solid sodium azide decomposes to produce nitrogen gas, which has a much greater volume than the solid.
3 Living things produced oxygen when they photosynthesised.
4 Venus's atmosphere contains a very high concentration of carbon dioxide, which keeps its surface very hot through the greenhouse effect. The concentration of carbon dioxide in Earth's atmosphere is increasing, although it will never become as great as that on Venus.

Page 46
1 It is a greenhouse gas, preventing heat from being lost into space.
2 A place where carbon is locked up; sedimentary rocks (especially limestone); plants (especially trees); fossil fuels.
3 There have been increases and decreases in global temperatures in the past, even when there were no humans and no fossil fuels were being burnt.
4 Producing less carbon dioxide to add to the atmosphere could help because the more we add, the more risk there is that temperatures will rise.

Page 48 Unit P1a Energy and electricity
1 The person is hotter than the trees, and so emits more thermal radiation, which is detected by the camera.
2 The total quantity of thermal energy depends on the amount of the material, not just its temperature.
3 Silver surfaces emit *and* absorb thermal radiation more slowly than black surfaces.
4 The bear's fur is an excellent thermal insulator, so not much infrared radiation escapes from its body.

Page 49
1 Hotter particles vibrate faster; they are in close contact with their neighbours, making them vibrate faster as well. Free electrons help transfer heat quickly along the rod.
2 It can absorb a lot of heat energy without its temperature going up too greatly. It is a fluid, so it can carry heat away by convection.
3 Their particles are very far apart and only rarely come into contact with each other.
4 It conducts heat so well that the heat moves quickly out of your hand and into the diamond, making your hand feel cold.

Page 50
1 10 million J
2 Movement of air (winds); movement of waves (caused by wind); rain (evaporation of water using Sun's heat, then condensation); growth of plants; you turning these pages over (plants produced food by photosynthesis, and now you are using the energy in it); etc.
3 Heat energy in the bike, the surrounding air and the ground.
4 The nuclei of hydrogen atoms fuse to form helium nuclei and energy; this is called nuclear fusion.

Page 51
1 150 J in the exhaust gases, and 550 J in the moving parts, so a total of 700 J.
2 They run on electricity, so do not burn petrol and release exhaust gases. However, the electricity used to charge their batteries has been produced in a power station, and that might have caused pollution.
3 The heat energy in the bath water is spread out through a much larger volume than the heat energy in the kettle water.
4 Normally, thermal energy would flow from the environment into the cold food. The refrigerator uses energy to make the thermal energy flow in the other direction.

Page 52
1 Each bag weighs $2 \times 10 = 20$ N. So work done $= 20 \times 20 \times 3 = 1200$ J. Power output $= 1200 \div 60 = 20$ W.
2 Less power is needed to pull the weight up the ramp. You use the same amount of energy, but over a longer period of time, so the power output is less.
3 $(500 \div 1000) \times 100 = 50\%$
4 Less energy will be lost as heat because of friction between the moving parts, so more of the energy will be transferred to useful energy.

Page 53
1 Fluorescent bulb; only loses 80% of its energy as heat.
2 Large amounts of solar energy do not often fall onto the roof, so only a fairly small amount of electricity is saved by using them.
3 Usually because there is no mains electricity supply available. (It may also be cheaper.)
4 Electrons in atoms are excited and raised to higher energy levels when electricity passes through; when they fall back to their normal level, they emit some of the energy as light.

Page 54
1 A thin wire has more resistance than a thick one; the more resistance, the more it heats up.
2 A metal (or other substance) that has almost no electrical resistance. They are used in MRI scanners, for example.
3 50 J.
4 Number of kWh = power (kW) × time (hours)
Number of kWh $= (50 \div 1000) \times 0.5 = 0.025$
Cost $= 8\text{p} \times 0.025 = 0.2\text{p}$.

Page 55
1 High voltage means low current, and this means less energy is lost as heat and the wires can be thinner.
2 The electricity can be easily changed to higher or lower voltages by using step-up and step-down transformers, which work using a.c.
3 If either the magnet or coil of wire are moved relative to each other, a current will be produced in the wire.
4 If a current is passed through the wire, it will cause the magnet or the coil of wire to move. (Movement is only produced while the current is *changing*, so you need to use a.c. because it keeps switching its direction.)

Page 56
1 35%
2 Less methane emitted to the atmosphere, which would contribute to climate change; less use of fossil fuels so less disturbance from mining.
3 The steam can turn a turbine connected to a generator; this is geothermal energy.
4 No fuel is used; fuel does not need to be transported to the power station, so fewer vehicle movements and less fossil fuel used; nothing is burnt, so no carbon dioxide released to the air. You may be able to think of others.

Page 57
1 So that the fuel can easily be brought to them by road in large vehicles.
2 Methane is a much more effective greenhouse gas than carbon dioxide, which is produced by a fuel-burning power station.
3 They rely on water flowing rapidly downhill; they need to be in areas where there is high rainfall.
4 It could be allowed to rot and release methane, which could be burned to generate electricity.

Page 59 Unit P1b Radiation and the Universe
1 Radar (speed detectors); telecommunications (satellites, mobile phones).
2 Our eyes cannot detect infrared, as it has too long a wavelength.
3 Short wavelength.
4 They are the same.

Page 60
1 Gamma, X-rays, ultraviolet, visible light, infrared, microwaves, radio.
2 Most go straight through, but some are absorbed.
3 A (long-wave) radio wave.
4 3×10^4 Hz (30 000 Hz)

Page 61
1 Ultraviolet.
2 It can ionise atoms and molecules, damaging DNA; this can cause cancer.
3 A satellite that always stays above the same spot on the Earth's surface.
4 UHF (ultra high frequency) and microwaves.

Page 62
1 They undergo total internal reflection when they hit the edge of the fibre.
2 It is less prone to interference; more information can be carried by a single cable at one time.
3 The radiation comes from the nucleus of an atom.
4 Alpha particles, beta particles and gamma rays.

Page 63
1 Alpha radiation is very ionising, but it cannot penetrate skin.
2 It is an electromagnetic wave, and has no charge.
3 The badge shows how much radiation they have been exposed to, so they can check that they stay within safe limits.
4 The first reading was the background radiation, so the source was emitting $80 - 9 = 71$ counts per second.

Page 64
1 The count rate will halve.
2 Carbon-14.
3 When the plant dies, it stops taking in carbon dioxide from the air. The carbon-14 already in the plant (e.g. in its wood) undergoes radioactive decay, so it steadily becomes less over time.
4 Alpha would not go through even the thinnest paper. Gamma would go through even the thickest paper.

Page 65

1 The waste is radioactive, so it could harm people, e.g. by damaging their DNA.

2 More light can enter a reflecting telescope.

3 The light that reaches it does not have to pass through the atmosphere, which can cause distortion.

4 Not all objects in space emit only visible light; many of them emit other types of radiation, and we can find out more about them by detecting these.

Page 66

1 They used the gravity of Jupiter and Saturn to speed them up as they flew past.

2 The pull of the Earth on their bodies is cancelled out by the outward force caused by their orbital velocity.

3 They become helium atoms and release energy.

4 In a supernova (exploding star).

Page 67

1 The gases were pushed further away than rocks when the early star exploded.

2 Charged particles in the solar wind interact with the Earth's magnetic field.

3 Some of their mass is converted to energy (in nuclear fusion reactions).

4 The outward force caused by radiation pressure falls as the nuclear fusion reactions stop; when it becomes less than gravity, gravity causes the star to collapse inwards.

Page 68

1 Hydrogen.

2 They are so far away that it takes 13 billion years for their light to reach us.

3 If something is moving away from us, the wavelength of the light coming to us appears to be greater. This 'shifts' visible light towards the red end of the spectrum. (Red light has a longer wavelength than other colours.)

4 It shows that the galaxies are moving away from each other, and that the further away they are, the faster they are moving; this could be explained if everything was spreading outwards from a single point after the Big Bang.

Page 70 Unit B2a Discover Martian living!

1 Ribosome.

2 Layers of membranes. (They contain chlorophyll, but this is not visible.)

3 In the linings of the alimentary canal and gas exchange system; they make mucus.

4 Root hairs, xylem, phloem.

Page 71

1 It decreases the rate, as particles move more slowly.

2 Their particles are moving freely and randomly. They bump into each other and the walls of the tube and bounce off in different directions. Eventually some of them meet each other part way along the tube.

3 The greater the concentration gradient, the faster the rate of diffusion.

4 They increase the surface area across which substances can diffuse.

Page 72

1 A dilute solution.

2 a Starch molecules were too large to diffuse through the membrane. **b** Iodine molecules could diffuse through the membrane; they mixed with the starch inside the tubing and produced a blue-black colour. **c** Water molecules diffused freely through the membrane in both directions; there was no net movement as the concentration was equal on each side of the membrane.

3 Cell wall.

4 Water moved out of the cells by osmosis, through the cell membrane which is a partially permeable membrane, because the contents of the cell were a more dilute solution than the sugar solution. This made the cells get smaller.

Page 73

1 Respiration.

2 It contains a lot of energy. It is insoluble, so does not affect osmosis in the cell.

3 They are near the top of the leaf, where they get sunlight. They are next to the spongy mesophyll layer, where there is carbon dioxide in the air spaces.

4 Nutrients from insects, including amino acids, fatty acids and glycerol (the same things that you would get from insects if you ate them).

Page 74

1 The plant is using as much light as it can; it may be limited by the amount of chlorophyll it contains, or by not having enough carbon dioxide to photosynthesise any faster.

2 Advantages: increases the temperature and increases carbon dioxide concentration, both of which may increase the rate of photosynthesis and therefore tomato production. Disadvantage: it is expensive.

3 Nitrogen in the ions combines with glucose to make amino acids, which join together to make proteins.

4 Fertiliser is expensive, so farmer only adds it where it is needed, not in parts of the field where there are already plenty of mineral ions. He does not want it to wash off the field into rivers, where it may cause pollution.

Page 75

1 Some sunlight does not fall onto plant leaves; some is reflected from their leaves; some passes through their leaves without being absorbed by chlorophyll.

2 18%

3 Mammals use a lot of food to produce heat energy to keep their body temperature constant. Reptiles just take on the heat of the surroundings.

4 Energy is lost at each transfer along the chain. There is not enough energy to support a population of organisms feeding so far along a food chain.

Page 76

1 The animals need less food to produce heat energy to keep themselves warm.

2 The fungus is near the beginning of a food chain (not much energy has been wasted in transfers between organisms before it reaches the fungus); it is grown on materials that may have no other uses and in controlled conditions in fermenters (little energy is wasted).

3 Farmers concentrated on producing as many eggs, in as small a space and at the lowest cost possible; this resulted in hens being kept in battery cages.

4 To get foods that we cannot grow in this country; to get foods out of their season in this country; to help support developing countries by buying their produce.

Page 77

1 Reactions happen faster at higher temperatures.

2 Bacteria and fungi that cause decay lose water by osmosis, into the concentrated sugar solution, across their partially permeable cell membranes. This prevents them from functioning normally, and may kill them.

3 Every organism, when it dies.

4 No light, so no photosynthesis. Food chains often begin with dead bodies that have sunk down from the water above, and the detritivores that feed on them.

Page 78

1 Carbohydrates; fats; proteins; chlorophyll.

2 From organisms that photosynthesise; the first ones evolved about 3 billion years ago, at which time there was no oxygen in the air.

3 By respiration.

4 They break down carbon-containing compounds in dead remains and waste products from plants and animals, so that the carbon becomes part of their bodies. They use some of these carbon compounds to provide energy by respiration, and this returns carbon dioxide to the air.

Page 80 Unit B2b Discover DNA!

1 The enzyme molecules are denatured – they lose their shape, so the substrate does not fit in the active site.

2 Around 40 °C; we need more results from temperatures between 30 and 50 °C to give a more precise answer.

3 It is alkaline, and neutralises the acidic contents arriving from the stomach. It contains bile salts, which emulsify fats so that lipase can digest them more easily.

4 Proteins (they are normally digested in the stomach).

Page 81

1 They have higher optimum temperatures, so clothes can be washed at higher temperatures, which may help detergents to work better.

2 Saliva is alkaline, and this helps to neutralise acid produced by bacteria in the mouth, which is the cause of tooth decay.

3 They break down proteins into amino acids, which babies can absorb.

4 Sucrase breaks down sucrose to glucose and fructose. These are more soluble than sucrose and form a thick, runny syrup.

Page 82

1 They keep their body temperature constant, usually above the temperature of their environment; this requires heat energy, which they release from food by respiration.

2 Aerobic respiration takes place inside the mitochondria, providing energy that a sperm uses for swimming.

3 The carbon dioxide diffuses from the muscle cell into the blood; it is transported to the lungs in the blood (dissolved in the blood plasma) and then diffuses from the blood into the alveoli.

4 There will be more carbon dioxide in the blood because the muscle cells will have been respiring more rapidly; there is therefore a bigger concentration gradient for carbon dioxide between the blood and the alveoli, which speeds up diffusion.

Page 83

1 Liver converts excess amino acids into urea; kidneys filter this out of the blood and excrete it in urine.

2 Urea contains nitrogen, because it has been made from amino acids which also contain nitrogen.

3 To conserve water, as water may have been lost in sweat.

4 The solution in the fish's cells has a higher concentration than the water around it, so they constantly absorb water by osmosis. This excess water is removed in the dilute urine.

Page 84

1 In the hypothalamus and in the brain; it monitors blood temperature and causes appropriate responses in the skin and muscles.

2 More sweat produced, which evaporates and cools the skin. Blood vessels supplying skin capillaries dilate, more blood flows close to skin surface, more heat lost.

3 Banting and Best injected fluid containing it into dogs with diabetes, which became healthy.

4 It would: monitor blood glucose all the time, rather than testing every now and then; respond instantly by releasing the right amount of glucose, rather than a person having to judge this for themselves.

Page 85

1 Mitosis.

2 It may damage a gene that controls cell division. If cell division is not properly controlled, then the cell may go on and on dividing, producing a tumour and perhaps spreading to other parts of the body.

3 They contain a random mixture of alleles of genes from both of their parents.

4 Four cells, each with one set of chromosomes.

Page 86

1 Red blood cells; white blood cells; cells to repair bone, cartilage, tendons or fat tissue.

2 The stem cells could develop into nerve cells that produce dopamine, replacing those that have stopped working.

3 Each new cell must have a complete copy of all the DNA.

4 The gene carries a code that determines how amino acids are joined together to make a particular protein.

Page 87

1 Homozygous. Organisms crossed with others the same as themselves, keep producing others like themselves, as there are no different alleles present.

2 About three-quarters of them, i.e. approx. 1725.

3 All eggs contain an X chromosome. Half of the sperm have an X, and half have a Y chromosome. So there is an equal chance of an egg being fertilised by an X sperm or a Y sperm, and therefore an equal chance of the zygote being XX or XY.

4 Changing the temperature at which the eggs are incubated affects the sex of the turtles that hatch. Temperature cannot change the X or Y chromosomes.

Page 88

1 Each parent could have a recessive allele for CF. These two alleles could come together in a zygote, and therefore in the child.

2 The child will definitely develop Huntington's disease. The parents need to decide whether to keep the child or whether the woman should have an abortion.

3 To check whether a person was present at a crime scene; to check whether a man could be a child's father.

4 If DNA is found at the crime scene and it does *not* match the suspect's DNA, then it is unlikely that he or she was present at the scene.

Page 90 Unit C2a Discover Buckminsterfullerene!

1 39. It has a total of 39 neutrons and protons in its nucleus. (Its atomic number is 19, so it has 19 electrons and 19 protons. So it must have 20 neutrons.)

2 A chemical reaction only involves the electrons in an atom. A nuclear reaction involves the nucleus (which contains protons and neutrons).

3 By losing one electron, leaving a full outer shell.

4 They both have two electrons in their outer shells, so they can both become stable by losing two electrons. They will therefore react in a very similar way.

Page 91

1 They all have the same number of electrons in their outer shell, giving them similar chemical properties.

2 Chlorine has two isotopes, which have different numbers of neutrons. The relative atomic mass of chlorine takes this into account, as well as the relative abundance of the two isotopes in a typical sample of chlorine, so the number is an average and works out not to be a whole number.

3 The total charge on the positive ions is exactly balanced by the total charge on the negative ions.

4 The charge moves because positive sodium ions move towards the negative electrode and negative chloride ions move towards the positive electrode.

Page 92

1 A compound made up of many ions all arranged in a regular lattice, held together by strong forces of attraction between the oppositely charged ions.

2 Each copper ion has a positive charge. It picks up electrons from the negative electrode and becomes an uncharged atom, which is deposited onto the electrode.

3 They each have a complete outer electron shell.

4 The forces of attraction between the molecules are weak, so the molecules stay far apart from each other, forming a gas.

Page 93

1 They do not contain any charged particles.

2 They are molecular substances with very weak bonds between the molecules, making it easy for the molecules to escape from one another.

3 The carbon atoms are arranged in layers so the layers can slide over one another.

4 One electron from each carbon atom is free to move throughout the structure.

Page 94

1 Their layers of atoms can easily slide over each other.

2 Alloys contain atoms of different sizes, and it is more difficult for electrons to move freely amongst them.

3 Rubidium, because metals become more reactive as you go *down* the Group.

4 Iron can be extracted from its ore by heating with carbon, which was a relatively easy thing to do. But magnesium is in Group 1 and is very reactive, so it could not be extracted in this way. Magnesium has to be extracted by electrolysis.

Page 95

1 They gain one electron to form negative ions.

2 Chlorine is more reactive than bromine, so bromine cannot displace chlorine from its compounds. All that will happen is that the orange-coloured bromine will dissolve in the solution, making it orange. No chlorine gas will be produced.

3 They could be used as new types of catalysts.

4 We have not yet had time to find out if nanoparticles are harmful or not. They are so small that they can easily enter the body, even through the skin.

Page 96
1 It darkens; it will then go back to its normal colour when the light is no longer present.
2 In a compound, elements are chemically combined and in fixed proportions. In a mixture, they are not chemically combined and can be in any proportion.
3 The lattice formed has one magnesium ion for every two chloride ions.
4 $N_2 + 3H_2 \rightarrow 2NH_3$
Page 97
1 $(12 \div 16) \times 100 = 75\%$
2 16 g
3 $CH_4 + 2O_2 \rightarrow CO_2 + 2H_2O$
4 One mole of methane is 12 + (1 × 4) = 16 g. The balanced equation shows us that 1 mole of methane produces 1 mole of carbon dioxide. The mass of 1 mole of carbon dioxide is 12 + (2 × 16) = 44. So we would get 44 g of carbon dioxide.
Page 98
1 The reaction may be reversible; some product may be lost when it is separated from other products; some of the reactants may react in unexpected ways.
2 $(210 \div 250) \times 100 = 84\%$
3 nitrogen + hydrogen $\rightleftharpoons$ ammonia
4 The reaction is reversible, so some of the ammonia that is formed changes back into nitrogen and hydrogen.
Page 99
1 No effect, but it does speed up the rate of the reaction.
2 The forwards reaction and backwards reaction happen at the same time, so there are always some reactants present.
3 Iron.
4 Making the reaction hot means that it happens more quickly. But a rise in temperature favours the reverse reaction rather than the forward one that makes ammonia, so you would not get as much ammonia formed.
Page 101 Unit C2b Discover electrolysis!
1 Collect the gas (hydrogen) produced during a measured time interval during the reaction and measure its volume.
2 There are 36.5 g in 1 mole, therefore 2 × 36.5 = 73 g in 1 dm^3 of the solution.
3 Measure the mass, and calculate how much mass is lost in a period of time.
4 If the gas dissolves in the mixture, then it is not collected in the gas syringe. The rate would therefore seem to be lower than it really is.
Page 102
1 It would double it (because particles would collide twice as frequently).
2 The number is the same.
3 It is twice as fast. (At 20 °C it took 500 s for the cross to disappear, and at 30 °C it took 250 s.)
4 The particles have more kinetic energy and collide more often; they collide twice as frequently when the temperature is raised by 10 °C.
Page 103
1 By that time, the reaction would be complete, and there is no difference in the *total* mass of gas lost at the end of the reaction.
2 The total mass of gas lost is the same, because there was the same quantity of acid.
3 There would be twice as many collisions between the particles, in the same amount of time.
4 There would be the same number.
Page 104
1 Transition metals.
2 The amount of energy required to start a reaction. Catalysts reduce activation energy.
3 The amount of energy taken in when the reaction goes one way is the same as the amount of energy given out when the reaction goes the other way.
4 It will feel hot, because the reaction between copper sulfate and water is exothermic.
Page 105
1 The calcium carbonate breaks apart (decomposition) when it is heated (thermal).
2 Increase the temperature.
3 Pressure 400 atmospheres; temperature 350 °C.
4 There are 4 molecules of gas on the left-hand side of the equation, and 2 on the right-hand side. So the gases on the right-hand side will occupy less volume than those on the left-hand side. Increasing the pressure pushes the equilibrium towards the side of the equation where the gases occupy less volume.
Page 106
1 Lead ions have a positive charge, so they are attracted to the negative charge on the negative electrode.
2 Sodium is more reactive than hydrogen, so it remains as an ion in solution. Hydrogen ions take electrons from the negative electrode in preference to sodium.
3 The negatively charged chloride ions each lose one electron, becoming uncharged atoms. (They then combine to form chlorine gas.) They are oxidised.
4 $Pb^{2+} + 2e^- \rightarrow Pb$ They are reduced (the ions *gain* electrons).
Page 107
1 As a fuel; making ammonia (for fertilisers); making margarine; in weather balloons.
2 Copper from the piece of impure copper forms positive ions, by giving up electrons to the positive electrode. These copper ions are attracted to the negative electrode, so they move through the solution and take electrons from the negative electrode, forming a deposit of pure copper on this electrode.
3 To get rid of any unreacted pieces of zinc.
4 $Mg + H_2SO_4 \rightarrow MgSO_4 + H_2$
Page 108
1 Zinc oxide and sulfuric acid.
2 So any base that has not reacted can be filtered off.
3 Using the indicator tells you exactly how much base and acid to react together. You can then use these quantities in another reaction, to get a sample of the salt that is not contaminated with indicator.
4 Add ammonia to nitric acid.
$NH_3(aq) + HNO_3 \rightarrow NH_4NO_3$
Page 109
1 hydrogen + hydroxide → water
2 $H^+(aq) + OH^-(aq) \rightarrow H_2O(l)$
3 The calcium ions react with the phosphate ions to form insoluble calcium phosphate, which precipitates out.
4 Add silver nitrate solution to the solution you are testing. Any chloride ions present will react with the silver ions to form a visible precipitate.
Page 111 Unit P2a Discover forces!
1 14 km/h
2 27 km/h
3 1 m/s^2
4 Its velocity is changing, because it is not moving in a straight line – it is being pulled towards the Earth and so it travels in a circle. It is accelerating, because acceleration is a change in velocity.
Page 112
1 The line starts off part way up the *x*-axis, then horizontal, then sloping downwards right to the bottom of the graph, then sloping upwards, then horizontal again at the same level at which it began.
2 Area under triangle = $\frac{1}{2}$base × height = $\frac{1}{2} \times 4 \times 10$ = 20 m.
3 Box does not move, because resultant force is 0.
4 Wind arrow: 10 mm = 40 N. Resultant force arrow = 20 mm, which represents (20 ÷ 10) × 40 = 80 N.
Page 113
1 …directly…
2 $F = ma$, so $F = 10 \times 10 = 100$ N
3 $a = F \div m$, so $a = 2 \div 1 = 2\ m/s^2$
4 The force of gravity is greater than air resistance, so there is a resultant downwards force on the sky diver.
Page 114
1 It increases air resistance, so the resultant downward force is smaller.
2 The surface area of his 'wings' is not as great as the surface area of the parachute, so air resistance is not so great. He falls at a faster speed before air resistance becomes large enough to equal the force of his weight (at which point he reaches terminal velocity).
3 Having a slower reaction time (e.g. because of drinking alcohol, taking drugs, being tired); driving faster.
4 Without treads, there are no grooves to channel water away from the tyre, so the tyre is in contact with a film of water, decreasing friction. Less tyre is in contact with the solid road surface, so when the brakes are applied, friction is not enough stop the car.
Page 115
1 The resistance against movement in water (drag) is greater than in air (air resistance) because water is denser than air.
2 Treacle is more viscous than water – its particles are more difficult to push apart. So drag is greater in treacle than in water.
3 **a** At the top of the swing; **b** at the bottom.
4 Electricity cannot be stored, so if a generator stops working there is no 'stored' electricity that can be supplied to consumers. A moving flywheel can store energy in the form of kinetic energy, and this can be used to generate electricity quickly.
Page 116
1 30 J (15 × 1 for each parcel).
2 He is using the formula work = force × distance, but he has forgotten that the force he is using to the push the toy is *not* the toy's weight, but a different, smaller force of friction (which he has not measured). His calculation would only be correct if he was lifting the toy up through a distance of 2 m.
3 The elastic potential energy stored in the elastic band.
4 $\frac{1}{2} \times 120\,000 \times 100^2 = 6 \times 10^8$ J
Page 117
1 Momentum has a direction (it is a vector quantity) whereas kinetic energy does not (it is a scalar quantity).
2 Momentum of the moving mass is 2 × 2 = 4 kg m/s. The momentum of the stationary mass is 0.
3 The total momentum of the two masses must be the same as it was before the collision, that is 4 kg m/s. So 4 = (2 × speed) + (4 × speed) = 6 × speed. The speed is therefore 4 ÷ 6 = 0.67 m/s.
4 At rest, the momentum of the balloons is 0. The momentum of the burst balloon and the momentum of the escaped air add up to 0 (they are moving in different directions).
Page 118
1 The seat belt slows down the rate of change of momentum when you suddenly slow down to a stop. The longer the time taken, the less force is used. Less force means less injury.
2 Momentum before braking = 2000 × 18 = 36 000 kg m/s
Momentum after braking = 2000 × 10 = 20 000 kg m/s
Change in momentum = 36 000 – 20 000 = 16 000 kg m/s
So force used = 16 000 ÷ 5 = 3200 N.
3 Electrons have moved from the shirt to the tent fabric.
4 Static electricity is caused by a build-up of electrons (or a lack of electrons). In a conductor, the electrons could flow away and would not stay together in one place.
Page 119
1 The strip gives electrons to the cap, and these move down the rod and the leaf. Now the rod and leaf have the same negative charge, so they repel each other.
2 When Suzanne stroked her cat, electrons transferred from its fur to her hand, Her hand acquired a negative charge, the cat's fur acquired a positive charge. When she touched the tip of the cat's ear, some of these electrons suddenly moved from her hand to the cat's ear because of the attraction between the positive and negative charges. (This only happens when the air is dry, and pointed parts of the cat, such as its ear, are most likely to cause a current.)
3 They all have a positive charge, so they repel each other.
4 The charge flows away to earth – the charged object is discharged.
Page 120
1 Any charge that builds up can be conducted safely away along the metal hose. If charge built up, e.g. on a non-conducting plastic hose, then a spark might jump and cause the oil to catch fire.
2 This prevents a static charge building up, which might cause an explosion of the flammable fluids or gases.
3 The drum had positive charge, but it only keeps its charge in the places corresponding to the dark parts of the object being photocopied. (The light parts make the drum able to conduct, so the charge flows away.)
4 The wires of the precipitator have a negative charge, and this causes the smoke particles to become negatively charged also. So they are repelled from the wires.
Page 122 Unit P2b Discover nuclear fusion!
1 Your circuit should look like the first circuit diagram on page 122, except that it contains another ammeter somewhere in the circuit, and only one lamp.
2 The sum of the currents in each branch of the circuit = the current from the cell.
3 They have free electrons that can move easily.
4 Potential difference (or voltage).
Page 123
1 The resistance increases.
2 $V = IR$ or $R = V \div I$. So $R = 6 \div 3 = 2\,\Omega$
3 Because its resistance is not constant.
4 As the voltage across the lamp increases, it gets hotter, and this makes its resistance increase. So at higher voltages, the greater resistance means the current is not as high as you would expect if it followed Ohm's Law, so the graph flattens out.
Page 124
1 Smoke detectors; automatic light controls; burglar alarms.
2 As the light intensity decreases (e.g. if the air becomes smoky, or if it gets dark), the resistance of the LDR increases. This affects the current in the circuit, which can switch on the smoke alarm or the lighting.
3 There is now more resistance in the circuit, so current decreases.
4 6 V (you add the p.d.s).
Page 125
1 12 A **2** 4.5 V
3 The fuse melts if too much current flows through it, breaking the circuit and preventing harm.
4 The earth wire provides the current with a safe route if other wires accidentally come into contact with the metal casing of an appliance and you touch the casing. If the outer casing is an insulator, there is no danger from this, so the earth wire is not needed.
Page 126
1 A straight, horizontal line. **2** 232 V
3 It breaks the circuit more quickly than a fuse.
4 This is an easy route for the current, so it will flow along the earth wire rather than through you (you have a higher resistance). Also, the high current flowing through the low-resistance earth wire melts the fuse and breaks the circuit.
Page 127
1 6.5 A
2 Current = charge ÷ time = 2 ÷ 4 = 0.5 A
3 Energy transferred = p.d. × charge = 1.5 × 2 = 3 J
4 5 protons; 6 neutrons; mass number 11.
Page 128
1 **a** Positive; **b** negative; **c** no charge.
2 Lead. **3** ${}^{214}_{84}Po \rightarrow {}^{210}_{82}Pb + {}^{4}_{2}He$
4 Radon gas. Radon comes from granite.
Page 129
1 Lots of the alpha particles went straight through the gold foil, meaning they must have gone through empty space rather than hitting anything.
2 They did not initially understand why the protons did not all repel each other, as they had the same charge.
3 A neutron.
4 ${}^{235}_{92}U + {}^{1}_{0}n \rightarrow {}^{144}_{56}Ba + {}^{90}_{36}Kr + {}^{1}_{0}n + {}^{1}_{0}n$
Page 130
1 They absorb excess neutrons, preventing the chain reaction getting out of control.
2 It is radioactive, so it is potentially dangerous to people and other living organisms.
3 In nuclear fission, the nucleus of an atom splits apart. In nuclear fusion, two atomic nuclei join together.
4 It does not produce any harmful gases such as carbon dioxide, methane or sulfur dioxide.

Glossary/index

acceleration The change of velocity in a given time; it is usually measured in metres per second per second (m/s^2). 111, 112, 113, 114
acid A chemical with a pH below 7. 103, 107, 108, 109
acid rain Rain with a very low pH resulting from the release of oxides of nitrogen and sulfur into the air. 22, 35, 104
activation energy The energy a molecule must have before it can take part in a particular chemical reaction. 102, 104
active site The part of an enzyme where the reaction occurs. 80
adaptation Characteristic that allows an animal or plant to survive in its environment. 15, 20, 73
addiction When a person is unable to manage without a particular drug. They are dependent on it. 9, 10
additives Chemicals put into food to keep it fresh or make it appetising. 42, 43
adrenaline A hormone produced by the adrenal gland that mobilises the body for action. 4
aerobic respiration The breakdown of glucose using oxygen to make energy available for chemical reactions in cells. 77, 82
air resistance The force air exerts on an object moving through it. 114, 115, 116
alcohol A family of compounds containing the –OH group. A general term for alcoholic drinks. 10, 38
alkali A substance with a pH above 7. 108, 109
alkali metal A metal which burns to form a strongly alkaline oxide. 94
alkane Chemicals containing only hydrogen and carbon with the general formula C_nH_{2n+2}. 34, 37
alkene Chemicals containing only hydrogen and carbon with the general formula C_nH_{2n}. 37, 38
allele A variety of a gene which determines different variations of a particular characteristic. 85, 87, 88
alloy A mixture of a metal element with one or more other elements. 30, 31, 32, 33, 94, 96
alpha particle A helium nucleus made of two protons and two neutrons. 62, 63, 128, 129
alpha radiation Alpha particles released from the nucleus of a radioactive isotope. 64, 128
alternating current Current that rapidly reverses in direction. 55, 59, 126
alveolus The sac-like end of an airway in the lungs. 82
amino acid The sub-unit making up protein molecules. 74, 80, 81, 82, 83, 86
ammeter A meter that measures the current in a wire, in amps (A). 55
ammonia An alkaline gas with the formula NH_3. 45, 71, 93, 98, 99, 104, 105, 108
ampere Or amp, is the unit used to measure electrical current (A). 127
anode The positive electrode in a circuit or battery. 31
antibiotic A substance produced by a microbe that kills other microbes. Some antibiotics can be purified and used to treat infections. 11, 12, 13
antibody Chemicals produced by white blood cells that destroy invading microorganisms. 12, 13
antigen Chemicals found in the cell membranes and cell walls of pathogens that trigger a reaction by the immune system. 12
antitoxin Chemicals produced by white blood cells that neutralise bacterial toxins. 12
antiviral drug A drug that acts against viruses. 12
arthritis Inflammation and restriction of the movement of a joint. 7
asexual reproduction Reproduction involving just one parent. The offspring are genetically identical to the parent (clones). 16, 85
atmosphere The layer of gases surrounding the Earth. 24, 45, 46, 65
atom The smallest part of an element, atoms consist of negatively charged electrons orbiting a positively charged nucleus. 26, 27, 29, 30, 31, 34, 37, 62, 63, 64, 65, 90, 91, 92, 94, 95, 96, 98, 122, 127, 129
atom economy The conversion efficiency of a chemical process in terms of all the atoms involved. In an ideal chemical process the amount of starting materials or reactants equals the amount of all products generated and no atom is wasted. 98
atomic mass The number of protons and neutrons in the nucleus of an atom. 26
atomic number The number of protons in the nucleus of an atom. 26, 90, 97, 127

background radiation Radiation from space and rocks that is around us all the time. 63, 68, 128
bacteria Microscopic single-celled living things that have no nucleus. 11, 12, 13, 77
balanced equation A symbol equation written to show that the number of atoms in the products equals the number of atoms in the reactants. 96
balanced forces When forces acting in opposite directions have the same value. 113
base A substance that reacts with an acid to form a salt, or dissolves in water to form hydroxide ions. 108
bauxite Aluminium ore, aluminium oxide. 31
beta particle A high-energy electron produced when a neutron changes to a proton. 62, 63
beta radiation Beta particles released from the nucleus of a radioactive isotope. 64, 128
Big Bang The event believed by many scientists to have been the start of the Universe. 68
bile An alkaline fluid produced by the liver that neutralises acid and breaks down fats. 80
biodegradable A substance which can be broken down by biological action in the environment. 39, 40
biodiversity All aspects of variety in the living world. 23, 31
biofuel A fuel that is produced from a living system. 38, 40
biological detergent A detergent containing enzymes. 81
biomass The mass of living material. 53, 56, 75
biosensor A sensor that uses biological materials, usually enzymes, to detect a particular substance. 95
bladder The organ where urine is stored before it is excreted. 83
blood pressure The pressure of blood in your arteries and veins. 7, 8
braking distance The distance it takes for the brakes to stop a moving vehicle. 114
bronchitis Inflammation and infection of the bronchi. 10

camouflage Natural colouring or characteristics that enable an animal to remain unseen by predators. 19
cancer A dangerous illness in which cells grow uncontrollably, and form a tumour. 10, 11, 21, 23, 61, 64, 65, 85
carbohydrase An enzyme that breaks down carbohydrates. 81
carbohydrate Chemicals found in all living things. They contain the elements carbon, hydrogen and oxygen. They are a source of energy. 7, 74, 75, 78
carbon cycle The constant cycling of carbon as carbohydrates, fat and proteins in living things and as carbon dioxide in the air. 23, 78
carbon sink A place where carbon is locked up, e.g. sedimentary rocks. 46
carcinogen A substance that causes cancer. 11
carnivore An animal that eats other animals. 75
carrier A person who is healthy but has a recessive allele for an inherited disorder. 88
catalyst A chemical that speeds up a reaction but is not changed or used up by the reaction. 37, 38, 42, 80, 95, 101, 102, 104
catalytic converter A device fitted to vehicle exhausts that reduces the level of nitrogen oxides and unburnt hydrocarbons in the exhaust fumes. 22, 33, 104
cathode The negative electrode in a circuit or a battery. 31
cell differentiation The change of an unspecialised cell into a particular type of cell. 86
cell membrane The thin layer around the outside of every cell, which controls what enters and leaves it. 7, 8, 70, 71, 72, 88
central nervous system The brain and spinal cord. 4
chain reaction A reaction where the products cause the reaction to go further or faster. 66, 129
charge A property of matter, charge exists in two forms (negative and positive) that attract each other. 94, 119
chemical energy Energy stored in chemicals. 50, 53, 73
chlorophyll A green pigment found in plants that traps energy from sunlight, which allows the plant to make glucose from carbon dioxide and water. 73, 74
chloroplast An organelle found in some cells in green plants that contains the chlorophyll needed for photosynthesis. 70, 73
cholesterol A chemical found particularly in animal fats; it is needed to make cell membranes, but too much in the blood can cause heart disease. 8, 41, 42
chromatography The separation of soluble substances in a mixture, producing chromatograms. The soluble substances move along paper or jelly-like films. 43
chromosome A thread-like body made of DNA and protein found in the nucleus of a cell. It only becomes visible during cell division. 16, 85, 86, 87
cilia Tiny hairs found on the surface of cells and some microorganisms. 70
circuit A complete path that electricity can flow along. 54, 59, 122
circuit breaker A device which breaks a circuit to switch off the power when a difference in the current is detected, e.g. residual current device. 126
climate change Warming of the Earth caused by a build-up of greenhouse gases. 24, 31, 35, 46, 57
clone Genetically identical organisms. 16, 17
community A group of animals or plants living or growing together in the same area. 77
competition Interaction between organisms needing a particular resource which is in short supply, e.g. plants competing for nutrients in the soil. 15, 20
compound Two or more atoms of different elements bound together, in fixed proportions, by chemical bonds. 27, 33, 34, 93, 94, 96, 97

concentration The amount of a substance in a given volume. 51, 71, 72, 74, 101, 102, 103
conduction Movement of energy as heat (or electricity) through a substance. 30, 49, 92, 93, 94
conductor A substance that will let heat or electricity pass through it, e.g. copper. 30, 32, 49, 119, 120, 122
consumer Organisms in an ecosystem that use organic matter produced by other organisms. 75
convection current Upwards movement of heated gases or liquids to float on top of the cooler, denser layers. 43, 44, 49
core The centre of the Earth. 43
cosmic ray Radiation from space which hits the atmosphere. Some passes through, while some is blocked. 63, 64, 128
count rate The number of nuclear events in a given time, often measured by a Geiger counter. 64
covalent bond A bond between atoms where electrons are shared. 27, 32, 34, 92, 93, 95, 96
cracking The breakdown of long-chain hydrocarbon molecules into shorter-chain ones using heat, pressure and sometimes catalysts. 37, 38
crude oil A mixture of hundreds of different compounds, mainly hydrocarbons. 20, 34, 38, 39
crust The Earth's outermost layer of solid rock. 43, 44
current The flow of electrical charge through an electrical circuit. 54, 55, 59, 119, 122, 123, 124, 125, 126, 127
cutting, plant Part taken from a plant that can be used to grow a new, genetically identical plant. 17
cystic fibrosis An inherited disorder that affects cell membranes. Caused by a recessive allele. 88
cytoplasm The material in the cell that is inside the cell membrane but outside the nucleus. 70, 72

decomposer An organism that breaks down dead organic matter. 78
deforestation The permanent clearing of forests. 23
denature To alter the shape of a protein molecule, e.g. by heating, so that it can no longer do its job. 80
density The mass of an object divided by its volume. 31, 94, 115
deoxyribonucleic acid See DNA.
detergent A chemical that makes grease soluble in water. 81
detritus feeder An animal that eats the dead and semi-decayed remains of living things, e.g. an earthworm. 77, 78
diabetes A disease in which blood sugar levels are not controlled because the pancreas cannot secrete insulin, or because body cells do not respond to insulin. 7, 18, 84
differentiation A process in which new cells develop specialised features in order to carry out a job. 86
diffusion The spreading of gases or liquids caused by the random movement of their molecules. 71, 72, 73, 82
digestion The breakdown of large complex chemicals into smaller simpler ones, e.g. starch into glucose. 77, 80, 81
digestive system The gut and the other body parts linked to it that help with digestion. 7, 80
digital signal A pulsed (on/off) signal. 61, 62
direct current A current that always flows in the same direction. 55, 126
distance-time graph A graph showing time along the *x*-axis and distance covered up the *y*-axis. 111, 112
distillation A process used to boil off a liquid from a mixture and then cool the vapours to produce a pure liquid. 34, 40

DNA Deoxyribonucleic acid. The molecule that carries the genetic information in all living organisms. 16, 18, 65, 86, 88
DNA fingerprint The sequence of bases in a DNA molecule unique to each individual. 88
dominant A dominant allele is one which produces a feature even when another different allele of the same gene is present. 87, 88
double insulated An electrical device in which there are at least two layers of insulation between the user and the electrical wires. 125
drag A force which slows down something moving through a liquid or a gas. 115
dynamo A device that transforms kinetic energy into electrical energy. 55, 115

earthing To be connected to the earth – a safety feature of many electrical appliances. Prevents the build-up of charge. 120, 125, 126
earthquake Sudden movements of tectonic plates against each other, causing vibrations that travel through the Earth. 44
effector An organ that responds to a stimulus. 4, 5
efficiency The ratio of useful energy output to the total energy input. 52, 56, 57, 76
egg A special cell (gamete) made by a female for sexual reproduction. 6, 16, 17, 87
electrical charge Electrical energy stored on the surface of a material, also known as a static charge. 30, 90, 93, 118, 119, 120, 127
electrode Bar of metal or carbon that carries electric current into a liquid. 92, 106, 107
electrolysis Using an electric current to split a compound, either in solution or in its molten state. 31, 33, 94, 106, 107
electromagnetic radiation Energy carried by a wave, which can pass through a vacuum. 48, 59, 60, 61, 63, 65, 67
electromagnetic spectrum The range of all the different types of electromagnetic waves in order of their energies. 59, 60, 65
electron A small negatively-charged particle that orbits around the nucleus of an atom. 26, 27, 30, 31, 45, 63, 68, 70, 90, 91, 92, 94, 118, 122, 123, 127, 128, 129
electron configuration Representation of the arrangement of electrons around an atom of element as a series of numbers, e.g. magnesium, 2, 8, 2. 26, 90
electroscope A device that detects if an object has an electrical charge (gold leaf electroscope). 119
element A substance that cannot be split into anything simpler by chemical means. 26, 33, 62, 90, 91, 93, 96, 97, 106
embryo An organism in the earliest stages of development, for example an unborn human baby or a young plant inside a seed. 6, 17, 86
embryo transplantation Moving an embryo from one mother, or laboratory vessel, to the uterus of another. 6, 17
emulsifier A chemical which can help to break down fats into small globules. 41
emulsion A mixture with one liquid dispersed in another. 41
endothermic reactions Reactions in which energy is taken in. 104, 105
energy The ability of a system to do something (work). We detect energy by the effect it has on the things around us, e.g. heating them up. 4, 7, 48, 49, 50, 51, 52, 53, 55, 59, 60, 63, 65, 66, 67, 68, 70, 71, 75, 76, 78, 82, 90, 104, 124
energy transfer When the same form of energy moves from one place to another. 48, 49, 50, 51, 52, 54, 61, 75, 115, 116
energy transformation When one form of energy is changed to another. 50, 51, 53, 56, 115, 116, 127

environment The conditions in which a plant or animal has to survive; everything that affects a living organism. 15, 19, 20, 39, 96
enzyme A biological catalyst that speeds up the rate of chemical reactions. 7, 12, 18, 77, 80, 81, 83, 88
epidemic An infectious disease affecting many people in a particular area. 12
equilibrium The point at which a forward reaction in a reversible reaction balances the reverse reaction. 99, 105
ethane An alkane with the formula C_2H_6. 37, 38
evolution The gradual change in living organisms over millions of years caused by random mutations and natural selection. 18, 19
exothermic reaction Reactions in which energy is given out. 99, 104, 105
extinction When all members of a species have died. 20, 23

fat Chemical made up of glycerol and fatty acids. It is used as a source of energy and to make cell membranes. 7, 8, 41, 42, 73, 78
fermentation The process by which anaerobic respiration in yeast changes sugar to alcohol. 38
fertilisation When male and female sex cells (gametes) fuse together. 16, 85
fertiliser A substance added to the soil to help plants to grow. 21, 74, 105, 108
fertility drug A drug used to make eggs mature in the ovary. 6
food chain A diagram using arrows to show how energy passes from one organism to another. 75, 76, 77
food web A diagram showing how all of the food chains in an area link together. 75
force A push or pull which is able to change the velocity or shape of a body. 52, 112, 113, 116, 118, 122, 129
fossil Preserved evidence of a dead animal or plant. 19
fossil fuel A fuel such as coal, oil and natural gas formed by the decay of dead organisms over millions of years. 22, 24, 35, 40, 46, 51, 56, 57
fractional distillation The separation of a mixture into its many components using their different boiling points. 34, 38
free electron An electron that is free to move between atoms in a substance. 49, 122, 123
frequency The number of waves passing a point in one second, measured in hertz (Hz). 59, 60, 126
friction The force that acts between two surfaces in contact with each other, resisting movement. 113, 114, 115, 116
fuel A substance that gives out energy, usually as light and heat, when it burns. 35, 38, 40
fuse A component in an electrical circuit containing a thin wire which heats up and melts if too much current flows through it, breaking the circuit. 54, 122, 125, 126, 127

gall bladder An organ associated with the liver that stores bile. 80
gamete Sex cell (egg or sperm) that joins to form a new individual during sexual reproduction. 16, 85
gamma rays High-energy, high-frequency radiation travelling at the speed of light. 60, 62, 63, 65
Geiger counter A device used to detect some types of radiation. 63, 64
gene A length of DNA on a chromosome in a cell that carries information about how to make particular proteins. 11, 12, 16, 18, 19, 85, 86, 87
generator A device for converting kinetic energy into electrical energy (current flow). 55, 56, 130

natural selection Factors in the environment affect animals and plants so that some survive to reproduce successfully and pass on their good combinations of genes. Others survive less well and do not pass on their poor combinations of genes as often. 13, 18, 19
nebula A huge gas cloud in space. 66, 130
negative ion An ion with a negative charge. 91, 92, 106
nervous system The central nervous system and nerves that control the actions and functions of the body. 4, 10, 88
neurone A nerve cell, specialised for conducting electrical impulses. 4, 5, 10
neutral A solution with a pH of 7 that is neither acidic nor alkaline. 108
neutralisation A reaction between an acid and an alkali to produce a neutral solution of a salt and water. 104, 108, 109
neutron A particle found in the nucleus of an atom; it has no electrical charge and a mass of 1 atomic mass unit. 26, 62, 68, 90, 91, 127, 128, 129, 130
newton The unit of force (N). 113, 116
nicotine An addictive drug found in tobacco. 11
nitrate A salt of nitric acid. 74, 82, 105
nitrogen A non-reactive gas that makes up most of the atmosphere. 27, 33, 45, 54, 74, 82, 99, 104, 105, 108
noble gas An element in Group 0 of the periodic table, which has a full outer electron shell and is unreactive. 45
non-renewable A resource that is being used up faster than it can be replaced, e.g. fossil fuels. 20, 35, 39
nuclear fission The splitting of an atom's nucleus to release nuclear energy. 90, 129, 130
nuclear fusion When nuclei of atoms fuse together giving out huge amounts of energy, e.g. when a star is born. 50, 66, 67, 68, 90, 130
nuclear power The energy obtained by transforming the energy from nuclear fission to electrical energy. 56, 57, 129, 130
nuclear radiation The energy released when the nucleus of an atom undergoes radioactive decay. 62, 63, 64, 65
nucleus: biology The control centre of the cell, the nucleus is surrounded by a membrane that separates it from the rest of the cell. 16, 17, 86
nucleus: chemistry The central part of an atom containing the protons and neutrons. 26, 62, 63, 64, 90, 127, 129, 130

oestrogen A hormone produced in females by the ovary. 6
Ohm's Law Current is directly proportional to potential difference across the resistor. 123, 124
optical fibre A very fine strand of silica glass, through which light passes by successive internal reflections. 62
optical telescope An instrument that uses light to form images of objects far away, e.g. in space. 65
optimum temperature The temperature range that produces the best reaction rate. 80
orbit A path, usually circular, of a smaller object around a larger object, e.g. a planet orbits the Sun; electrons orbit the nucleus in an atom. 66
organelle An internal part of a cell, e.g. ribosome, mitochondrion. 70
oscilloscope A device that displays a line on a screen showing regular changes (oscillations) in something. An oscilloscope is often used to look at sound waves collected by a microphone. 126
osmosis The diffusion of water molecules from a dilute solution to a more concentrated solution through a partially permeable membrane. 72, 73
ovary Female reproductive organ. 4, 6, 85
oxidation Loss of electrons from an atom. 29, 31, 104, 106
ozone A substance in the atmosphere formed from oxygen (O_3), which acts as a shield against ultraviolet radiation. 45

pancreas An organ in the abdomen that produces enzymes to break down food, and the hormone insulin to reduce blood sugar level. 5, 80, 84
pandemic A world-wide epidemic. 12, 13
Pangaea The supercontinent thought to be the first landmass to form. 44
parallel circuit An electrical circuit in which the current divides to take different routes then joins up again. 122, 125
partially permeable membrane A membrane through which only some molecules can pass. 72
pathogen An organism that causes a disease. 11, 12, 13
payback time How long it takes for savings to equal costs. 53
periodic table A chart grouping the elements according to their number and electron configuration. 26, 90, 91
pesticide A chemical used to kill a pest. 21
pH The level of acidity or alkalinity of a substance measured as a scale from 1 to 14: neutral, 7; acidic, below 7; alkaline, above 7. 80, 108, 109
phagocyte A white blood cell that ingests bacteria in a process called phagocytosis. 12
photochromic A material that darkens on exposure to sunlight. 96
photosynthesis The production, in green plants, of glucose and oxygen from carbon dioxide and water using light as an external energy source. 21, 22, 23, 45, 46, 73, 74, 75, 78, 80
pollutant A chemical that causes pollution. 31
pollution Human activity that causes damage to the environment, e.g. dumping raw sewage at sea. 20, 21, 22, 23, 33, 35, 39
pollution indicator An organism that is sensitive to a particular level of environmental pollution. The presence or absence of a particular pollution indicator can be used to assess the degree of pollution in an environment. 23
polymer A molecule made of many repeating subunits, e.g. polythene or starch. 37, 38, 39
polymerisation The process of forming large polymers from smaller monomer molecules. 38
polyunsaturates Fats with more than one carbon-carbon double bond. 8, 41
population All the organisms of the same species that live in a particular place. 20
positive ion An ion with a positive charge. 91, 92, 94, 106, 123
potential difference The difference between two points which will cause current to flow in a closed circuit, measured in volts (V). 120, 122, 123, 124, 125, 126
potential energy Stored energy in chemicals or objects. 50, 52, 53, 115, 116
power The rate that a system transfers energy, i.e. does work, usually measured in watts (W). 52, 55
power rating The rate at which an appliance transforms electrical energy. 54, 127
power station A place that generates electricity. 22, 35, 56, 57
precipitate To fall out of solution. An insoluble solid. 109
predator An animal that hunts and kills other animals. 15, 19, 20
pressure The force acting on a surface divided by the area of the surface, measured in newtons per square metre (N/m^2). 99, 102, 105
prey Animals that are hunted by other animals. 75
producer An organism that makes organic material from inorganic substances. Green plants are primary producers because they use energy in sunlight to make sugar. 75
product Something made by a chemical reaction. 98, 101, 107
protease An enzyme that digests proteins. 81
protein A complex molecule that contains carbon, hydrogen, oxygen, nitrogen, and sometimes sulfur. It is made of one or more chains of amino acids. It is used for growth, repair and energy. 7, 73, 74, 76, 78, 80, 81, 82, 83, 86
protein synthesis To make protein in a cell from amino acids. 80
proton A particle found in the nucleus of an atom with a charge of +1 and a mass of 1 atomic mass unit. 26, 62, 63, 68, 90, 91, 127, 128, 129
pyramid of biomass A diagram to show the masses of living organisms present at each step in a food chain. 75

radiation Energy that travels as light or other forms of electromagnetic waves. 50, 59, 60, 61, 62, 63, 64, 65, 127, 128
radio wave A form of electromagnetic radiation used to carry radio signals. 50, 60, 61, 65
radioactive decay The breakdown of an unstable nucleus of an atom, releasing energy. 43, 62, 64, 128
radioisotope A radioactive isotope. 62, 63, 64, 128
radiotherapy Using radiation (gamma rays) to treat certain types of disease, e.g. cancer. 64
radon A naturally occurring radioactive gas, found near granite rocks. 63, 128
reactant A chemical taking part in a chemical reaction. 98, 101, 102
reaction time: biology/physics The time taken to respond to a stimulus. 114
reaction time: chemistry The time taken for a reaction to finish. 101, 102, 103, 105
reactivity A measure of how easily a chemical will react. 31, 32, 33, 37, 94, 95, 106, 107
reactivity series A series of metals in order of their ability to react with water and acids. 107
receptor A special cell that detects a stimulus. 4
recessive An allele which produces a feature only when the dominant allele of the same gene is not present. 87, 88
redox reaction A chemical reaction involving the transfer of electrons – reduction and oxidation. 31
reduction Gain of electrons by an atom. 31, 106
reflex action A rapid automatic response to a stimulus. 5
reflex arc The route of the electrical impulse in a reflex action along a sensory, a relay and a motor neurone. 5
relative atomic mass The mass of an atom (A_r) is the sum of its protons and neutrons (same as its mass number). 91, 97
relative formula mass The mass of a molecule (M_r) in comparison to the mass of hydrogen, which is taken as 1. 97
relative mass The mass of the particles (proton, neutron or electron) in an atom relative to one another. 26, 90
renewable A resource that can be replaced, e.g. wind power, wave power and solar power are all renewable sources of energy. 33, 38, 40, 56, 57
resistance: biology When an organism's genetic make-up changes so that it is unaffected by an agent used to destroy it. 12, 13, 18
resistance: physics The amount by which a conductor prevents the flow of electric current. 54, 115, 122, 123, 124

resistor A conductor that reduces the flow of electric current. 122, 123, 127

respiration The chemical process that releases energy from food in all cells. All living things must respire. 7, 23, 70, 71, 73, 75, 77, 78, 80, 82, 83

resultant force The sum of all the different forces acting on an object. 112, 113, 114

reversible reaction A reaction whose direction can be changed by a change in conditions. 96, 98, 99, 104, 105

ribosome An organelle in the cell in which protein synthesis takes place. 70

salt A compound made when an acid reacts with an alkali. Common salt is sodium chloride. 8, 74, 91, 94, 95, 107, 108, 109

Sankey diagram A diagram that shows the input and output of energy in a device. 51, 52

satellite A body orbiting around a larger body. Communications satellites orbit the Earth to relay messages from mobile phones. 61, 65

saturated Fats or oils containing only single bonds between carbon atoms. 8, 34, 41

secretion The production of useful substances by organs or cells, e.g. hormones by glands. 4, 5

series circuit An electrical circuit where components are connected in one continuous loop. 122, 124

sewage Waste matter and rubbish carried in sewers. 21, 23

sex chromosomes Chromosomes that carry the genes that determine sex. 87

sexual reproduction Reproduction involving gametes and fertilisation. There are usually two parents, each contributing half the genes of the offspring. 16

shape memory When a smart alloy that can be deformed is able to return to its original shape when heated. 33, 39, 96

shell A grouping of electrons in an energy level around an atom. 26, 27, 30, 90, 92, 94, 95

silicon dioxide Sand. Covalently bonded silicon and oxygen atoms. 93

skin graft Surgical replacement of damaged skin by undamaged skin. 17

smart alloy An alloy that has a shape memory. 33

smart material A material that responds to a change in the environment by changing in some way that is reversible. 38, 96

smoke Tiny particles (e.g. carbon particles) suspended in air, produced by burning materials. 22, 64

solar cell A device that converts light energy into electrical energy. 57

solar energy Energy from the Sun. 53, 56, 57, 75

Solar System The collection of planets and other objects orbiting around the Sun. 66, 67

species A group of living things of the same type. Humans belong to the species *Homo sapiens*. 18, 19, 20

spectral pattern The pattern formed when an element is heated and emits light, as viewed by a spectroscope. 68

spectroscope A device that splits light into bands of different colours. 68

speed The distance travelled by an object in a certain time. 111, 112, 113, 114, 115, 116

sperm A special cell (gamete) made by a male for sexual reproduction. 16, 70, 85, 87

star A large gas body in space that emits light from internal nuclear reactions. 65, 66, 67, 130

starch An insoluble carbohydrate used as an energy store in plant cells. 73

static electricity Electric charge on the surface of a non-conductor, i.e. it is charge that cannot flow easily. 118, 119, 120

stem cell An undifferentiated cell found in embryos and certain parts of adults that can develop into many other types of cell. 86

stimulus A change inside or outside of an organism that brings about a response, e.g. altered blood sugar levels; a flash of light. 4, 5

stoma Tiny hole in the surface of a leaf which allows gases to diffuse in and out. 73

stopping distance The distance required to bring a moving vehicle to a halt. The sum of the thinking distance and the braking distance. 114

subduction zone A region where one tectonic plate moves under another. 44

substrate Something that an enzyme causes to react to make a product. 88

sucrase An enzyme that digests sucrose. 81

surface area The area of a surface. Surface area has a significant effect on the rate of many chemical reactions. 15, 95, 102

sustainable development Development that does not make life more difficult for future generations. 24, 98, 105

synapse The gap between the ends of two neurones. 5, 10

target organ Organs affected by the presence of a particular hormone, e.g. adrenaline from the adrenal gland affects the heart. 4

tectonic plate Cracked sections of the lithosphere that ride on top of the fluid mantle. 44

telecommunications Communication over a distance by cable, telephone or broadcasting. 61

temperature A measure of how hot something is. 46, 48, 71, 74, 77, 80, 83, 84, 87, 99, 101, 102, 105, 123, 124

terminal velocity The maximum velocity reached by a falling object, when weight is balanced by air resistance. 114, 115

temperature receptor Cell in the brain that monitors blood temperature. 84

testis Male reproductive organ. 85

thermal decomposition Breaking down a chemical using heat. 28, 37, 104, 105

thermal energy *See* heat.

thermal radiation The transfer of heat energy by rays. Also called infrared radiation. 48, 68

thermochromic Material that changes colour when it gets hotter or colder. 96

thermoregulatory centre An area of the brain containing temperature receptors that monitors blood temperature. 84

thinking distance The distance travelled by a moving vehicle before the brakes are applied. The product of speed and reaction time. 114

tidal power The energy obtained by transforming the energy in the rise and fall of the tides into electrical energy. 56, 57

tissue A group of cells of the same type, e.g. nervous tissue contains nerve cells. 17, 70, 73, 75

total internal reflection When light hits a glass/air interface at an angle of more than 42°, it is reflected back at the boundary. 62

toxin A chemical that harms living things, a poison. 9, 11, 21

transformer A device that alters the voltage of an alternating current. 55

transition element A metal belonging to the transition group in the periodic table. 30, 33

turbine A device that transfers kinetic energy into circular movement, usually to drive a generator. 55, 56, 57, 130

ultraviolet radiation Radiation just beyond the blue end of the spectrum of visible light. UV light causes tanning and some sorts of skin cancer. 59, 60, 61, 65

Universe Everything, everywhere. 68

upthrust An upward force, e.g. of water on a floating object. 113, 115

uranium A radioactive metal used in nuclear power stations and bombs. 56, 129, 130

urea A substance made in the breakdown of proteins in the liver. 5, 83

urine A solution of urea and other wastes in water, made in the kidneys for excretion by the bladder. 4, 5, 83

uterus The organ in the female where the baby grows during pregnancy; also known as the womb. 6

vaccine A chemical developed to immunise people against a particular disease. 13

vaccination Injection of a vaccine into the bloodstream. 13

vacuum A space with no particles in it. 48, 59

Van de Graaff generator A device used to generate static electrical charges. 119, 120

variation The existence of different characteristics in the same species. 16, 19, 23, 85

vegetable oils Oils obtained from the fruits, seeds and nuts of some plants, e.g. olive, rape, coconut. 40

velocity The speed of an object in a given direction (*v*). 111, 112, 113, 114, 115, 117, 118

villi Tiny finger-like structures lining the small intestine. 71

virus A microorganism (a thousand times smaller than a bacterium) that invades cells, reproduces and bursts cells, making people ill. 11, 12, 13

vitamin A substance found in foods that is essential to maintain health. 40

volcano The result of upthrust of the crust of the Earth at a point of weakness, allowing magma from the mantle to be forced out. 44

volt The international unit of electrical potential (V). 126

voltage The potential difference across a component or circuit (*V*). 55, 120, 122

voltmeter A meter that measures the voltage between two points. 122, 124, 125

watt A unit of power (W), 1 watt equals 1 joule of energy being transferred per second. 52, 54, 127

wave power The energy obtained from transforming the energy of waves into electrical energy. 56, 57

wavelength The distance between two identical points on a wave. 59, 60, 68

weight The force of gravity acting on a body measured in newtons (N). 52, 114

wind farm A group of wind turbines used to generate electricity from the kinetic energy in moving air. 56

withdrawal symptoms The combination of physical and psychological symptoms produced when an addictive drug is withheld for a period of time. 9, 10

work Work is done when a force moves. The greater the force or the larger the distance, the more work is done. 52, 115, 116

X-ray Electromagnetic radiation used for imaging or to destroy some types of cancer cells. 59, 60, 61, 65, 88

XX chromosomes The sex chromosomes present in a human female. 87

XY chromosomes The sex chromosomes present in a human male. 87

yeast A unicellular fungus used extensively in the brewing and baking industries. 38

yield The ratio of product to starting materials; a high yield means that most of the starting material is converted to useful products. 98, 105

Collins

Collins Revision

GCSE Higher Science

Exam Practice Workbook

FOR AQA A + B

Coordination

D–C

1 a The diagram shows a neurone from the human body.

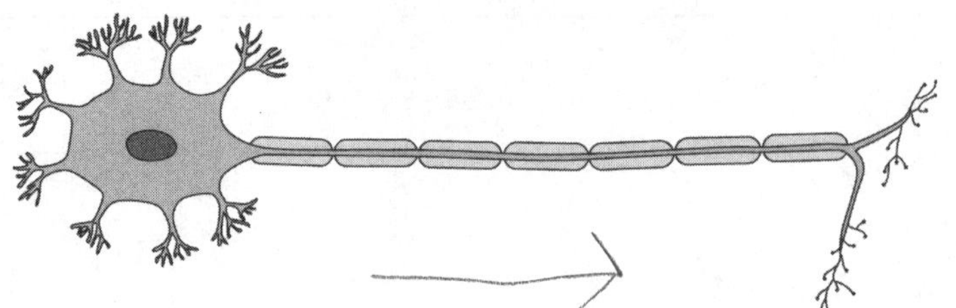

i What type of neurone is this? Motor Neurone ✓ [1 mark]

ii Draw an arrow on the diagram to show the direction in which the nerve impulse travels along the neurone. [1 mark]

B–A*

b In what form does the nerve impulse travel along the neurone?

electrical impulses. [1 mark]

Receptors

D–C

2 If you touch a very hot object, you react by moving your hand away very quickly.

a Name the receptor and effector in this response.

Receptor Skin Effector Motor Neurone Muscle in arm [2 marks]

B–A*

b Receptors transform one form of energy into another. Name the **two** forms of energy in this response.

The receptor transforms Heat energy to electrical energy. [2 marks]

D–C

3 a Adrenaline is secreted by the adrenal glands when you get a sudden fright. It makes the heart beat faster.

i Name the target organ of adrenaline.

Heart [1 mark]

ii Suggest how adrenaline could help you to escape from danger.

It helps you breath faster to get more oxygen to the muscles [1 mark]

B–A*

b Describe **two** ways in which transmitting information using hormones differs from transmitting information using neurones.

1 The Neurones send the information to the CNS

2 Hormones use chemicals to pass the information. [2 marks]

Grades

Reflex actions

D–C

1 a The diagram shows a synapse.

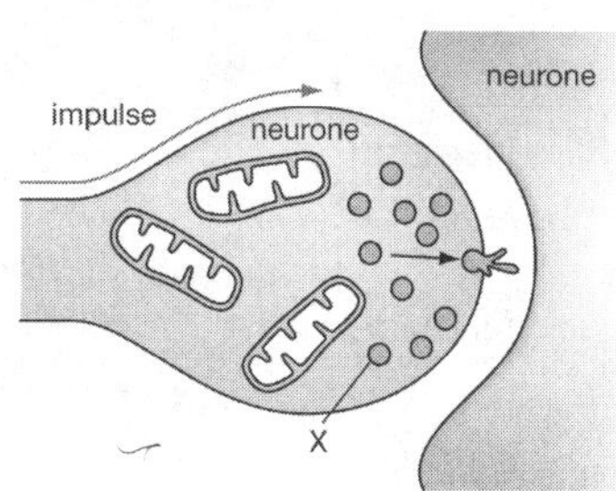

i What is contained in the structure labelled X? information. [1 mark]

ii Describe how the nerve impulse travels across the synapse.

When the impulse reachs a synapse it secrets a chemical which travel over the gap to the next neurone [2 marks]

B–A*

b Adam puts his hand onto a pin. A sensory neurone carries an impulse to his central nervous system, where the impulse is passed on to a motor neurone.

i How many synapses are there in this pathway? 2 [1 mark]

ii Suggest **one** advantage of having synapses in pathways such as this.

It can get over obsticles such as bones [1 mark]

In control

D–C

2 a Sports drinks usually contain glucose, sodium ions and chloride ions. Explain why an athlete might need to take in extra sodium and chloride ions.

____

______________________________ [2 marks]

b If you have too much glucose in your blood, your pancreas secretes a hormone called insulin. Describe how insulin helps to make the blood glucose level go back to normal.

____

______________________________ [2 marks]

B–A*

c i Jill has diabetes. Her pancreas does not secrete insulin. Suggest why Jill's body may run out of glucose if she does not eat carbohydrate-containing foods at regular intervals.

____

______________________________ [2 marks]

ii Explain why it is dangerous for the body to run out of glucose.

____

______________________________ [2 marks]

Grades

Reproductive hormones

D–C

1 a The flow diagram shows how three hormones help to control the menstrual cycle.

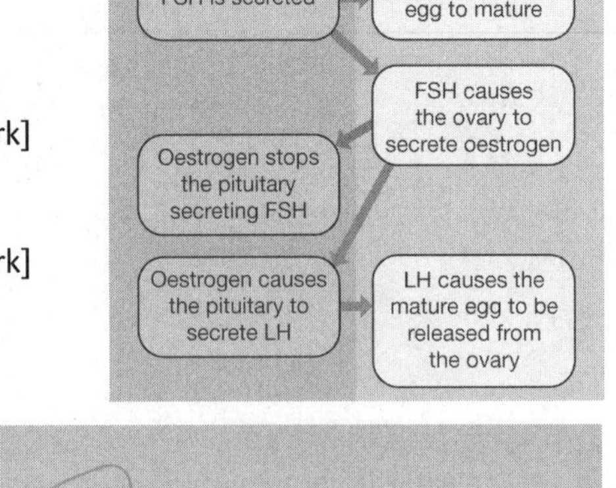

i Name the gland that secretes FSH and LH.

________________________ [1 mark]

ii Name the gland that secretes oestrogen.

________________________ [1 mark]

B–A*

b The graph shows how the levels of FSH, LH and oestrogen change during the menstrual cycle.

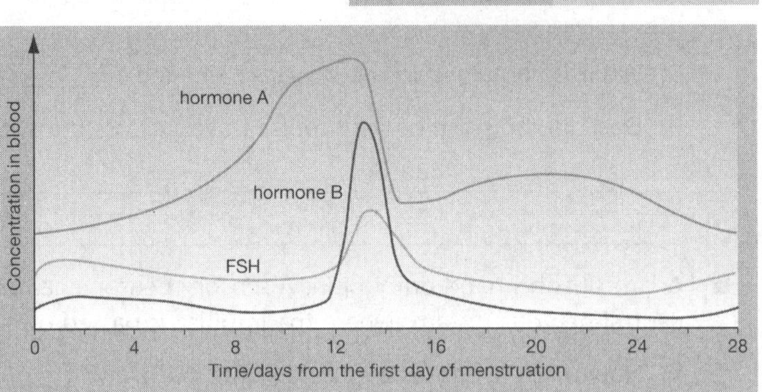

i On the graph, use an arrow to show when ovulation is most likely to occur. [1 mark]

ii During which days is menstruation most likely to occur? [1 mark]

iii Name hormones A and B.

A ________________

B ________________ [2 marks]

Controlling fertility

D–C

2 This table shows the success rate for one IVF treatment in women of different ages.

Age of woman in years	Chance of having a baby after IVF
23–35	More than 20%
36–38	15%
39	10%
40 and over	Less than 6%

a Describe how a woman's age affects the success rate for IVF treatment.

________________________ [1 mark]

b Using the data in the table, and your own ideas, suggest arguments supporting the idea that women over 55 years of age should not be given IVF treatment on the National Health Service.

____

________________________ [2 marks]

B–A*

3 FSH may be given to women whose eggs do not mature naturally, to help them to conceive. This treatment often results in multiple births. Suggest an explanation for this.

____

________________________ [2 marks]

Diet and energy

D–C

1 a Katie has a high metabolic rate. Olivia's metabolic rate is lower. Olivia is jealous of Katie because Katie can eat more food than Olivia without putting on weight.

i What is **metabolic rate**?

______________________________ [1 mark]

ii Explain why Katie's high metabolic rate means that she can eat more food than Olivia without putting on weight.

____

______________________________ [2 marks]

iii Suggest what Olivia could do to increase her metabolic rate.

______________________________ [1 mark]

B–A*

b Carbohydrate, fat and protein are the three nutrients that provide us with energy.

i Which of these three nutrients contains the most energy per gram?

______________________________ [1 mark]

ii Describe how body cells release energy from carbohydrate.

____

______________________________ [2 marks]

Obesity

D–C

2 The graph shows the percentage of obese men and women in five different countries.

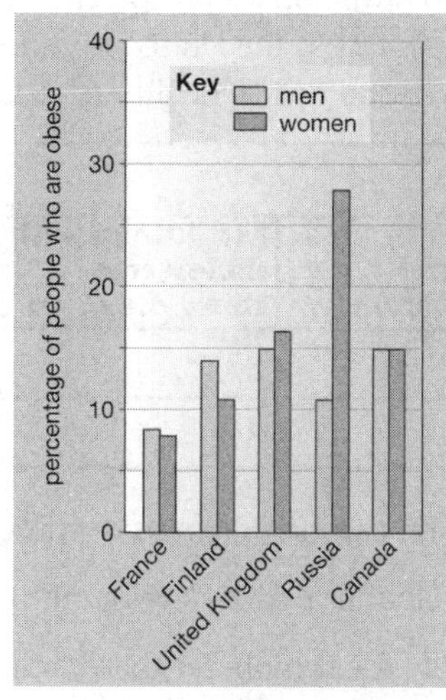

a In what way does the pattern of obesity in Russia differ from the pattern in Canada?

______________________________ [1 mark]

B–A*

b Lisa says that the pattern of obesity in Russia suggests that it is caused by a person's eating and exercise habits, rather than by their genes. Explain how the data supports this view.

____

______________________________ [2 marks]

c Explain why obese people are more likely to suffer from arthritis than people of normal weight.

____

______________________________ [2 marks]

Not enough food

Grades

D–C

1 a A slimming programme advertises that you can lose 18 lbs (8 kg) in four days. It says:

> The 18-in-4 programme identifies common, but specific FRUITS, VEGETABLES, CHICKEN AND BEEF that need to be eaten in specific combinations. These combinations of specific foods work with the body's chemistry and result in accelerated weight loss.

i Explain why the **combination** of food that a person eats, rather than the **quantity** of food that they eat, is unlikely to help them lose weight.

______________________________ [1 mark]

ii Suggest **two** reasons why it is not good for a person to lose this much weight so quickly if they are trying to improve their health.

1 ______________________________

2 ______________________________ [2 marks]

B–A*

iii In 2005, a slimming diet that involved eating mostly proteins and fats, with no carbohydrate, was very popular. Explain why this diet could increase the risk of developing heart disease.

____

______________________________ [2 marks]

Cholesterol and salt

D–C

2 This table shows information about the cholesterol concentration in the blood of men and women in two cities in the United Kingdom. A blood cholesterol concentration above 6.6 units is dangerous to health.

City	Men		Women	
	Average cholesterol concentration	Percentage with cholesterol above 6.6 units	Average cholesterol concentration	Percentage with cholesterol above 6.6 units
A	5.9	27	5.9	31
B	6.1	35	6.1	36

a What conclusion can be drawn from this data?

A Men have a higher risk of developing heart disease than women.

B People in City A eat more fat than people in City B.

C People in City B have a higher risk of developing heart disease than people in City A.

D Women eat more fat than men. [1 mark]

B–A*

b Suggest how this data should have been collected in order to make a fair comparison between City A and City B.

____

______________________________ [2 marks]

c The liver makes cholesterol when blood cholesterol levels are low. Drugs called statins inhibit the enzyme that makes cholesterol. Suggest why taking statins can reduce blood cholesterol levels more than can be done by changing your diet.

____

______________________________ [2 marks]

Grades

B1 A HUMAN BIOLOGY

Drugs

D–C

1 a Explain what is meant by being **addicted** to a drug.

____

____ [2 marks]

B–A*

b Cocaine is an illegal and very addictive drug. This graph shows the percentage of people between 16 and 29 and between 30 and 59 who told researchers that they took cocaine in 2006. In each age group, people are grouped according to how often they visit nightclubs.

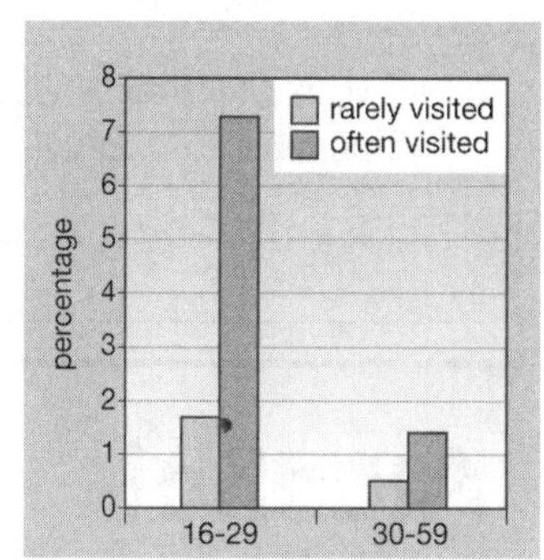

i Compare the percentages of the frequent nightclub visitors in the two age groups who reported taking cocaine in 2006.

____

____ [2 marks]

ii Describe the relationship between visiting nightclubs and taking cocaine, as shown by this data.

____

____ [2 marks]

Trialling drugs

D–C

2 This table shows the results of a trial to test a new drug to help people to recover from flu.

		Given zanamivir	**Given a placebo**
A	Mean age of subjects in years	19	19
B	Number of days until their temperature went down to normal	2.00	2.33
C	Number of days until they lost all their symptoms and felt better	3.00	3.83
D	Number of days until they felt just as well as before they had flu	4.5	6.3
E	Average score the volunteers gave to their experience of the five major symptoms of flu	23.4	25.3

Answer these questions using information in the table.

a Give the letter of the row in the table that shows how the researchers kept one variable constant in their trial. ______________ [1 mark]

b Give the letters of **all** the rows in the table that show that the drug helps people to get over flu more quickly.

____ [1 mark]

c The trial was a **double-blind** trial. What does this mean?

A Both the researchers and the subjects know who is being given the drug or placebo.

B Neither the subjects nor the researchers know who is being given the drug or a placebo.

C Only the researchers know whether which subjects are being given the drug or a placebo.

D Only the subjects know whether they are being given the drug or a placebo. [1 mark]

B–A*

d Despite extensive trials such as this one, some new drugs have unexpected side effects when they are in general use. Suggest why these side effects are not discovered during the trials.

____

____ [2 marks]

Illegal drugs

1 a Explain how injecting heroin into the blood can increase the risk of getting HIV/AIDS. (D–C)

___

___ [2 marks]

b Explain why people who have begun taking cocaine may find it almost impossible to stop taking the drug.

___

___ [2 marks]

Alcohol

2 The table shows the number of people who were admitted to hospital in one area of the USA in one year because they had been taking drugs. (D–C)

Drug they had taken	Number of men admitted to hospital	Number of women admitted to hospital
Alcohol	380	140
Heroin and similar drugs	90	50
Cocaine	40	30
Cannabis	100	20

a What does this data show?

A Alcohol causes more hospital admissions than heroin.

B Heroin is more dangerous than cocaine.

C More men than women drink alcohol.

D More people use cannabis than heroin. [1 mark]

b Zachary says that the table shows that more men than women take these drugs. Is he right? Explain your answer.

___

___ [2 marks]

c Excessive drinking of alcohol damages the liver. Why does this happen?

___ [1 mark]

d Explain how this data supports the statement that the overall impact of legal drugs on health is greater than the impact of illegal drugs. (B–A*)

___

___ [2 marks]

Grades

Tobacco

D–C

1 This graph shows information about the numbers of men who smoked and the numbers of men who died of lung cancer between 1911 and 2001.

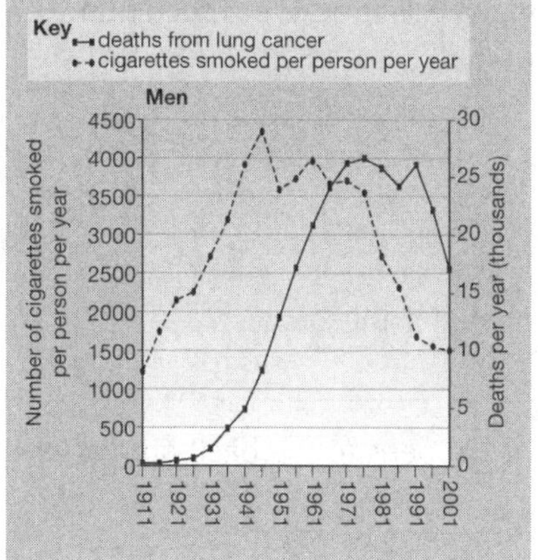

a Using the information on the graph, suggest why it was not until the 1950s that it was recognised that smoking could be causing cancer.

____

____ [2 marks]

b The data in the graph shows a correlation between smoking and lung cancer. Explain why this does not prove that smoking causes lung cancer.

____ [1 mark]

B–A*

c Tobacco smoke contains carbon monoxide. Explain how this can reduce the birth weight of a baby born to a woman who smokes during pregnancy.

____

____

____ [3 marks]

Pathogens

D–C

2 a What is a pathogen?

A An infectious disease.

B A microorganism that causes disease.

C An organism that transmits disease.

D A gene that causes inherited disease. [1 mark]

b Ignaz Semmelweis worked in labour wards in Austria in the 1840s. Describe how he helped to reduce the death rate of women who gave birth there.

____

____ [2 marks]

B–A*

c Even today there are still serious problems with the spread of infections in hospitals. Suggest why these problems are greater in hospitals than in other places.

____

____ [2 marks]

Body defences

Grades

1 The diagrams show a cell attacking a bacterium. D–C

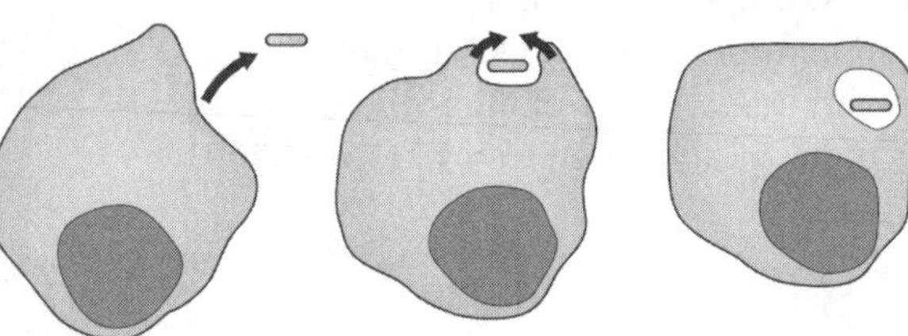

a What is the name for this process?

____ [1 mark]

b Describe what happens to the bacterium after the cell has surrounded it.

____

____ [2 marks]

2 This table shows the total number of people in the world who suffered from the new infectious disease, SARS, in February and March 2003. B–A*

Country	Number of cases of SARS
China	95
Viet Nam	40
Singapore	20
Canada	2
Switzerland	1
Thailand	1

a How could SARS have spread from China, Viet Nam and Singapore to Canada and Switzerland?

____

____ [2 marks]

b Explain how the table suggests that this outbreak of SARS was an epidemic, rather than a pandemic.

____

____ [2 marks]

Drugs against disease

3 a Drug companies have not been very successful in finding effective antiviral drugs. What helps to explain this? D–C

A Viruses are too small to see.

B Viruses are too dangerous for researchers to work with.

C Viruses reproduce inside body cells.

D Viruses reproduce very slowly. [1 mark]

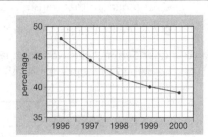

b This graph shows the percentage of children between the ages of 0 and 4 who were prescribed antibiotics each year between 1996 and 2000 in the USA.

i By how much did antibiotic prescriptions for this group of children fall between 1996 and 2000?

____ [1 mark]

ii Explain why it is important to try to reduce the number of antibiotic prescriptions. B–A*

____

____ [2 marks]

Arms race

D–C

1 Most deaths from MRSA happen in hospitals. Which **cannot** help to explain this?

A A lot of antibiotics are used in hospitals, increasing the chance of antibiotic resistance developing.

B It is easy for bacteria such as MRSA to spread from one person to another in a hospital ward.

C Many people in hospital are already ill and weak.

D MRSA is only found in hospitals. [1 mark]

B–A*

2 *Escherichia coli* is a bacterium that can cause infections of the blood and cerebrospinal fluid (the fluid inside the brain and spinal cord). This graph shows the percentage of these infections in the United Kingdom that were caused by strains of *E. coli* that were resistant to the antibiotic ciprofloxacin in the years from 1990 to 2004.

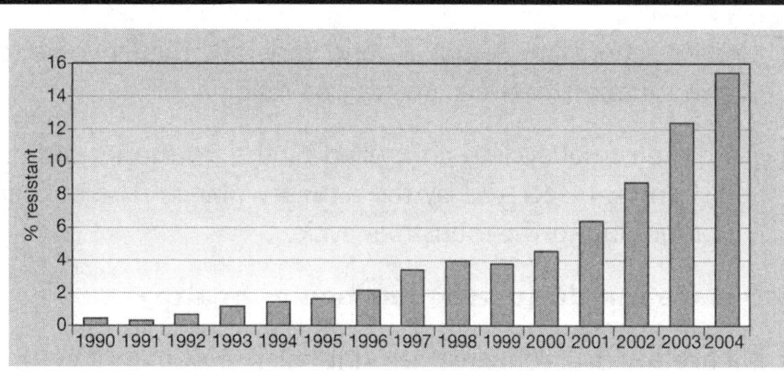

a Describe the trend in antibiotic resistance in *E. coli* between 1990 and 2004.

____

____ [2 marks]

b Suggest how this antibiotic resistance has come about.

____

____ [2 marks]

Vaccination

D–C

3 TB is an infectious disease caused by a bacterium. A vaccine called BCG was introduced in 1951. It is given to most school pupils between the ages of 10 to 14. The table shows the number of new cases of TB in the United Kingdom in some of the years between 1950 and 1990.

Year	Number of new cases of TB
1950	50 000
1960	25 000
1970	16 000
1980	9000
1990	6000

a Do the figures in the table prove that the BCG vaccination has been effective? Explain your answer.

____

____ [2 marks]

b Suggest why the number of new cases of TB took many years to drop as low as 6000 cases per year.

____

____ [2 marks]

c The TB vaccine contains an extract made from the bacteria that cause TB. Explain how the vaccine can make a person immune to TB.

____

____

____

____ [4 marks]

B1a revision checklist

I know:

how human bodies respond to changes inside them and to their environment

- ☐ nerves and hormones coordinate body activities and help control water, ion and blood sugar levels and temperature
- ☐ receptors transform energy from stimuli into electrical impulses, enabling information to travel rapidly along the nerves
- ☐ reflex actions are automatic and very fast; they involve sensory, relay and motor neurones, with synapses between them, in a reflex arc
- ☐ the menstrual cycle is controlled by the hormones oestrogen (secreted by the ovaries), FSH and LH (secreted by the pituitary gland). The concentrations of these hormones change during the menstrual cycle

what we can do to keep our bodies healthy

- ☐ a healthy diet contains the right balance of the different foods you need and the right amount of energy
- ☐ obesity increases the risk of arthritis, diabetes, high blood pressure and heart disease
- ☐ too much cholesterol in the form of LDL in the blood increases the risk of heart disease
- ☐ the liver makes cholesterol if intake is too low

how we use/abuse medical and recreational drugs

- ☐ drugs affect people's behaviour and can damage the brain, and some hard drugs (heroin) are addictive. Some drugs are legal (alcohol) and some are illegal (cocaine, heroin)
- ☐ the substances in tobacco smoke cause many diseases, e.g. cancer and bronchitis
- ☐ alcohol is a depressant and hinders the activity of parts of the brain
- ☐ medical drugs are beneficial, but must be thoroughly tested before use

what causes infectious diseases and how our bodies defend themselves against them

- ☐ pathogenic microorganisms cause infectious diseases
- ☐ some white blood cells (phagocytes) ingest pathogens and kill them; and others (lymphocytes) make antibodies
- ☐ antibiotics kill bacteria inside the body but do not kill viruses; however, some bacteria can develop resistance to antibiotics
- ☐ immunisations and vaccinations offer protection from various diseases

Grades

Hot and cold

D–C

1 The diagram shows the blood flow in the leg of a caribou. Caribou live in very cold places.

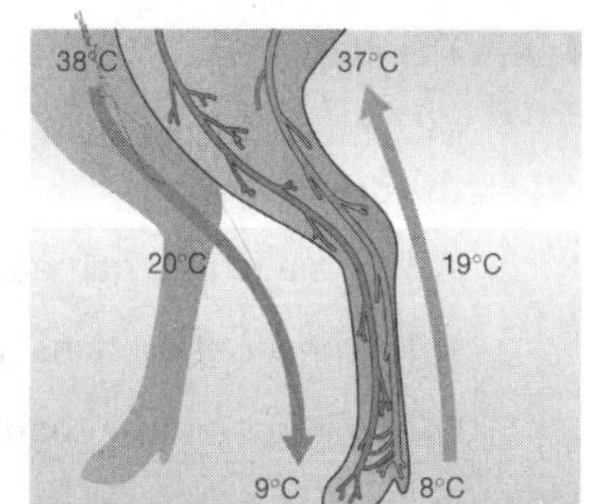

a What is the temperature of the blood in the **artery** as it flows into the top of the caribou's leg?

37 °C [1 mark]

b Explain why the blood in the caribou's foot has a lower temperature than this.

further away from all other parts of the body [1 mark]

c How does the arrangement of blood vessels shown in the diagram help to stop the temperature deep inside the caribou's body from dropping too low?

They spread as much as possible. [1 mark]

B–A*

2 Sun bears live in hot places. They are quite small for a bear, with large ears and brown fur. Polar bears are much larger, with small ears and white fur.

Explain how each of these features is an adaptation to the conditions where the bears live.

a Size of body and ears

The ears are small to reduce heat loss.
The size of the body is large to have a small SA:V ratio [2 marks]

b Colour of fur

So it can camouflage in the surroundings [1 mark]

Adapt or die

D–C

3 Complete the sentences by using the **correct** words from the box.

adapted	**camouflaged**	**competitors**	**poisons**
	predators	**thorns**	

Plants and animals are adapted for survival in their habitat. Some plants have thorns in their leaves, which harm insects that eat the leaves. Some plants have warning colours to deter predators. [3 marks]

B–A*

4 In Lee's garden, there is a large tree at the edge of the lawn. Much less grass grows under the tree than on other areas of the lawn. Suggest why there is less grass under the tree.

It is deprived of light and water. [2 marks]

Two ways to reproduce

D–C

1 a The diagram shows what happens during sexual reproduction.

Match statements, **A**, **B**, **C** and **D**, with the labels **1** – **4** in the diagram.

A This is a female gamete.

B The new cell contains genes from both parents.

C Two gametes fuse together.

D This is a male gamete.

1	2	3	4
D	A	C	B

[4 marks]

B–A*

2 Explain why asexual reproduction produces clones, but sexual reproduction does not.

It produces clones as it only has one pair of genes from 1 parent So it is geneticaly identical.

[2 marks]

Genes and what they do

D–C

3 a Where in a cell is DNA found?

Nucleus.

[1 mark]

b Describe the function of DNA in a cell.

DNA Contains all the genetic data

[2 marks]

B–A*

4 Snuppy was the first cloned dog. Some egg cells were taken from a female Afghan hound and their nuclei were removed. Some other cells were taken from a male Afghan hound's ear. An egg cell was fused with an ear cell. This formed a 'zygote', which grew into Snuppy.

a Would Snuppy be a clone of the female or the male dog? Explain your answer.

Male as it has the genetic data of the male in the egg cell

[2 marks]

b Dogs have 78 chromosomes.

i How many chromosomes would there have been in the egg cell taken from the female dog?

39

[1 mark]

ii How many chromosomes would there have been in the ear cell taken from the male dog?

39 78

[1 mark]

iii How many chromosomes are there in each of Snuppy's cells?

2 78

[1 mark]

Grades

Cuttings

D–C

1 Oak trees reproduce sexually. They produce seeds called acorns. Two acorns from an oak tree grow into new trees. One of the trees is much taller than the other.

a Suggest **two** reasons why one tree is taller than the other.

1 The Soil may be better in one patch than other

2 May derrived of light.

[2 marks]

b Gardeners can also grow new trees from cuttings. They take a small piece of stem from a tree and place it in soil. The piece of stem grows roots and becomes a new plant. Mia wants to grow an oak tree exactly like one she already has in her garden. Explain why using cuttings would be better than planting acorns.

As it is already an established plant and has a better chance at survial.

[2 marks]

B–A*

2 Banana plants are unable to reproduce sexually.

a Suggest how banana growers can produce new banana plants.

A by sowing seeds

B by taking cuttings (circled)

C by crossing one plant with another

D by introducing insects to pollinate the flowers

[1 mark]

b A disease is killing banana plants. Explain why it is difficult for banana growers to produce a new variety of banana plants that are resistant to this new disease.

As they are all genti cally identical.

[2 marks]

Clones

D–C

3 This diagram shows how new cattle can be produced using a technique called **embryo transplants**.

a Explain why the embryos that are transplanted into the replacement mothers are clones.

As the ball of cells have the same gentic data but have been split.

[2 marks]

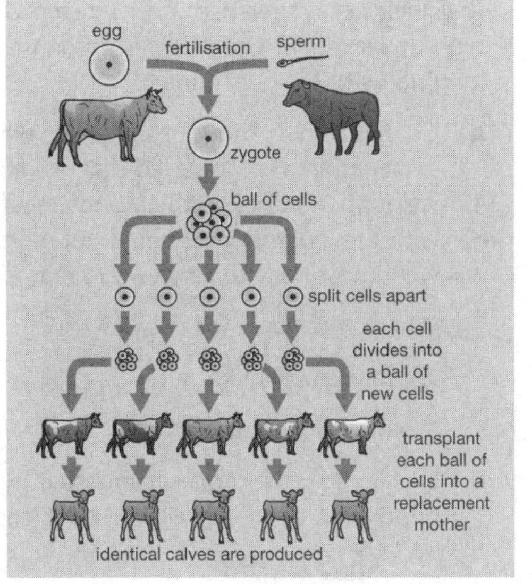

B–A*

b Suggest how this technique could help to save a rare species of animal from extinction.

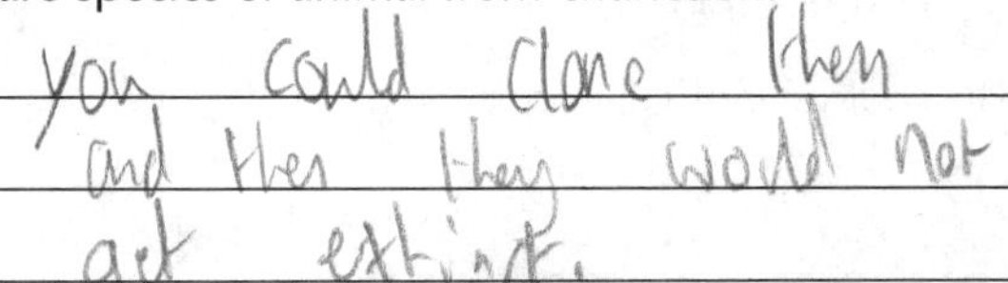

you could clone them
and then they would not
get extinct.

[2 marks]

Genetic engineering

D–C

1 Read this article about genetically engineered maize.

> Maize seeds are a protein- and carbohydrate-rich food. Maize is grown in many parts of the world. Insect pests reduce yields of maize. Many farmers spray the maize plants with pesticides to kill the pests.
>
> Now some new varieties of maize plants have been produced using genetic engineering. A gene for making a toxin has been transferred into the maize cells. The plants make a toxin (poison) in their leaves, which kills any insects that eat them.

a Explain how the genetically engineered maize could be an **advantage** to each of the following:

i a farmer

They would not have to buy the pesticide ✓ [1 mark]

ii a person buying food made with the maize

It would be cheaper. [1 mark]

B–A*

b Suggest why some people may not want to eat foods made from genetically modified maize.

____

______________________________ [2 marks]

D–C

2 The diagram shows how genetic engineering has been used to produce insulin from bacteria.

Explain why the insulin that the bacteria make is exactly the same as the insulin made by human cells.

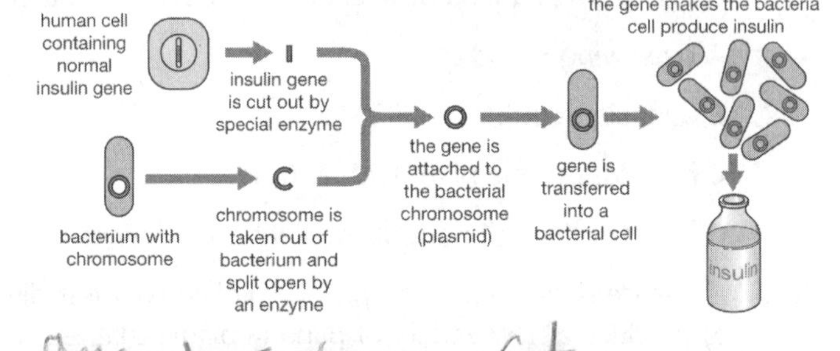
human cell containing normal insulin gene
insulin gene is cut out by special enzyme
bacterium with chromosome
chromosome is taken out of bacterium and split open by an enzyme
the gene is attached to the bacterial chromosome (plasmid)
gene is transferred into a bacterial cell
the gene makes the bacteria cell produce insulin
insulin

As the insulin gene has been cut out the human cell. [2 marks]

Theories of evolution

D–C

3 The drawings show how Lamarck thought that giraffes evolved their long necks. He thought that their necks stretched as they reached for high-up leaves. The giraffes passed on their long necks to their offspring.

a Charles Darwin thought that there was a different explanation. He suggested that there was natural variation in neck length amongst giraffes. The ones with the longest necks could get more food, so they were most likely to survive and reproduce.

What is the name for this process?

A acquired characteristics **B** extinction

C natural selection (circled) **D** speciation [1 mark]

b Explain how the process suggested by Darwin could eventually produce a population of giraffes all with long necks.

The shorter neck giraffes would not be able to reach the food so die whereas the taller [2 marks] will survive and breed.

Grades

Natural selection

D–C

1 Peppered moths have pale wings with dark speckles. This camouflages them against tree bark so they are less likely to be eaten by birds. In the industrial revolution, air pollution made tree bark darker. A variety of peppered moths with dark wings became more common than the pale moths.

a What is the term for the relationship between the birds and the moths?

A camouflage **B** competition **C** evolution **D** predation [1 mark]

b What can explain the dark moths becoming more common?

A Any moths that had dark wings were more likely to survive.

B The moths changed their colour so that they were better camouflaged from birds.

C The pale peppered moths flew away to less polluted areas.

D The polluted air made the moth's wings darker. [1 mark]

B–A*

c In the 1950s, the air became much cleaner. Suggest what would have happened to the peppered moth population and explain your suggestion.

They would get caught by the birds more as they were less camoflaged.

[3 marks]

D–C

2 Species of organisms that live on remote islands are often different from related species that live on the mainland. Suggest why.

as the organism have adapted to their local surroundings

[2 marks]

Fossils and evolution

D–C

3 The diagram shows some horses that lived long ago.

Years ago	What it looked like and idea of size	Bones of its front leg	How it lived
10 million	height: 1.0 m		lived in very dry places; it was a fast runner
40 million	height: 0.6 m		lived in dryer conditions; needed to be able to run away from predators
60 million	height: 0.4 m		lived on soft ground near water; its feet could support its weight without sinking into the mud

a How do we know about the structure of the legs of horses that lived millions of years ago?

[1 mark]

b Use the theory of natural selection, and the information in the diagram, to suggest why the feet of horses that lived 40 million years ago had three toes but the feet of horses that lived 10 million years ago had only one toe.

[2 marks]

B–A*

c Explain why we cannot conclude that the first or second species of horses shown in the diagram are ancestors of the third species.

[2 marks]

Grades

Extinction

D–C

1 Crayfish live in streams and rivers in England. A larger species of crayfish has been introduced from America. The introduced species eats the same food as the English crayfish. It carries a disease that does not affect it but that kills English crayfish.

Explain why the English crayfish may become extinct.

____

____

______________________________ [3 marks]

B–A*

2 Dodos were huge, flightless birds that lived on Mauritius, in the Indian Ocean. There were no predators on Mauritius. Dodos became extinct after people arrived on the island.

Suggest why the dodos became extinct.

____

______________________________ [2 marks]

More people, more problems

D–C

3 a The graph shows how the human population has changed since 1750.

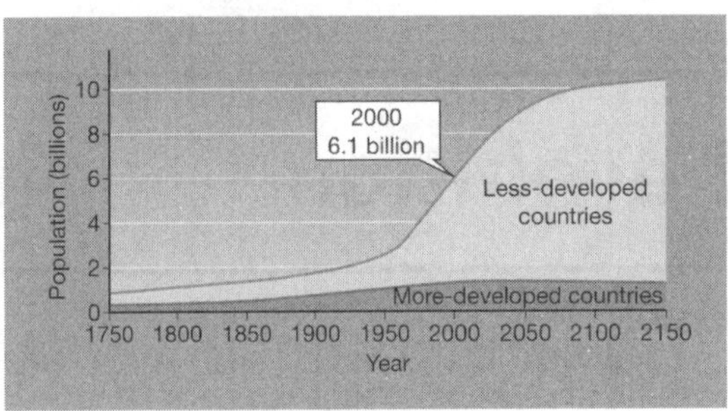

i Which part of the graph is a **prediction**?

______________________________ [1 mark]

ii How many people lived on Earth in 1920?

______________________________ [1 mark]

iii By how many times did the world population increase between 1920 and the year 2000?

A 2 **B** 3 **C** 4 **D** 5 [1 mark]

b List **three** ways in which an increase in the human population reduces the amount of land available for other species.

1______________________________

2______________________________

3______________________________ [3 marks]

B–A*

c Explain why the human population has grown so rapidly during the 19th and 20th centuries.

____

______________________________ [2 marks]

Land use

D–C

1 The graph shows an area of land in the United Kingdom that was used for different types of agriculture between 1968 and 2005.

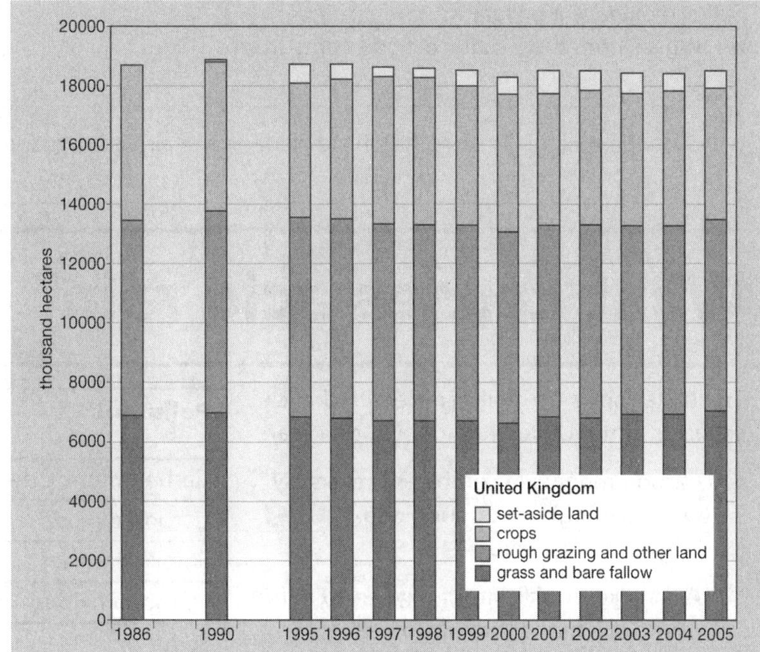

a What can be concluded about changes in the total area of agricultural land between 1986 and 2005?

A It decreased greatly.

B It increased a little.

C It stayed approximately the same.

D We cannot tell because the data does not give enough information.

[1 mark]

B–A*

b Set-aside land is land on which farmers do not grow crops. Farmers are given grants to leave some of their land as set-aside. This is to avoid having too many food crops produced in the European Union which would reduce their price.

i Describe the changes in the area of set-aside land between 1986 and 2005.

____

____ [2 marks]

ii Use the graph to suggest which type of land has provided most of the set-aside land.

____ [1 mark]

iii Suggest how set-aside land could increase biodiversity on farmland.

____

____ [2 marks]

Pollution

D–C

2 DDT is a persistent insecticide. It does not break down inside the bodies of animals and plants. The diagram shows what happens as DDT passes along a food chain.

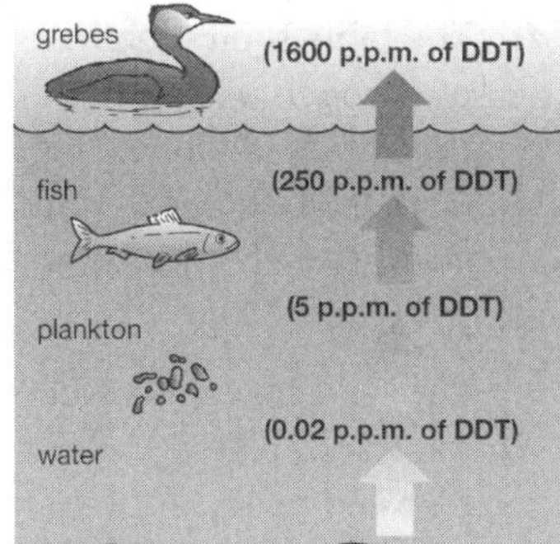

a How does the information in the diagram explain why animals at the top of food chains are more likely to be poisoned by DDT than other animals?

____

____

____

[2 marks]

b DDT is now banned in Europe and America but DDT is still found in small amounts in animals' bodies. Suggest why this is so.

____

____ [2 marks]

Grades

Air pollution

D–C

1 Sulfur dioxide is a pollutant.

a Where does most sulfur dioxide come from?

______________________________ [1 mark]

b Describe **one** way in which sulfur dioxide in the air can damage a person's health.

______________________________ [1 mark]

What causes acid rain?

D–C

2 The table shows the pollutants emitted in car exhausts in the United Kingdom in one year.

Pollutant	Emissions in thousands of tonnes
Carbon monoxide	3300
Carbon dioxide	31 200
Smoke particles	130
Nitrogen oxides	714
Sulfur dioxide	12

a Carbon monoxide combines irreversibly with haemoglobin in our blood. Why is this dangerous?

A It makes it difficult for oxygen to diffuse into the blood from the lungs.

B It makes it difficult to breathe.

C It makes it difficult to supply enough oxygen to respiring cells.

D It makes it difficult to remove carbon dioxide from the blood. [1 mark]

b Explain how breathing in smoke particles can damage health.

______________________________ [1 mark]

c Nitrogen oxides and sulfur dioxides both contribute to the formation of acid rain. Describe how acid rain is formed.

____

______________________________ [2 marks]

d Describe **two** harmful effects of acid rain on living organisms.

1______________________________

2______________________________ [2 marks]

e Explain how adding a base, such as limestone, to acidified lakes can help to reduce the harmful effects on wildlife.

____

______________________________ [2 marks]

B–A*

f The table shows the mass of nitrogen oxides, in thousands of tonnes, that were emitted in exhausts from road vehicles between 1970 and 2005.

Year	**1970**	**1980**	**1990**	**2000**	**2003**	**2004**	**2005**
Thousands of tonnes of NO_2 from cars	765	989	1324	818	636	597	549

Catalytic converters were first introduced into Europe in 1985. Legislation requiring them to be fitted to vehicles was introduced in 1993. Use this information, and your own knowledge, to suggest reasons for the changes in NO_2 emissions shown in the table.

____

____

____

______________________________ [4 marks]

Grades

Pollution indicators

D–C

1 The table shows some freshwater animals and the level of oxygen that they need to live.

Animal	**Oxygen level needed**
Mayfly larva	High
Cased caddis fly larva	Fairly high
Snail	Fairly low
Blood worm	Low
Rat-tailed maggot	Very low

a Explain how a scientist could use this information to quickly find out how much oxygen there is in a river.

____

____ [2 marks]

b A stream has raw sewage flowing into it. Which animal would you be most likely to find in the water in the stream? Explain your answer.

____ [2 marks]

B–A*

c **i** Blood worms get their name because they contain the red pigment haemoglobin. Suggest how this helps them to live in their environment.

____ [2 marks]

ii Blood worms are able to live in conditions where there is plenty of oxygen, but they are not normally found in these areas. Suggest a reason for this.

____ [2 marks]

Deforestation

D–C

2 Some parts of the rainforest in Sarawak have been cut down so that the wood can be sold. The table shows the numbers of some species of animals that lived in an area before the forest was logged, immediately after logging and two years after logging. After logging, the forest slowly regenerated.

	Mean number of animals per km^2		
Mammal	**Before**	**Immediately after logging**	**Two years after logging**
Marbled cat	Present	0	0
Oriental small-clawed otter	Present	0	0
Giant squirrel	5.18	1.48	3.75
Smaller species of squirrel	15.82	24.08	103.50
Tree shrews	10.28	4.92	10.04
Barking deer	2.91	0.79	10.22

B–A*

a Which animals have been worst affected by the logging? Explain your answer.

____

____ [2 marks]

b Suggest reasons for the results for tree shrews and barking deer.

____

____ [2 marks]

c Suggest reasons for the results for smaller species of squirrel.

____

____ [2 marks]

d Apart from the loss of animal species, give **two** other reasons why we should try to reduce the amount of logging of rainforests.

1_______________________________________

2_______________________________________ [2 marks]

The greenhouse effect – good or bad?

Grades

1 The graphs show the changes in average world temperature and carbon dioxide concentration in the air between 1700 and 2000.

D–C

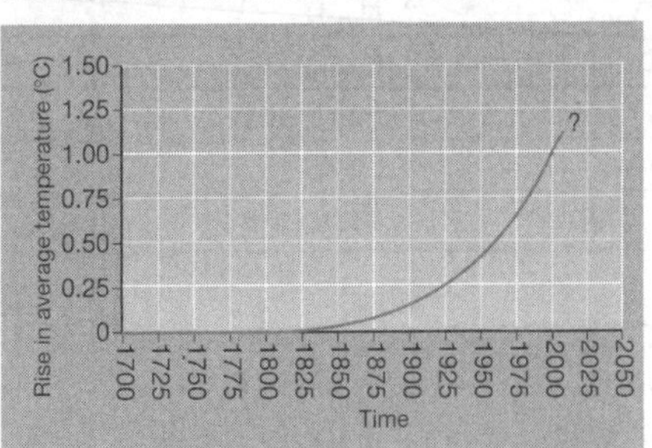

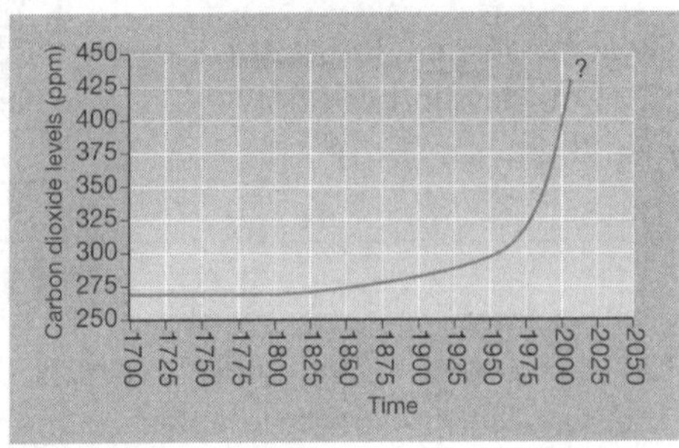

a What can we definitely conclude from these graphs?

A Both carbon dioxide and temperature have increased since 1700.

B Increasing carbon dioxide causes an increase in temperature.

C Increasing temperature causes an increase in carbon dioxide.

D By 2050, the average temperature will be about 1.5 °C higher than in 1700. [1 mark]

b Suggest reasons for the especially rapid increase in carbon dioxide concentration since 1950.

B–A*

____

____ [2 marks]

Sustainability – the way forward?

2 a Jack likes to cycle to school rather than being taken in the car by his father. He says it keeps him fit and it is a small contribution towards sustainable development.

D–C

i Explain what is meant by **sustainable development**.

____

____ [2 marks]

ii Suggest how Jack's decision about his journey to school makes a small contribution to sustainable development.

____

____ [2 marks]

b The table shows the U-values for different structures in buildings.

B–A*

Structure	U-values
Tiled roof without insulation	2.0
Tiled roof with insulation	0.3
Cavity wall without insulation	1.4
Cavity wall with foam insulation	0.4
Single-glazed window	5.7
Double-glazed window	2.9

The U-value can be used to calculate the heat energy lost from that part of the building, using the formula:

heat energy loss (watts) = U-value × area (m^2) × temperature difference between inside and outside

A house has a tiled roof with no insulation. The family who live in the house like to keep the temperature at 15 °C. Calculate by how much they could reduce the rate of heat loss when the temperature outside is 5 °C, if they insulated their roof.

____

____ [2 marks]

B1b revision checklist

I know:

what determines where particular species live

- ☐ animals and plants are adapted to live in different habitats, to compete for resources and to survive attack from predators
- ☐ they may have adapted (e.g. thorns, poisons, warning colours) to cope with specific features of their extreme environment

why individuals of the same species are different from each other; and what new methods there are for producing plants and animals with desirable characteristics

- ☐ DNA is the genetic material that controls inherited characteristics
- ☐ reproduction can be sexual (genetic variation in offspring) or asexual (identical offspring)
- ☐ cloning and genetic engineering can be used to produce plants and animals with desirable characteristics
- ☐ concerns about GM organisms include: their safety for use, genes jumping to other species, and long-term consequences of eating them
- ☐ cross-species embryo transplantation may help to preserve endangered wild species

why some species of plants and animals have died out and how new species of plants and animals develop

- ☐ theories of evolution have changed over time; Darwin proposed the theory of evolution by natural selection
- ☐ fossils tell us how present-day species have evolved and how they compare to prehistoric species
- ☐ life on Earth may have originated: as a result of early light/air/water conditions; from a meteorite; or in deep oceans
- ☐ species can become extinct due to environmental change and human impact

how humans affect the environment

- ☐ increases in human population use up more resources and produce more waste and pollution
- ☐ human action contributes to acid rain, air pollution, water pollution, over-use of land and loss of diversity in rainforests
- ☐ living organisms such as lichens, invertebrates and fish can be used as indicators of pollution
- ☐ increasing the greenhouse effect can lead to climate change
- ☐ sustainable development, e.g. using renewable energy resources and recycling can help to safeguard the environment for future generations

Elements and the periodic table

Grades

D–C

1 This question is about the periodic table.

a Name an atom that is in the same group as sodium but has one more shell of electrons.

____________________ [1 mark]

b Name an atom that has a relative atomic mass that is double the relative atomic mass of carbon. ____________________ [1 mark]

c Which atom contains half as many protons as neon?

____________________ [1 mark]

B–A*

d When scientists first tried to put elements in order, they put them in order of their relative atomic masses. Mendeleev changed the order of tellurium iodine (relative atomic mass 128) and iodine (relative atomic mass 127) so that iodine was in Group 7. Explain why this was a better way to classify iodine.

____

____________________ [2 marks]

Atomic structure 1

D–C

2 Bromine is an element in the periodic table.

80 35 **Br** bromine

a Fill in the gaps to show the numbers of the different types of particles in an atom of bromine.

An atom of bromine contains __________ protons, __________ electrons and __________ neutrons. [3 marks]

b When bromine reacts with other elements, it gains electrons. Gaining electrons does not affect the mass of the bromine atoms. Explain why.

____________________ [1 mark]

c **i** Give the symbol of the element that contains atoms with one more proton than bromine.

____________________ [1 mark]

ii Give the symbol of an element that is in the same group as bromine but has fewer protons.

____________________ [1 mark]

B–A*

3 This question is about electronic configurations. The electronic configurations of some atoms are shown here:

2, 8, 2	**2**	**2, 8, 1**	**2, 8**
atom A	**atom B**	**atom C**	**atom D**

a **i** Which atom is from an element in Group 2? ____________________ [1 mark]

ii Explain how you can tell.

____________________ [1 mark]

b **i** Which atom is a noble gas? ____________________ [1 mark]

ii Explain how you can tell.

____________________ [1 mark]

c Write the electronic configuration of the element that is directly above atom C in the same group. ____________________ [1 mark]

Bonding

D–C

1 Match formulae, **A**, **B**, **C** and **D**, with the sentences **1 – 4** in the table.

A NaCl

B N_2

C H_2O

D KNO_3

1	2	3	4

1	This is a compound that contains three different types of atom.
2	This is a molecule with two identical atoms joined together.
3	This compound contains a covalent bond between hydrogen and oxygen atoms.
4	This compound contains chloride ions.

[4 marks]

B–A*

2 This is a diagram of an oxygen molecule. O═O

Which of the following statements about the bonding in an oxygen molecule is correct?

A Atoms are held in a sea of mobile electrons.

B There are four shared electrons between the oxygen atoms.

C There is a single covalent bond between the oxygen atoms.

D The oxygen atoms are held together by ionic attraction. [1 mark]

Extraction of limestone

B–A*

3 A company sells large quantities of limestone.

a Give **two** large-scale uses of limestone.

____

______________________________ [2 marks]

b The company wants to open a new quarry. Local people protest against the quarry. This is one of the protesters.

Give **three** reasons why local people might protest against a new limestone quarry.

____

____

____

____

[3 marks]

c Suggest **one** argument that the company could use to persuade people that the quarry would benefit local people.

______________________________ [1 mark]

Thermal decomposition of limestone

Grades

D–C

1 a When calcium carbonate is heated it reacts to make a solid compound and a gas. Complete the word and symbol equation for the reaction by filling in the boxes.

calcium carbonate → [ ] + [ ]

$CaCO_3$ → CaO + [ ] [3 marks]

b Eve does an experiment. She weighs a lump of limestone before and after heating it. She finds that the lump is lighter after heating. Which sentence gives the best explanation for this?

A The atoms get smaller when they are heated.

B A gas is given off.

C The limestone has been burned.

D The lump got wet when it was heated. [1 mark]

B–A*

2 The equation shows what happens when carbon dioxide is bubbled through a solution of calcium hydroxide (limewater).

$CO_2(g)$ + $Ca(OH)_2(aq)$ → $CaCO_3(s)$ + ____________

a Complete the equation by filling in the missing formula. [1 mark]

b Explain why limewater goes cloudy when carbon dioxide is bubbled through it.

____

____________ [2 marks]

Uses of limestone

D–C

3 Cement and concrete are both made using limestone.

a Describe **one** difference between cement and concrete.

____________ [1 mark]

b Cement is used to hold bricks together; concrete is used for making structures such as bridges. Explain why the two materials have different uses.

____

____________ [2 marks]

B–A*

4 How is toughened glass made?

A By adding metal atoms to the glass when it is made.

B By heating the molten glass to a higher temperature.

C By cooling the molten glass more quickly.

D By using less limestone in the glass mixture. [1 mark]

Grades

The blast furnace

D–C

1 Complete the sentences by choosing the **correct** words from the box.

carbon monoxide	**silicon dioxide**	**oxidation**	**oxygen**	**reduction**

In the blast furnace, iron oxide reacts to form iron by reacting with ______________________.

During the reaction, the iron oxide loses ______________________.

This type of reaction is called ______________________. [3 marks]

B–A*

2 In the blast furnace, carbon from coke reacts with carbon dioxide to form carbon monoxide.

$$C + CO_2 \rightarrow 2CO$$

What is happening in this reaction?

A thermal decomposition

B oxidation only

C combustion

D both oxidation and reduction [1 mark]

Using iron

D–C

3 The diagrams show the arrangement of atoms in some different materials.

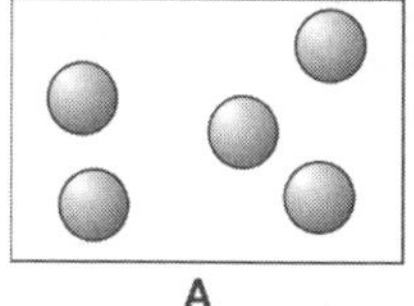
A

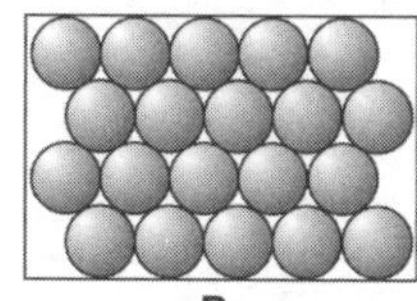
B

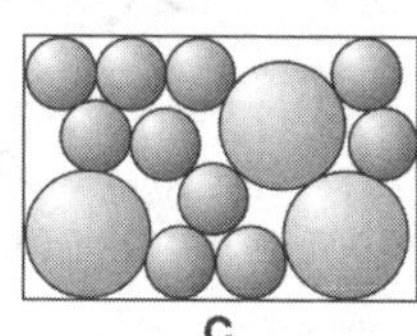
C

a **i** Which diagram shows the arrangement of atoms in steel?

______________________ [1 mark]

ii Explain your reasoning.

____

______________________ [2 marks]

B–A*

b Why is wrought iron easily shaped?

A The atoms are in layers that can slide over each other.

B The bonds between atoms are weak.

C The atoms are free to move around freely.

D The atoms are far apart from each other. [1 mark]

Grades

Using steel

D–C

1 The diagrams show how the atoms are arranged in wrought iron and stainless steel.

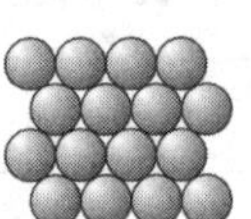

wrought iron

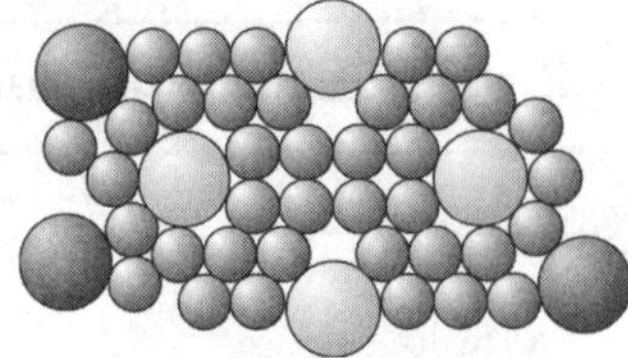

stainless steel

a Describe the difference in the **structure** of wrought iron and stainless steel.

____

____

____ [3 marks]

b Explain why stainless steel is more suitable than wrought iron for making cutlery.

____

____ [2 marks]

Transition metals

D–C

2 The diagram shows the structure of a metal.

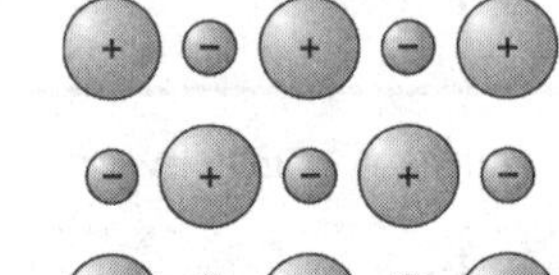

a Name the **two** types of particles shown in the diagram.

____________________ and ____________________ [2 marks]

b Use the diagram to explain why metals conduct electricity.

____

____ [2 marks]

B–A*

3 The table shows some information about some metals.

Metal	**Transition element?**	**Radius of one atom (nm)**	**Melting point (°C)**	**Strength of metal**
Sodium	No	0.2	98	Very weak
Iron	Yes	0.1	1535	High
Calcium	No	0.2	850	Weak
Cobalt	Yes	0.1	1492	High

a Which **two** metals in the table are transition metals?

____________________ and ____________________ [2 marks]

b How do the properties of transition metals differ from the other metals in the table?

A They have higher melting points but are weaker.

B They have smaller atoms and lower melting points.

C They have smaller atoms and are weaker.

D They have higher melting points and are stronger. [1 mark]

Aluminium

D–C

1 a Aluminium is extracted from aluminium oxide using electrolysis.
Complete the sentences about electrolysis.

Aluminium cannot be extracted by reacting with carbon because it is too ____________________.

During electrolysis the positive aluminium ions are attracted to the ____________________ electrode.

The ions form aluminium atoms when they ____________________ electrons. [3 marks]

B–A*

b During electrolysis, oxygen ions gain electrons to form oxygen.

i Balance the equation.

$$2O^{2-} \rightarrow \_\_\_\_\_\_ e^- + O_2$$ [1 mark]

ii Does this reaction involve oxidation or reduction? Explain your reasoning.

____

____ [2 marks]

B–A*

2 In the process of electrolysis of aluminium, cryolite is added. Why does adding cryolite lower the cost of the process?

A cryolite acts as a cheap fuel

B adding cryolite means less aluminium oxide needs to be used

C cryolite reduces aluminium ions

D molten cryolite acts as a solvent for aluminium oxide [1 mark]

Aluminium recycling

D–C

3 Here is some information about aluminium.

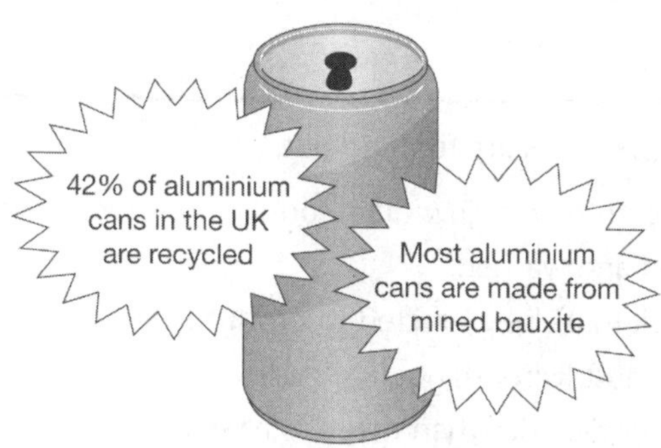

a Bauxite is the main ore that is used to make aluminium. Give **two disadvantages** of mining bauxite.

____

____ [2 marks]

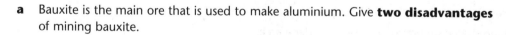

b Suggest reasons why **less than half** the aluminium cans we use are recycled.

____

____ [2 marks]

Grades

Titanium

D–C

1 The table shows some information about titanium, aluminium and iron.

	Titanium	**Aluminium**	**Iron**
Density in g/cm³	4.5	2.7	7.9
Melting point °C	1675	660	1535
Corrosion resistance	Does not corrode	Does not corrode	Corrodes
Main oxide in ore	TiO_2	Al_2O_3	Fe_2O_3
Bonding in oxide	Covalent	Ionic	Ionic
Cost of extraction	Very high	High	Low

a Titanium and aluminium are used to make bicycle frames. Use information from the table to explain why they are better metals to use than iron.

____

____

______________________________ [3 marks]

b i Aluminium is extracted by electrolysis of molten aluminium oxide.
Use information from the table to explain why molten aluminium oxide conducts electricity but molten titanium oxide does not.

____

______________________________ [2 marks]

ii Aluminium is electrolysed at 900 °C. At this temperature, the aluminium leaves the electrolysis tank as a liquid.
Explain why titanium is not a liquid at this temperature.

______________________________ [1 mark]

Copper

D–C

2 The supplies of high quality copper ores are running out.

a Why does extracting copper from very low grade ores cause more environmental damage?

A The ores must be crushed and washed.

B Very large volumes of rock must be extracted to produce a small amount of copper.

C Copper ions are toxic to most living things.

D Copper ores produce sulfur dioxide when they are heated. [1 mark]

b Copper can be extracted from old mine waste heaps by leaching. What does the term **leaching** mean?

A Electrolysing a dilute solution of copper.

B Crushing the waste ore to increase its surface area.

C Dissolving copper ores in dilute acid.

D Roasting the waste ore in air. [1 mark]

B–A*

3 Bronze is used to make coins. Which of the following statements about bronze is true?

A Bronze is an alloy of non-metals.

B Bronze is a transition metal element.

C Bronze contains a mixture of copper and tin.

D Bronze is a compound of copper and zinc. [1 mark]

Smart alloys

D–C

1 One use of smart alloys is for making braces to straighten teeth.
The diagram shows some information about smart alloys.

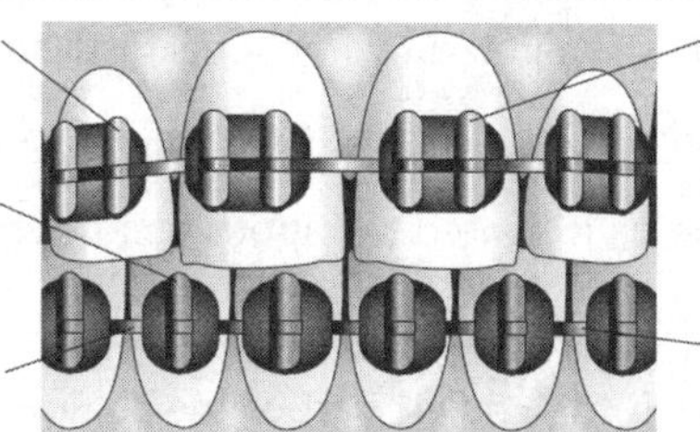
This smart alloy is made from nickel and titanium
Smart alloys are mixtures of different elements
Smart alloys have a 'shape' memory
Smart alloys can be bent and stretched into different shapes
Smart alloys contain mainly metal atoms

a Complete the sentences about smart alloys.
Smart alloys are **similar** to normal alloys because ______________________

____

____

Smart alloys are **different** from normal alloys because ______________________

____

____ [4 marks]

b What do you have to do to a smart alloy to return it to its original shape?

____

____ [2 marks]

Fuels of the future

D–C

2 Car exhaust emissions release pollutant gases into the atmosphere.

a Which exhaust gases are a result of incomplete combustion in the car engine?

A sulfur dioxide and carbon dioxide

B nitrogen and carbon dioxide

C petrol vapour and carbon monoxide

D nitrogen dioxide and sulfur dioxide [1 mark]

b Which of the following gases is **not** produced by reactions in a catalytic converter?

A carbon dioxide

B water vapour

C nitrogen

D hydrogen [1 mark]

B–A*

3 Claudia and Saul are discussing whether electric trams cause less pollution than buses.

The trams give out much less pollution than buses – they are a real benefit to the environment.

I'm not so sure – surely using electricity must be harmful to the environment?

a Explain why using electric trams gives out less pollution than buses in the city.

____

____

____ [2 marks]

b Why is using electricity harmful to the environment?

____

____ [2 marks]

Crude oil

Grades

D–C

1 Crude oil is separated by fractional distillation.

a Match descriptions, **A**, **B**, **C** and **D**, to the numbers **1 – 4** on the diagram.

A the oil is heated to vapour point here

B condensation happens here

C this is the coolest part of the column

D molecules with the highest boiling point leave the column here

1	2	3	4
D			A

[4 marks]

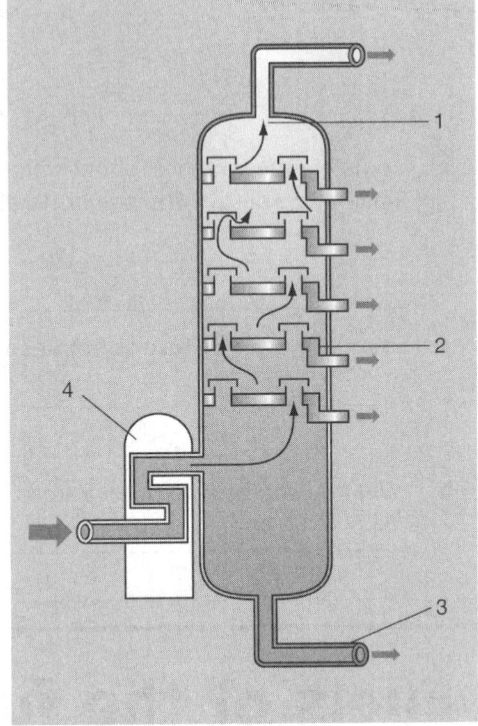

B–A*

b Which fraction leaving the column has the strongest intermolecular forces between its molecules?

A the fraction leaving the top of the column

B the fraction containing the largest molecules

C the fraction with the lowest boiling point

D the fraction containing the lowest number of carbon atoms [1 mark]

Alkanes

B–A*

2 The diagrams show the structures of some alkane molecules.

```
                               H
                               |
                             H-C-H
   H H H H H              H    |   H          H H H         H H H H
   | | | | |              |    |   |          | | |         | | | |
 H-C-C-C-C-C-H          H-C----C---C-H      H-C-C-C-H     H-C-C-C-C-H
   | | | | |              |    |   |          | | |         | | | |
   H H H H H              H    H   H          H H H         H H H H

   A                       B                   C              D
```

a Which alkane has the highest boiling point? __________ [1 mark]

b Which **two** alkanes are isomers of each other? __________ [1 mark]

c Which alkane **cannot** form isomers? __________ [1 mark]

d Which alkane is called propane? __________ [1 mark]

Pollution problems

D–C

1 Petrol mainly contains hydrocarbons but also contains small amounts of sulfur as an impurity.

a Complete the equations to show what happens when petrol burns completely.

hydrocarbon + ________________ → ________________ + ________________

sulfur + ______________________________ → ______________________________ [4 marks]

B–A*

b Burning petrol adds to the acid rain problem. Fish caught in lakes affected by acid rain are usually older and larger. Explain why fewer young, small fish are found.

____

____ [2 marks]

D–C

2 An article about pollution from burning petrol contains this information.

> **PETROL TO PUT US IN THE DARK?**
> We need to find ways to use less petrol. Don't forget that petrol is a non-renewable fuel. Apart from all the well-known environmental problems from burning petrol, it might also lead to global dimming – a worrying thought!

a What does the article mean when it says that petrol is **non-renewable**?

A Burning petrol gives off carbon dioxide that may cause climate change.
B We use huge quantities of petrol worldwide.
C Supplies of petrol cannot be replaced when they are used.
D Not all pollution from petrol can be removed by a catalytic converter.

[1 mark]

B–A*

b Which of the following is a correct explanation of **global dimming**?

A Water from melting ice caps cause clouds that cut out sunlight.
B Pollution reduces the light that can pass through the ozone layer.
C Particles of carbon from burning fuels block out the sunlight.
D Increased carbon dioxide absorbs energy from the Sun.

____ [2 marks]

Reducing sulfur problems

D–C

3 All petrol garages in the United Kingdom sell low sulfur petrol.

Which of these statements about sulfur in petrol are true?

A Petrol companies used to mix sulfur compounds into petrol.

B Sulfur is now removed from petrol using a solvent.

C Sulfur in petrol harms the ozone layer.

D Sulfur in petrol is oxidised when the petrol burns.

________________________ and ________________________

[2 marks]

B–A*

4 Many scientists think that using aeroplanes is very damaging to the environment because each journey burns very large amounts of fuel and produces large amounts of pollutant gases. Some people who are very concerned about the environment choose to go on holiday by ship.

a Why does burning large amounts of fuel lead to environmental damage?

____

____ [2 marks]

b Explain how environmental damage can also be caused by ship travel.

____

____ [2 marks]

C1a revision checklist

I know:

how rocks provide building materials

- ☐ limestone, calcium carbonate, can be used to make cement, mortar, concrete and glass which are used as building materials
- ☐ an element consists of one type of atom; two atoms of the same element can join together to form a molecule; a compound consists of atoms of two or more elements joined together
- ☐ atoms are held together in molecules and lattices by chemical bonds, which involves giving, taking or sharing electrons
- ☐ the electronic configuration of an atom shows the number of its electrons and their arrangement in shells; atoms with a full outermost shell of electrons are stable
- ☐ how to interpret chemical equations in symbol form and balance equations in terms of numbers of atoms

how rocks provide metals and how metals are used

- ☐ metals are extracted from their ores, often oxides, by reduction with carbon (iron), electrolysis (aluminium and copper) or other chemical reactions (titanium)
- ☐ pure metals are soft and easily shaped because the atoms form a regular arrangement – the layers of atoms can slide easily over each other
- ☐ metals are mixed together to make alloys (e.g. iron and other metals or carbon make steel)
- ☐ aluminium is expensive to produce, is often too soft on its own, but forms strong alloys with other metals
- ☐ copper is a hard, strong, good conductor and can be used for wiring and plumbing
- ☐ aluminium and titanium are resistant to corrosion and have a low density

how we get fuels from crude oil

- ☐ crude oil is a mixture of hydrocarbon compounds that can be separated by fractional distillation; some fractions can be used as fuels
- ☐ most of the compounds in crude oil are saturated hydrocarbons called alkanes, which have the general formula C_nH_{2n+2}
- ☐ burning fossil fuels releases useful energy but also harmful substances, e.g. sulfur dioxide causes acid rain; carbon dioxide causes climate change; smoke particles cause global dimming

Cracking

D–C

1 This molecule is made in a cracking reaction.

$$\mathrm{H_2C{=}CH{-}CH_3}$$

Which of the following statements is **not** true for this molecule?

A It is called propene.

B It is a hydrocarbon.

C It is a saturated molecule.

D It contains some single bonds. [1 mark]

B–A*

2 Which of the following molecules **cannot** be made by cracking decane $C_{10}H_{22}$?

A C_2H_4 **B** C_3H_8 **C** C_3H_{10} **D** $C_{12}H_{24}$ [1 mark]

B–A*

3 Novane, C_9H_{20}, is an alkane. Novane can be cracked to form hexane and one other molecule, molecule A.

$$\mathrm{H{-}(CH_2)_9{-}H \rightarrow H{-}(CH_2)_7{-}H} + \boxed{\text{molecule A}}$$

a **i** Draw the structure of molecule A in the box. [1 mark]

ii What is the name of molecule A?

______________________________________ [1 mark]

Alkenes

D–C

4 When alkanes in crude oil are cracked some alkenes are formed.

a Butene is an alkene. It has the formula C_4H_8. Pentene has one more carbon atom than butene. What is the formula for pentene?

______________________________________ [1 mark]

B–A*

b Which of the following statements is true for alkenes?

A contain only single bonds

B are saturated

C contain the same number of hydrogen atoms as carbon atoms

D have the general formula $C_nH_{(2n+2)}$ [1 mark]

B–A*

5 Ethene gas is used by suppliers of very large supermarkets to make fruit ready for sale.

a Explain why fruit can be made ready for sale by being treated with ethene gas.

______________________________________ [1 mark]

b Explain why fresh flowers are kept **away** from ethene gas.

______________________________________ [1 mark]

Making ethanol

Grades

D–C

1 This information appeared in a newspaper article about ethanol.

BIOFUELS FOR THE FUTURE

Fuels such as petrol and diesel could soon be replaced by renewable biofuels that contain ethanol made from sugar beet. These new biofuels could solve our problems of relying on finite crude oil reserves.

Many farmers already use biofuels for farm vehicles.

a Explain how sugar beet can be used to produce ethanol.

____

____ [2 marks]

b Explain why ethanol made in this way is a 'renewable biofuel'.

____

____ [3 marks]

B–A*

c Ethanol, C_2H_5OH, can also be made by passing a mixture of ethene and steam over a catalyst.

i Write a symbol equation for the reaction.

____ [2 marks]

ii Explain why ethanol is **not** a hydrocarbon.

____ [1 mark]

Plastics from alkenes

B–A*

2 The diagrams show two compounds that contain fluorine.

```
F       F            F   F
 \     /             |   |
  C = C          H – C – C – H
 /     \             |   |
F       F            F   F
```

compound A   compound B

a Compound A forms a polymer. Draw a diagram to show the structure of the polymer that forms from compound A.

[2 marks]

b Explain why compound B cannot form a polymer.

____ [1 mark]

B–A*

3 Which of the following is most likely to be made from a **smart polymer**?

A plastic lenses for spectacles that darken in bright light

B metal electrical wiring in a computer

C plastic food containers that can be thrown away

D metal car bumpers that can be beaten back to shape after accidents [1 mark]

Polymers are useful

D–C

1 Look at the structures of polymers **A**, **B**, **C** and **D**.

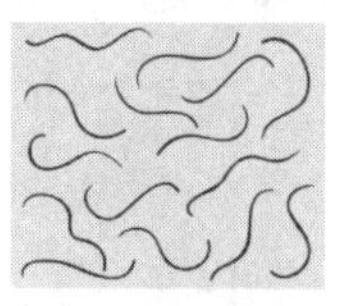
A

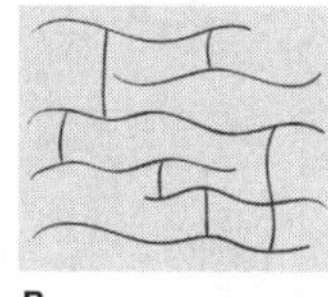
B

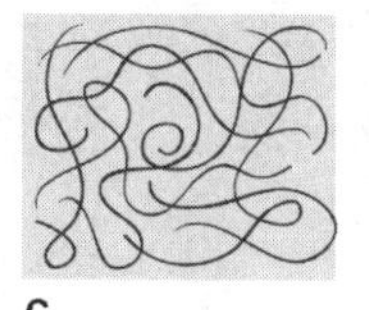
C

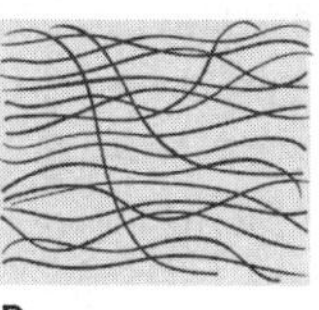
D

a Which polymer contains **cross links**?

______________________________ [1 mark]

b Which polymer has the highest **density**?

______________________________ [1 mark]

c Polymers A and C are heated until they go soft and runny. Explain why polymer A is softer and runnier than polymer C.

____

______________________________ [2 marks]

B–A*

2 Which one of the following statements is true for shape memory polymers?

A All shape memory polymers are biodegradable.

B An irreversible change happens when shape memory polymers are heated.

C At higher temperatures the structure of shape memory polymers returns to its original state.

D Shape memory polymers are elastic materials that bounce back to their original shape. [1 mark]

Disposing of polymers

D–C

3 The table shows some methods of disposing of waste plastics.

Method of disposal	**How it is carried out**	**Advantage**
Burning	Waste is burned in an incinerator	Energy can be used to make electricity
Landfill	Waste is buried in the ground	No need to sort the waste, very cheap
Recycling	Waste is processed to make new polymers	New polymers can be used to make new products such as coat padding
Making biodegradable polymers	Polymers are designed to rot away after being used	Waste is not a long term problem

Which **two** methods reduce the need to use fossil fuels? Explain your reasoning.

1 ______________________________

____

2 ______________________________

______________________________ [2 marks]

B–A*

4 Pins made from both metals and polymers can be used to hold two broken leg bones together while they heal. The patient's leg is cut open and the pins are put through the bone. Explain why polymers are more suitable than metals for use inside the body.

____

______________________________ [2 marks]

Grades

Oil from plants

D–C

1 Vegetable oil made from crushed seeds contains a lot of impurities. The diagram shows an apparatus that can be used to separate pure oil from the mixture.

a Label the diagram by filling in the empty boxes.

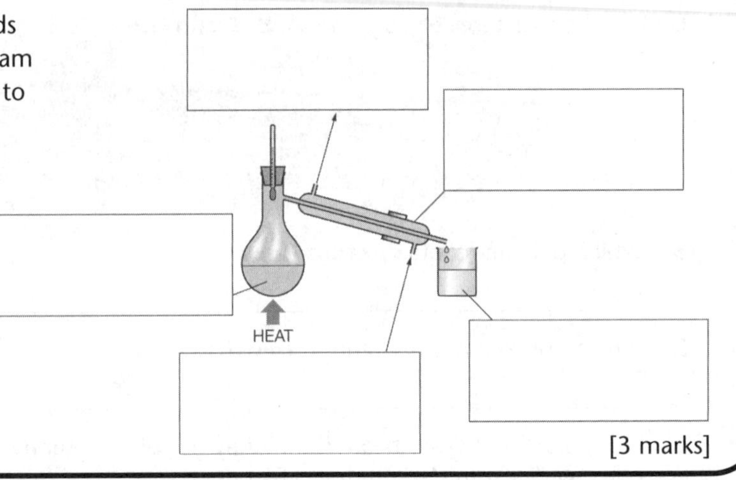
HEAT

[3 marks]

B–A*

2 Why are omega-3 fatty acids beneficial to people's health?

A Walnuts are very high in omega-3 fatty acids.

B Omega-3 fatty acids are lower in fat than other fatty acids.

C Omega-3 fatty acids raise blood pressure.

D Omega-3 fatty acids reduce levels of fat in the blood. [1 mark]

Green energy

D–C

3 Harriet is not sure that biofuels are a good idea. This is what she has to say.

Growing plants for biofuels uses up a lot of land that we need to grow food.
Some oils that come from plants are far too thick to be used for most engines.
They cost a lot more to produce than petrol.

a Harriet says that some oils are too thick to be used in engines. How is this problem overcome?

____________________ [1 mark]

b Harriet talks about the disadvantages of using biofuels. Give **two advantages** of using biofuels.

____

____________________ [2 marks]

c In the future, petrol will become more expensive than biofuels. Explain why.

____

____________________ [2 marks]

B–A*

d Biodiesel is one type of biofuel. Biodiesel is a renewable fuel. Give **two** other advantages of using biodiesel rather than standard diesel as a fuel.

____

____________________ [2 marks]

Grades

Emulsions

D–C

1 The diagrams show what some different types of emulsions contain.

a Complete the table about these emulsions. Use the words in the box.

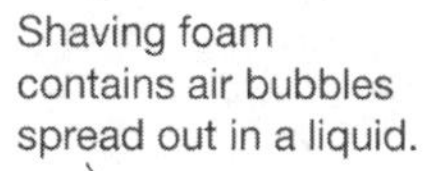
Shaving foam
contains air bubbles
spread out in a liquid.

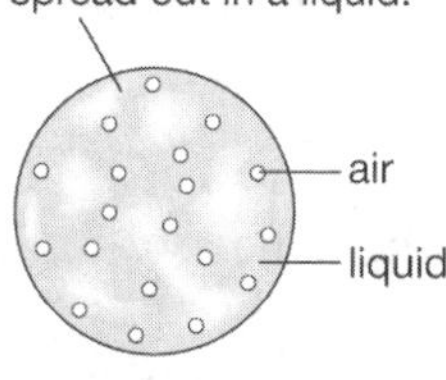
air
liquid

shaving foam

After shave cream has water droplets spread out in an oil.

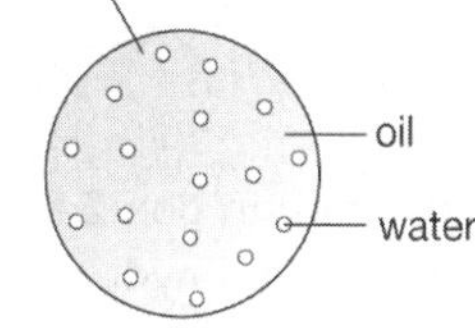
oil
water

after shave cream

air	**oil**	**liquid**	**water**

Type of emulsion	Dispersed phase	Continuous phase
Shaving foam		
After shave cream		

[2 marks]

b The ingredients in these emulsions are immiscible. What does **immiscible** mean?

______________________________ [1 mark]

B–A*

2 Which of the following statements about mayonnaise is true?

A Egg yolk in mayonnaise makes the ingredients immiscible.

B Oil and water in mayonnaise forms layers because they are miscible.

C The continuous phase in mayonnaise is water.

D Vinegar is used in mayonnaise as an oil. [1 mark]

Polyunsaturates

D–C

3 Olive oil is a **monounsaturated oil**. It contains one double bond.

a How is the bonding in a **saturated fat** different from olive oil?

______________________________ [1 mark]

b How is the bonding in a **polyunsaturated oil** different from olive oil?

______________________________ [1 mark]

c Complete the table to show the observations you would expect when each oil or fat reacts with **bromine water**.

Type	Olive oil	Saturated fat	Polyunsaturated oil
Observations	Orange to colourless		

[2 marks]

B–A*

4 Tim's doctor tells him that he has high blood cholesterol levels. The doctor advises Tim

A to eat fats from vegetable oils rather than animal fats.

B to eat more saturated fat to lower cholesterol.

C that monounsaturated fats are better at lowering cholesterol than polyunsaturated fats.

D that high cholesterol is worrying because it widens the arteries. [1 mark]

Making margarine

Grades

D–C

1 The diagram in box 1 shows part of a molecule of vegetable oil.

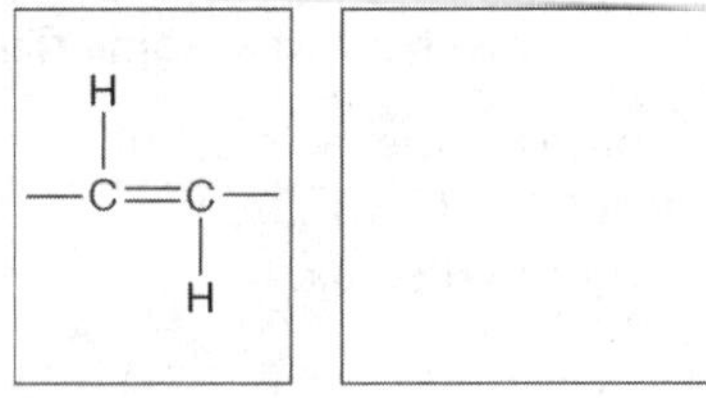

a Draw a similar diagram in box 2 to show the same molecule after it has been hydrogenated. [2 marks]

b Give **two** ways hat his process can be speeded up when it is carried out on an industrial scale.

____

____ [2 marks]

B–A*

2 Match words, **A**, **B**, **C** and **D**, with the sentences **1 – 4** in the table.

A HDL cholesterol

B LDL cholesterol

C hydrogenated vegetable oil

D trans free fat

1	**2**	**3**	**4**

1	this can be used to make softer margarines
2	eating this can lead to increased blood pressure
3	this compound in the blood is linked to high blood pressure
4	eating trans fats reduces the concentration of this compound in the blood

[4 marks]

Food additives

D–C

3 The table shows some information about food additives with E-numbers.

Purpose	**Number**
Colouring	E100-181
Preservative	E200-285 and 1105
Antioxidant	E300-340
Emulsifier	E400-499

a A cake contains the additive E220 and E340. Explain why manufacturers add these additives to foods such as cake.

____

____

____ [3 marks]

B–A*

b The manufacturers used to use additive E102. This additive has since been banned. Explain why it is sometimes necessary to ban food additives.

____

____ [2 marks]

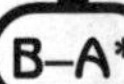

Grades

CIB OILS, EARTH AND ATMOSPHERE

Analysing chemicals

B–A*

1 One method of testing dyes used in foods is by using chromatography. This chromatogram shows four known food dyes, **A**, **B**, **C** and **D** and an unknown dye used to colour sweets.

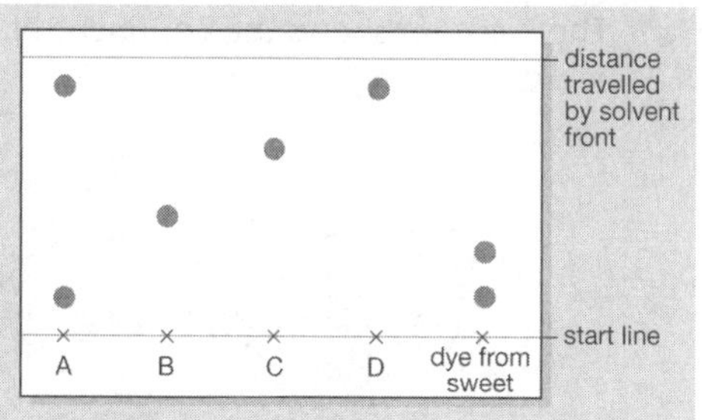

a Food dye **A** is banned because one of its ingredients is harmful. Food dye **D** is not banned. Does the sweet contain the harmful ingredient? Explain your reasoning.

____

____

____ [3 marks]

b Ruby wants to calculate the retention factor of dye **B**. Write a set of instructions to tell her how to do this.

____

____

____ [3 marks]

c Dye **C** has a retention factor of 0.75. The solvent front is 4 cm above the start line. How far above the start line did the spot from dye **C** travel?

____

____

____ [3 marks]

The Earth

D–C

2 The diagram shows the structure of the Earth.

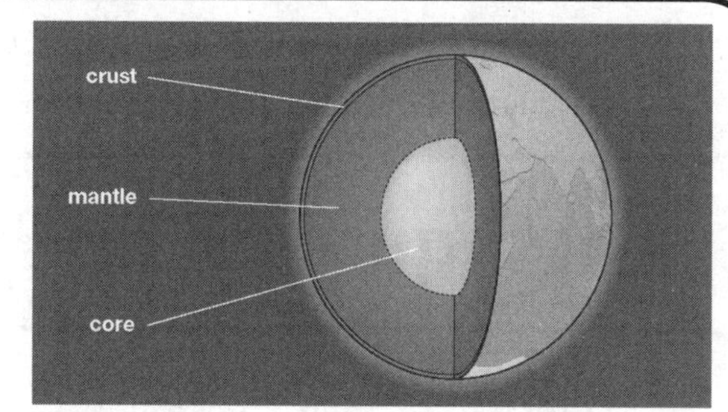

Match words, **A**, **B**, **C** and **D**, to the sentences **1 – 4** in the table.

A mantle

B inner core

C outer core

D crust

1	this is the furthest from the Earth's surface
2	a thin layer of solid rock
3	convection currents here reshape the Earth's surface
4	this contains a liquid mixture of iron and nickel

1	2	3	4

[4 marks]

B–A*

3 The magnetic field of the Earth

A has a particular strength that remains constant.

B protects us from harmful UV light from the Sun.

C repels electrically charged particles.

D creates an effect called the solar wind. [1 mark]

Earth's surface

D–C

1 a The diagram shows the continent of Pangaea 200 million years ago.

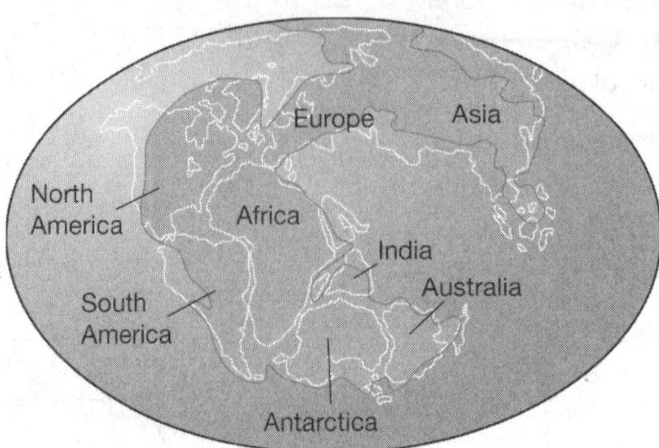

The sentences describe how the continents moved to their current positions. Match words, **A**, **B**, **C** and **D**, to the numbers **1 – 4** in the sentences.

A lithosphere **B** convection currents **C** tectonic plates **D** mantle

The upper mantle and crust form the Earth's ______ **1** ______ .

The continents are carried on large masses of rock called ______ **2** ______ .

Heat from the Earth's core causes the rocks to move by ______ **3** ______ .

This happens in the ______ **4** ______ .

1	2	3	4

[4 marks]

B–A*

b Europe and North America were joined 200 million years ago. Nowadays, they are separated by the vast ocean of the Atlantic. This change happened because

A new rock formed at a mid-ocean ridge.

B magma rose to the surface near a subduction zone.

C a continental plate rose over an oceanic plate.

D earthquakes caused a sudden movement of rock. [1 mark]

Earthquakes and volcanoes

D–C

2 a Look at the diagram that shows where two plates meet. Match sentences, **A**, **B**, **C** and **D**, with the numbers **1 – 4** on the diagram.

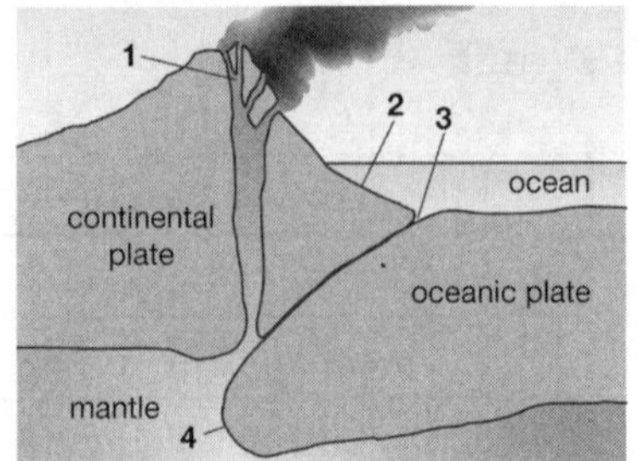

A earthquakes are most likely to happen here

B rock is melting here

C rock is being pushed up here

D lava is cooling down most quickly here

1	2	3	4

[4 marks]

b Which of the plates, the continental plate or the oceanic plate is the **least** dense? Explain your reasoning.

____ [1 mark]

c Explain why this diagram shows an example of a subduction zone.

____

____ [1 mark]

The air

D–C

1 The gases in the atmosphere

A are all pure elements.

B are all pure compounds.

C contain molecules that are a mixture of metals and non-metals.

D contain a mixture of molecules of non-metals. [1 mark]

B–A*

2 Airbags work because the crash triggers the following reaction.

$2NaN_3$ g ________ N_2 + ________ Na

a Balance the equation. [1 mark]

b Give the names of the products of the reaction.

________________________________ [1 mark]

c Explain why this reaction causes the airbag to inflate.

________________________________ [2 marks]

Evolution of the air

D–C

3 This table shows gases coming from a modern active volcano on a Pacific island. The **temperature** of the volcano is over **1000 °C**.

Name of gas	**Percentage of gas**
Water	37.1
Carbon dioxide	48.9
Sulfur dioxide	11.8
Hydrogen	0.49
Carbon monoxide	1.51

a **i** Why is water a gas when it comes out of the volcano?

________________________________ [1 mark]

ii The Earth's early atmosphere contained large amounts of water vapour. Describe what happened to all this water vapour.

____

________________________________ [3 marks]

b Scientists think that our early atmosphere came from volcanoes.

i Which gas is present in the **largest quantity** in the gases from this volcano?

________________________________ [1 mark]

ii What **two** processes removed most of this gas from the early atmosphere?

________________________________ [2 marks]

c Give **two** other important differences between the gases from this volcano and our modern atmosphere.

____

________________________________ [2 marks]

B–A*

4 The Earth and the Moon are both the same distance from the Sun. The Earth has an average surface temperature of 17 °C. The average surface temperature of the moon is – 33°C. Use ideas about the atmosphere to explain why the surface temperatures are so different.

____

________________________________ [3 marks]

Atmospheric change

Grades

D–C

1 Some friends are discussing the environment. This is what they have to say.

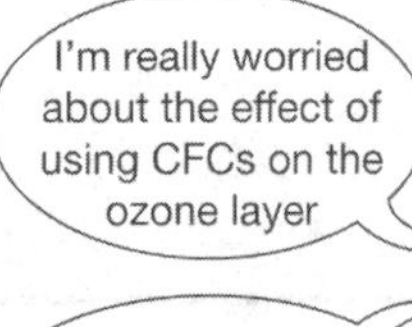

I'm really worried about the effect of using CFCs on the ozone layer
I think we should plant more trees in forests
I think we should all burn biofuels
I think it is ridiculous the amount of fossil fuels we burn

Put ticks (✓) in the table to show whether each activity increases, decreases or makes no difference to the amount of carbon dioxide in the air.

Activity	**Decreases amount of carbon dioxide in air**	**Increases amount of carbon dioxide in air**	**Does not affect amount of carbon dioxide in air**
Using CFCs			
Planting more trees			
Burning biofuels			
Burning fossil fuels			

[4 marks]

D–C

2 The Earth stores carbon dioxide in carbon sinks. Plants are one type of carbon sink.

a Choose the names of two other types of carbon sinks. Draw a ring around **two correct** answers.

crude oil **rainwater** **limestone** **photosynthesis** **respiration** [2 marks]

b Plants act as carbon sinks. Trees are much better carbon sinks than short-lived plants, such as grass. Explain why.

____

____ [2 marks]

c The seas contains large amounts of dissolved carbon dioxide. More carbon dioxide can dissolve in cold water than in warm water. Scientists are worried that if the climate gets warmer, the amount of carbon dioxide in the **air** will increase very quickly. Explain how this could happen.

____

____ [2 marks]

B–A*

3 Most people think that the average surface temperature of Earth is increasing due to increased concentrations of carbon dioxide in the atmosphere. Some scientists do not agree. They think that the increase in temperature is due to changes in the Sun's activity. Which statement supports the scientists' views?

A Different planets in the solar system have different surface temperatures.

B Not all planets in the solar system have an atmosphere.

C The surface temperature of other planets in our Solar System has risen over the last 100 years.

D Not all planet atmospheres contain carbon dioxide. [1 mark]

C1b revision checklist

I know:

how polymers and ethanol are made from oil

- ☐ crude oil is made from long-chain hydrocarbons that can be cracked by thermal decomposition to form shorter-chain alkanes and alkenes
- ☐ alkenes are unsaturated hydrocarbons, they contain double carbon-carbon bonds and have the general formula C_nH_{2n}
- ☐ alkenes can be made into polymers, which are long-chain molecules created when lots of small molecules called monomers are joined together in polymerisation
- ☐ polymers can be used to make useful substances, e.g. waterproof materials and plastics, but many are not biodegradable

how plant oils can be used

- ☐ vegetable oils can be hardened to make margarine in a process called hydrogenation
- ☐ biodiesel fuel can be produced from vegetable oils
- ☐ oils do not dissolve in water; they can be used to produce emulsions, e.g. in salad dressings
- ☐ processed foods may contain additives to improve appearance, taste and shelf-life; E-numbers identify permitted additives and must be listed; some additives can be harmful and may be banned
- ☐ chemical analysis can be used to identify additives and food colouring in foods, e.g. by paper chromatography

what the changes are in the Earth and its atmosphere

- ☐ the Earth has three main layers: the crust, mantle and core
- ☐ the Earth's atmosphere has changed over millions of years; many of the gases that make up the atmosphere came from volcanoes
- ☐ for 200 million years, the proportions of different gases in the atmosphere have been much the same as they are today
- ☐ human activities have recently produced further changes, e.g. the levels of greenhouse gases are rising

Heat energy

Grades

D–C

1 **Thermal radiation** is a type of **electromagnetic** wave.

a Write down **three** properties of thermal radiation.

[3 marks]

b Emily has just moved to a new house which has large south-facing windows. During the summer the house gets very hot.

i Explain why the house stays cooler if she draws the curtains in the daytime.

[2 marks]

ii Her friend suggests that she could paint the outside of her house a different colour. Explain how this would help the house feel cooler in summer.

[3 marks]

B–A*

2 Craig has made some biscuits.

a Explain why the uncooked biscuits have less internal energy compared with the biscuits when they are cooking in the oven.

[1 mark]

b Write down how temperature and internal energy are connected.

[2 marks]

Thermal radiation

D–C

3 Match words, **A**, **B**, **C** and **D**, with the numbers **1 – 4** in the sentences.

A reduce **B** absorb **C** reflect **D** increase

White surfaces ____ **1** ____ heat well and ____ **2** ____ heat badly.

One company painted the outside of its refrigerated lorries white to ____ **3** ____ heat transfers and keep the food cool. Painting the lorry black would ____ **4** ____ heat transfers.

1	**2**	**3**	**4**

[4 marks]

B–A*

4 **a** Write down **two** factors that affect the amount of thermal energy an object has.

[2 marks]

b Explain why an ice cube has less thermal energy than the drink it is in.

[2 marks]

c Explain why a thermal imaging camera will detect the drink more easily before the ice cube is added.

[2 marks]

Conduction and convection

D–C

1 One end of a metal rod is placed in the flame of a Bunsen burner. After a while, the other end of the rod becomes hot.

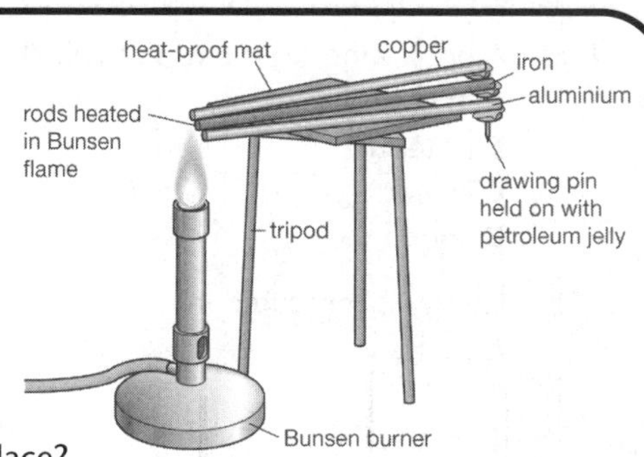

a What is the name of the type of heat transfer that takes place?

____ [1 mark]

b Explain how heat is transferred from one end of the metal rod to the other.

____

____

____ [3 marks]

c The experiment is repeated with a metal rod and a glass rod. Explain why the glass rod does not conduct the heat as well as the metal rod.

____

____ [2 marks]

B–A*

2 Match words, **A**, **B**, **C** and **D**, with the numbers **1 – 4** in the sentences.

A heat capacity **B** temperature **C** thermal energy **D** warm

A hot water bottle stores more ____ **1** ____ than a bottle of hot air at the same ____ **2** ____. The water has a higher ____ **3** ____, and stays ____ **4** ____ for longer.

1	2	3	4

[4 marks]

Heat transfer

B–A*

3 a A diamond feels very cold to touch. This means that
A it is a bad heat conductor.
B it is cooler than its surroundings.
C it has a low specific heat capacity.
D it is a good heat conductor. [1 mark]

b A diamond has no free electrons. It conducts heat because its atoms
A are very closely linked.
B are so loosely linked that they can change places.
C vibrate easily.
D have a high heat capacity. [1 mark]

D–C

c Convection currents occur in liquids because
A heated fluids are more dense than cool fluids.
B heated fluids are less dense than cool fluids.
C energy is transferred when the atoms vibrate.
D they have free electrons. [1 mark]

d The particles in gases are not close together which means that
A they are poor convectors and good heat conductors.
B they are poor convectors and poor heat conductors.
C they are good convectors and poor heat conductors.
D they are good convectors and good heat conductors. [1 mark]

Grades

Types of energy

D–C

1 **a** A book lying on the top of a wall has ________________ energy.
A chemical
B potential
C kinetic
D sound [1 mark]

b Energy is measured in
A joules
B watts
C metres
D kilograms [1 mark]

c Energy can be stored in these forms
A light and kinetic
B kinetic and potential
C chemical and potential
D electrical and chemical [1 mark]

B–A*

d Fossil fuels were created as a result of
A global warming.
B the Sun's radiation over recent years.
C human activity.
D the Sun's radiation over millions of years. [1 mark]

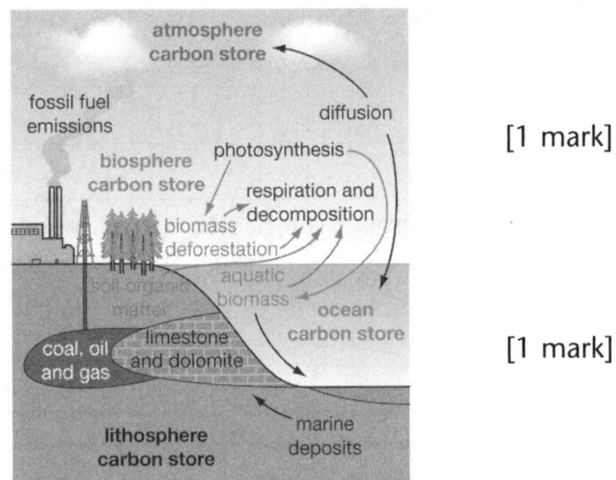

Energy changes

D–C

2 George drives his car along a level road.

a Some of the energy is wasted. Write down **two** forms of wasted energy produced by the car.

____ [2 marks]

b Explain what happens to the wasted energy.

____

____ [2 marks]

B–A*

c Describe how the energy from the fuel is transformed as George drives his car up a hill and then stops it.

____

____

____

____ [4 marks]

D–C

3 Match words, **A**, **B**, **C** and **D**, with the numbers **1 – 4** in the sentences.

A energy **B** mass **C** fusion **D** conservation

1	2	3	4

The law of Energy ____ **1** ____ says that energy cannot be created or destroyed.

B–A*

In the Sun, energy is given out due to nuclear ____ **2** ____. The ____ **3** ____ of the nuclei changes into energy directly. In the Sun, both mass and ____ **4** ____ are conserved. [4 marks]

Grades

Energy diagrams

D–C

1 The picture shows a Sankey diagram for a petrol engine.

a Complete this equation:

energy input = energy __________ = useful energy

+ __________ energy [2 marks]

b The energy transformation takes place in two stages in the car. State where most of the energy wastage takes place.

____ [1 mark]

c An improved engine has been designed. It changes 1000 J of chemical energy into 500 J of kinetic energy. The rest is wasted heat. Draw a Sankey diagram to show this.

[4 marks]

B–A*

2 Match words, **A**, **B**, **C** and **D**, with the numbers **1 – 4** in the sentences.

A rechargeable **B** electricity **C** energy **D** pollution

An electric car uses a ____ **1** ____ battery. The ____ **2** ____ in the battery comes from power stations. The car does not pollute the roads, but the power stations providing ____ **3** ____ to charge the battery cause ____ **4** ____.

1	**2**	**3**	**4**

[4 marks]

Energy and heat

D–C

3 a Wasted energy is transferred to the surroundings, which
A concentrate the energy for future use.
B store the energy.
C heat up.
D cool down. [1 mark]

b When a plate heats up, its particles
A vibrate more vigorously.
B start vibrating.
C change places.
D spread out. [1 mark]

c Useful sources of energy are normally
A diluted
B concentrated
C spread out
D hard to find [1 mark]

B–A*

d A fridge transfers heat in the opposite direction to normal. A power supply is needed because
A it provides cold energy to the fridge.
B heat energy is needed for the fridge to work properly.
C energy is needed to spread the heat out.
D energy is needed to concentrate and transfer the heat. [1 mark]

Energy, work and power

Grades

1 The weight of Molly and her skis is 500 N. A ski lift carries her to the top of the slope, which is a height of 60 m. D–C

a How much **work** is done by the ski lift?

__________ joules [3 marks]

b If this took 75 seconds, work out the **power** of the ski lift.

__________ watts [3 marks]

c The ski lift is replaced by a more **efficient** model. Explain why the energy used by the new ski lift will be less than the energy used by the first lift. B–A*

____

______________________________ [2 marks]

d Explain why the force needed to lift Molly vertically up a mountain slope would be greater than the force needed to lift her along a slope to the same point.

____

______________________________ [2 marks]

Efficiency

2 a Efficiency compares D–C
A energy usefully transferred with work done.
B energy usefully transferred with total energy supplied.
C wasted energy with total energy supplied.
D percentage of wasted energy with work done. [1 mark]

b The power of a kettle is 200 kW. The energy used by the kettle in one minute is
A 12 000 kJ
B 200 kJ
C 12 000 kW
D 200 kW [1 mark]

c A kettle uses 170 kJ to heat water. 200 kJ of electrical energy is supplied to the kettle. The kettle's efficiency is
A 118%
B 60%
C 85%
D 8.5% [1 mark]

d A perpetual motion machine will never work because B–A*
A science is not advanced enough.
B they use up too much energy.
C energy cannot be transformed in two directions.
D some energy is always wasted with any energy transfer. [1 mark]

Using energy effectively

D–C

1 Luke is worried that he is wasting energy at home.

a How can he reduce the energy spent on lighting?

[2 marks]

b Why is it a good idea for Luke to install insulation?

[1 mark]

D–C

2

Type of insulation	Installation cost (in £)	Annual saving (in £)	Payback time (in years)
Loft insulation	240	60	4
Cavity wall insulation	360	60	6
Draught-proofing doors and windows	45	15	3
Double glazing	2500	25	100

a Explain what is meant by **payback time**.

[2 marks]

b The Brown family are planning to stay in their house for ten years and would like to install some insulation. Use the table to explain why draught proofing is not the best insulation method for them. Include a calculation for full marks.

[3 marks]

B–A*

c Government grants are available for homeowners to install solar panels, but this may not save energy nationally. Write down **two** disadvantages of installing solar panels.

[2 marks]

Why use electricity?

D–C

3 a Mains electricity is very useful because
A it can be stored.
B it is an instant, convenient supply of energy.
C using electricity does not contribute to global warming.
D there are lots of uses for energy in the form of electricity. [1 mark]

b Batteries are useful because
A they are cheap to use.
B they can store larges amounts of energy.
C they are portable and can supply electricity.
D they store electricity. [1 mark]

c When there is no source of electricity
A batteries can be recharged using solar energy or clockwork springs.
B batteries cannot be charged up.
C batteries cannot be used.
D batteries can only be recharged by plugging them into a socket. [1 mark]

d LEDs are being used in torches instead of bulbs. Batteries in these torches last longer because
A LEDs use a larger current than bulbs.
B LEDs use electrons.
C LEDs are different colours.
D LEDs use less energy than bulbs do. [1 mark]

Electricity and heat

D–C

1 a Write down **two** factors that affect the resistance of a wire.

____ [2 marks]

b Explain why a radio uses a thinner flex than an electric cooker.

____ [1 mark]

c Fuse wires have different ratings. Explain why the cooker needs a higher rated fuse than the radio.

____ [2 marks]

B–A*

2 An MRI scanner uses superconductors.

a How does the resistance of a superconductor change with temperature?

____ [1 mark]

b Explain why a superconductor has low heat losses due to the currents flowing through it.

____ [2 marks]

c Write down **one** reason why superconductors are not used often in everyday activities.

____ [1 mark]

The cost of electricity

3

Appliance	Power rating	Appliance	Power rating
Laptop computer	100 W	Television	50 W
Electric kettle	2.4 kW	Microwave oven	1.2 kW
Hairdryer	1.5 kW	Electric fire	2.0 kW
Radio	4 W	Lawn mower	1.3 kW

a What is meant by the 'power rating of the electric kettle is 2.4 kW'?

____ [2 marks]

b Calculate the total energy used in when the hairdryer is switched on for 15 minutes.

______________________________________ kWh [3 marks]

D–C

4 a The reading on an electricity meter is 32 456 in December. The previous reading in September was 31 667. How much electricity was used in those three months?

A 789
B 64 123
C 31 667
D 32 456 [1 mark]

b Electricity costs 10p per kWh. The cost of electricity when 347 kWh are used is

A 347p
B 34.7p
C £3470
D £34.70 [1 mark]

c The power of a DVD player is 600 W. It is switched on for 60 seconds. The energy used in Joules is

A 10
B 36
C 36 000
D 600 [1 mark]

B–A*

d Digital TV set-up boxes use less energy on stand by than when they are in use. This means that when the box is on stand by

A no energy is being wasted.
B the amount of wasted energy is reduced.
C the wasted energy does not change.
D the box cannot overheat. [1 mark]

Grades

The National Grid

D–C

1 Match words, **A**, **B**, **C** and **D**, with the statements **1 – 4** in the table.

A step-up transformer **B** National Grid **C** power station **D** substation

1	a network of power stations and cables carrying electricity across the country
2	place where electricity is generated
3	part of the National Grid that reduces the voltage of electricity
4	device that increases the voltage

1	**2**	**3**	**4**

[4 marks]

D–C

2 Match numbers, **A**, **B**, **C** and **D**, with the spaces **1 – 4** in the table.

A 6 **B** 3 **C** 2 **D** 4

1	**2**	**3**	**4**

Item	**Power (watts)**	**Current (amps)**	**Voltage (volts)**
Electric blanket	920	**1**	230
Remote controlled car	12	2	**2**
Television	460	**3**	230
Torch	**4**	0.5	6

[4 marks]

B–A*

3 a What type of current is supplied from a battery?

____ [1 mark]

b Explain how this current is different from the current supplied by power stations.

____ [2 marks]

D–C

c Explain why energy losses are reduced if electricity is transported at a very high voltage.

____ [2 marks]

Generating electricity

D–C

4 a Anna's bicycle has a **dynamo** attached to its lights. A dynamo is used to change kinetic energy into electrical energy.

i Why do the lights shine when the wheels are turning?

____ [1 mark]

ii Write down the energy change that takes place in the dynamo.

________________ energy g ________________ energy [1 mark]

b The diagram shows a coil of wire attached to a sensitive ammeter. When a current flows, the needle moves.

A
N S
coil

i Why does the needle flick if the magnet moves into the coil?

__________________________________ [1 mark]

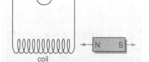

ii Explain why a larger current is registered when the magnet moves faster.

____ [1 mark]

Power stations

Grades

D–C

1 Match words, **A**, **B**, **C** and **D**, with the statements **1 – 4** in the table.

A biomass **B** coal **C** hydroelectricity **D** uranium

1	a non-renewable fuel
2	a nuclear fuel
3	a renewable energy source
4	a renewable fuel

1	2	3	4

[4 marks]

D–C

2 The diagram shows how electricity is generated in a power station.

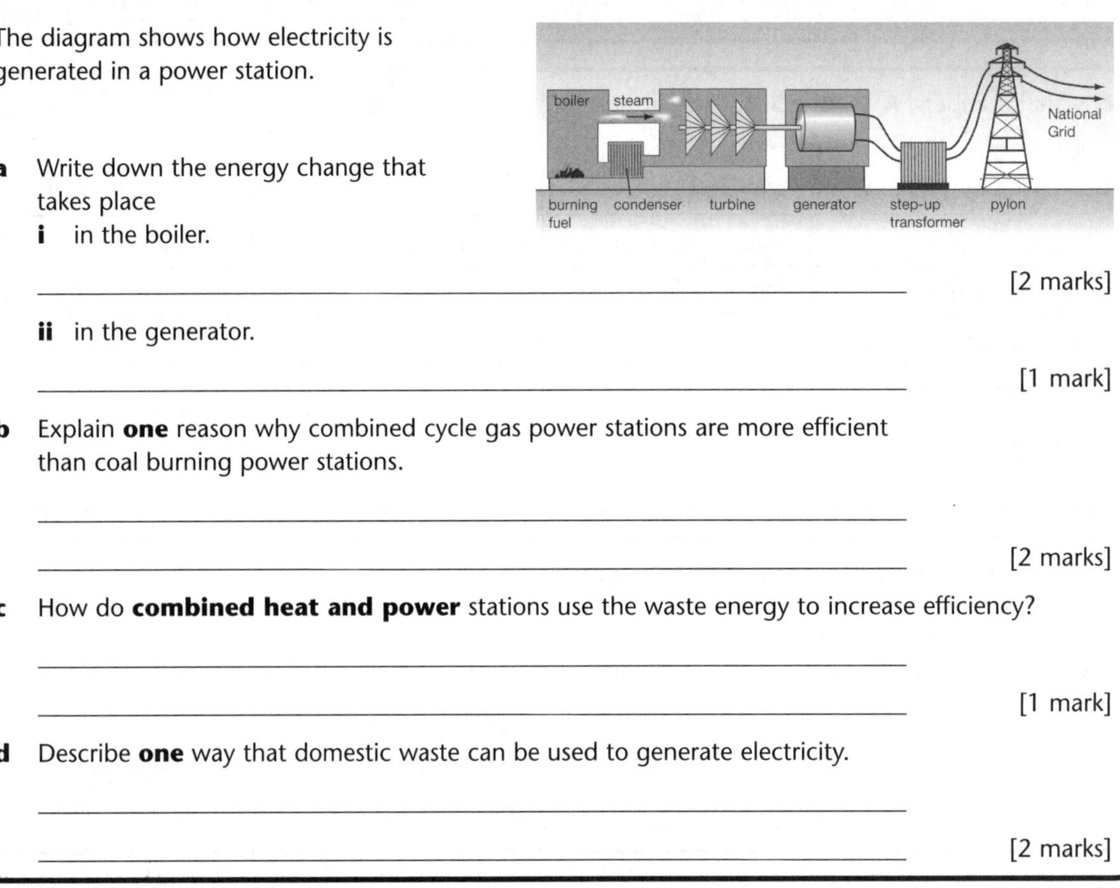

a Write down the energy change that takes place

i in the boiler.

[2 marks]

ii in the generator.

[1 mark]

b Explain **one** reason why combined cycle gas power stations are more efficient than coal burning power stations.

[2 marks]

c How do **combined heat and power** stations use the waste energy to increase efficiency?

[1 mark]

B–A*

d Describe **one** way that domestic waste can be used to generate electricity.

[2 marks]

Renewable energy

D–C

3 France has a large tidal power station.

a Explain how the tidal power station generates electricity.

[3 marks]

b Write down **one disadvantage** of using tidal power stations.

[1 mark]

B–A*

c Explain **one** way that the Sun's heat can be used to generate electricity.

[3 marks]

Grades

Electricity and the environment

D–C

1 a Write down **two** harmful effects of burning fossil fuels.

____ [2 marks]

b Explain why wind turbines are useful in remote areas.

____

____ [2 marks]

Making comparisons

D–C

2 a Sienna is comparing the effects on the environment of coal-fired power stations and nuclear power stations.

i Which of these power stations release greenhouse gases?

____ [1 mark]

ii Which of the fuels used in these power stations, will run out one day?

____ [2 marks]

b Power stations need to be located on suitable sites. Write down **three factors** that a company may consider before choosing a site for a coal-fired power station.

____

____ [3 marks]

B–A*

c Sometimes there are unwanted effects when siting hydroelectric power stations. Explain why rotting vegetation caused when a site repeatedly floods and then experiences drought may contribute to greenhouse gases.

____

____ [3 marks]

D–C

3 Match words, **A**, **B**, **C** and **D**, with the numbers **1 – 4** in the sentences.

A cheap **B** efficient **C** flexible **D** reliable

Fossil fuel power stations are designed to waste less energy, so they are more ____ **1** ____.

At the moment, fossil fuels are ____ **2** ____ but their price will rise when supplies get low.

Solar cells cannot generate electricity when it is dark or cloudy because they are not ____ **3** ____.

In some places hydroelectric stations can provide electricity in seconds, making it very ____ **4** ____.

1	2	3	4

[4 marks]

D–C

4 a What **two** things affect the total cost of electricity produced?

____ [2 marks]

b Explain why a new gas power station may cost more to build in future, but have lower running costs than existing power stations.

____

____ [3 marks]

B–A*

c Describe **one** other method of generating electricity direct from organic matter.

____

____ [3 marks]

P1a revision checklist

I know:

how heat (thermal energy) is transferred and what factors affect the rate at which heat is transferred

- ☐ heat energy can be transferred by conduction, convection and thermal radiation
- ☐ thermal conductors (e.g. metals) transfer heat energy easily; thermal insulators (e.g. plastic, glass) do not
- ☐ dark, dull surfaces emit and absorb thermal radiation better than shiny, light surfaces
- ☐ the bigger the temperature difference between an object and its surroundings, the faster the rate at which heat is transferred

what is meant by the efficient use of energy

- ☐ energy is never created nor destroyed; some energy is usually wasted as heat
- ☐ the greater the percentage of the energy that is usefully transformed in a device, the more efficient the device is
- ☐ how to calculate the efficiency of a device:

$$\textbf{efficiency} = \frac{\textbf{useful energy output}}{\textbf{total energy input}}$$

why electrical devices are so useful

- ☐ they transform electrical energy to whatever form of energy we need at the flick of a switch
- ☐ the National Grid transmits energy around the country at high voltages and low current to keep energy losses low
- ☐ dynamos produce electricity when coils of wire rotate inside a magnetic field
- ☐ how to work out the power rating of an appliance (the rate at which it transforms electrical energy)
- ☐ how to calculate the amount of energy transferred from the mains:

energy transferred = **power** × **time**

- ☐ how to calculate the cost of energy transferred from the mains:

total cost = **number of kilowatt-hours** × **cost per kilowatt-hour**

how we should generate the electricity we need

- ☐ we need to use more renewable energy sources, including wind, hydroelectric, tidal, wave and geothermal power
- ☐ most types of electricity generation have some harmful effects on people or the environment; there are also limitations on where they can be used

Uses of electromagnetic radiation

Grades

D–C

1 a Write down **three properties** that all electromagnetic waves share.

___

___ [3 marks]

b The diagram shows the members of the electromagnetic spectrum. Fill in the names missing on the diagram.

Radio waves	**1**	Infrared	**2**	Ultraviolet	**3**	Gamma rays

Increasing frequency →

Increasing energy →

1 ___ **2** ___ **3** ___ [3 marks]

B–A*

c Match words, **A**, **B**, **C** and **D**, with the numbers **1 – 4** in the sentences.

A infrared **B** ultraviolet **C** visible light **D** X-rays

___ **1** ___ are used to check baggage at airports for dangerous and illegal items. Forged bank notes show up under ___ **2** ___ light. Trespassers can be detected if they walk through ___ **3** ___ beams in buildings. Security cameras use ___ **4** ___ to take pictures in high security areas.

1	2	3	4

[4 marks]

D–C

2 Match words, **A**, **B**, **C** and **D**, with the numbers **1 – 4** in the sentences.

A gamma rays **B** infrared **C** radio waves **D** ultra violet

___ **1** ___ are used for broadcasts. ___ **2** ___ are used to detect parts of the body that are warmer due to inflammation. Our skin develops a suntan when exposed to ___ **3** ___. ___ **4** ___ can be used to kill cells, and sterilise surgical instruments.

1	2	3	4

[4 marks]

Electromagnetic spectrum 1

D–C

3 a Add a label to the diagram to show one wavelength. [1 mark]

b Under the wave, add a diagram of another wave which carries more energy. [2 marks]

c Write down the relationship between frequency and energy.

___ [1 mark]

B–A*

d An electromagnetic wave is absorbed by a radio antenna, creating an alternating current. How are the frequency of the radiation and the alternating current connected?

___ [1 mark]

Electromagnetic spectrum 2

D–C

1 Match words, **A**, **B**, **C** and **D**, with the numbers **1 – 4** in the sentences.

A absorb **B** carry **C** reflects **D** transmits

All electromagnetic waves ____ **1** ____ energy. A mirror ____ **2** ____ light so we can see an image. A window ____ **3** ____ light helping us see during the day. Black material ____ **4** ____ light so it looks dark.

1	2	3	4

[4 marks]

D–C

2

Wave type	Wavelength	Sources	Detectors
Gamma rays	10^{-12} m	Radioactive nuclei	Geiger-Müller tube
X-rays	10^{-10} m	X-ray tubes	Geiger-Müller tube
Ultraviolet	10^{-7} m	Sun, very hot objects	Skin, photographic film, fluorescent material
Visible light	0.0005 mm	Hot objects, Sun, lasers, LEDs	Eyes, photographic film
Infrared	0.1 mm	Warm or hot objects	Skin, thermometer
Microwaves	1–10 cm	Radar, microwave ovens	Aerial, mobile phone
Radio waves	10– 1000+ m	Radio transmitters	Aerial, TV, radio

Complete these sentences using the words **absorbed**, **transmitted** or **reflected**.

a Satellites use microwaves communications because these waves are ________ by the atmosphere.

b Ultraviolet waves are ________ by our skin, causing suntan.

c White cars stay cool in the sun because infrared is ________ by the light colour.

Waves and matter

D–C

3 a What type of electromagnetic radiation is used to take shadow pictures of bones?

________________________ [1 mark]

b Explain how a picture of a person's bones can be taken using electromagnetic radiation.

____

____

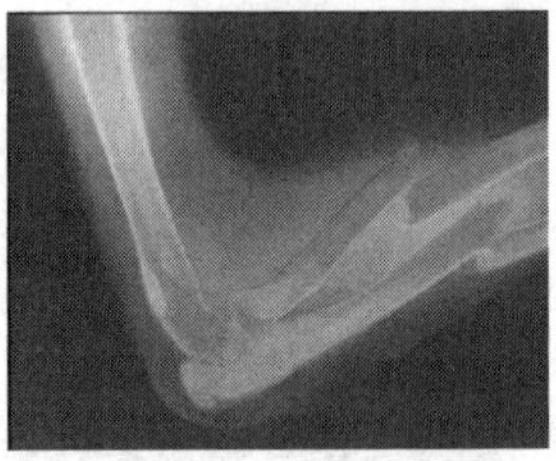

________________________ [3 marks]

c Suggest **one** reason why these pictures are taken only when necessary.

________________________ [1 mark]

B–A*

4 a How fast does light travel? ________________________ [1 mark]

b What is the equation that links speed, frequency and wavelength?

________________________ [1 mark]

c Calculate

i the wavelength of radio waves if their frequency is 300 000 Hz.

____

________________________ [2 marks]

ii the frequency of microwaves if their wavelength is 0.03 m. Include the units.

____

________________________ [3 marks]

Grades

Dangers of radiation

D–C

1 Match words, **A**, **B**, **C** and **D**, with the numbers **1 – 4** in the sentences.

A absorb **B** burn **C** ionise **D** kill

Cells ____ **1** ____ many types of radiation which can damage them.

High doses of gamma rays ____ **2** ____ cancer cells.

Low doses of X-rays and gamma rays ____ **3** ____ cells, damaging them.

Infrared radiation and microwaves ____ **4** ____ cells.

1	2	3	4

[4 marks]

D–C

2 Rachel has put sun-block on to protect her skin before she goes out in the Sun.

a What type of electromagnetic radiation does sun-block protect her from?

______________________________ [1 mark]

b Write down **three** risks from too much exposure to this type of radiation.

____

____

______________________________ [3 marks]

c Explain why darker skins are less likely to be damaged by the Sun's radiation.

____

____

______________________________ [3 marks]

B–A*

d Explain what effect the shorter wavelength electromagnetic waves have on the molecules in cells and why this may cause cancer.

______________________________ [1 mark]

Telecommunications

D–C

3 a What is the name given the type of signal shown in figure A?

____

[1 mark]

A

B

b Explain how digital signals carrying a cable TV programme are sent.

____

____

____

____

______________________________ [2 marks]

c Suggest **one** reason why underground telephone cables are no longer being laid.

______________________________ [1 mark]

d Many communication links are now made using satellite links.
Write down **two advantages** of using satellite communications.

____

______________________________ [2 marks]

Fibre optics: digital signals

Grades

D–C

1 a Explain how an **optical fibre** is used to send digital signals.

____

____ [2 marks]

b The diagram shows light travelling down an optical fibre.

i Continue the path of the light ray until it emerges from the optical fibre. [3 marks]

ii What is the **name** given to the process that occurs inside the optical fibre?

____ [1 mark]

B–A*

c State **two advantages** of using digital technology to store information on car driving licences.

____

____ [2 marks]

Radioactivity

D–C

2 Match words, **A**, **B**, **C** and **D**, with the arrows **1 – 4** in the diagram.

A electron

B neutron

C nucleus

D proton

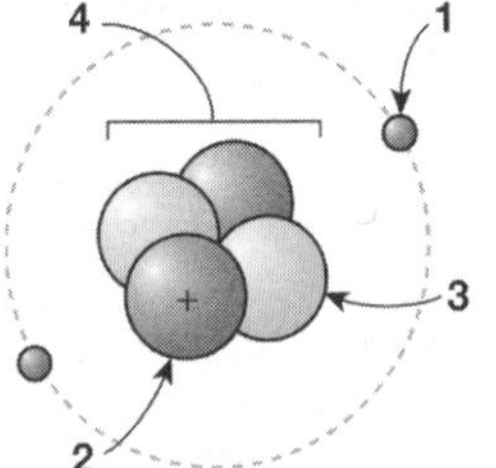

1	**2**	**3**	**4**

[4 marks]

D–C

3 The table gives some information about different atoms.

Element	**Number of protons**	**Number of neutrons**
A	6	6
B	6	7
C	7	7
D	8	8

a Which elements are isotopes of each other?

____ [1 mark]

b How many electrons are orbiting the nucleus of element A? _______________ [1 mark]

c What is the mass number of element C? _______________________________ [1 mark]

d Explain what is meant by **radioactive decay**.

____

____ [2 marks]

Alpha, beta and gamma rays 1

D–C

1 a Explain what is meant by an **alpha particle**.

______________________________ [2 marks]

b Alpha particles are very ionising. Inside the body, they damage cells. Explain why alpha particles are unlikely to cause damage to people in most cases.

______________________________ [1 mark]

c Write down **two** uses for alpha radiation.

______________________________ [2 marks]

d Why are gamma rays **less ionising** than alpha particles?

______________________________ [2 marks]

B–A*

2 a Which **two** types of radiation are deflected by electric fields?

______________________________ [2 marks]

b Give **one** reason why these types of radiation are deflected in different directions.

______________________________ [1 mark]

Background radiation 1

D–C

3 a Background radiation is

A very rare.
B found only in a few places.
C found all around us.
D only caused by natural sources. [1 mark]

b Radon gas is found in some areas of the United Kingdom. It is

A not a health risk.
B a health risk causing raised levels of lung cancer in these regions.
C caused by man-made sources only.
D not possible to change your exposure to radon gas. [1 mark]

c Cosmic radiation is caused by the Sun. It is

A mainly absorbed by the atmosphere.
B a significant health risk on all flights.
C not a health risk.
D the same intensity at sea level and on high altitude flights. [1 mark]

d Radiation badges are worn

A to protect people from exposure to radioactivity.
B as a source of radioactivity.
C to give an instant read-out of exposure to radioactivity.
D to monitor exposure to radioactivity over a period of time. [1 mark]

B–A*

4 Lindi is using a Geiger counter to monitor a radioactive source. Explain why

a there is a reading even when the source is stored in a lead lined container.

______________________________ [1 mark]

b the readings from the source fluctuate.

______________________________ [1 mark]

Half-life

Grades

1 a After two half-lives, in a radioactive sample
A all of the atoms will decay.
B none of the atoms will decay.
C exactly half of the atoms will decay.
D about half of the atoms will decay. [1 mark]

b The length of a sample's half-life depends on
A the type of material that is decaying.
B the temperature the sample is kept in.
C the pressure the sample feels.
D the country the sample was found in. [1 mark]

c The half-life of a sample is the time taken for
A half of its mass to disappear.
B its original activity to fall to half.
C half of the samples measured to decay completely.
D its count rate to change. [1 mark]

d A sample of cobalt has a half-life of 5 years, and its count rate is 1200 counts per second. After 15 years, the count rate is
A 600 **B** 300 **C** 150 **D** 0 [1 mark]

D–C

2 a Explain what is meant by **half-life**.

____

_______________ [3 marks]

D–C

b What types of materials can be dated using **radiocarbon dating**?

_______________ [1 mark]

B–A*

c The proportion of radioactive carbon in a wooden arrow has been measured. Explain whether the proportion of radioactive carbon in an arrow 10 000 years old will be more or less than that found in an arrow made from modern wood.

____

_______________ [2 marks]

d The proportion of radioactive carbon in a wooden arrow has been measured. Explain whether the proportion of radioactive carbon in an arrow 10 000 years old will be more or less than that found in an arrow made from modern wood.

____

_______________ [3 marks]

Uses of nuclear radiation

3 The diagram shows one method of controlling the thickness of paper in a factory.

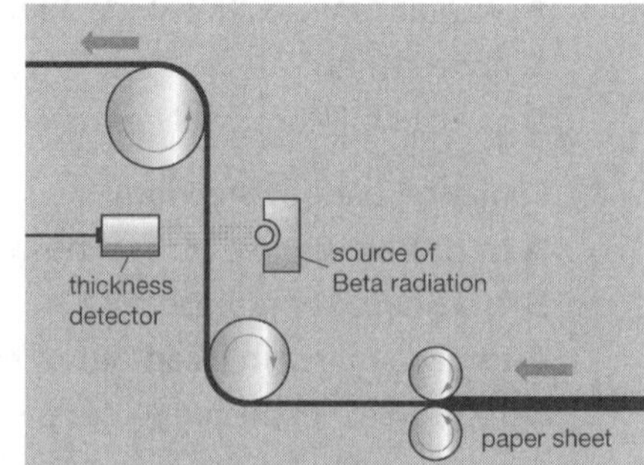

D–C

a Write down **one** reason why is it important to monitor the thickness.

____

_______________ [1 mark]

b Why is alpha radiation not used to detect the thickness of the paper?

_______________ [1 mark]

c Explain how the machine should adjust the rollers if the amount of radiation reaching the detector gets smaller.

____

_______________ [2 marks]

Safety first

D–C

1 a A person is working with a radioactive sample. Which precaution will make no difference to her exposure?

A using tongs
B putting the sample back in its case after use
C wearing a lead apron
D reducing the temperature of the room [1 mark]

b Special badges are worn by people who work with radioactivity so that

A the radioactivity is neutralised.
B the amount of radioactivity received can be measured.
C their colleagues know that they are authorised.
D they do not develop cancer. [1 mark]

c When radioactivity is absorbed, which of these effects never occurs?

A infection
B cell mutation
C absorption of energy
D ionisation [1 mark]

d One effect of exposure to radiation is to

A selectively repair previous cell damage.
B ionise cells.
C ionise DNA molecules in cells.
D make cells radioactive. [1 mark]

Searching space

D–C

2 Telescopes are used to examine the night sky.

a Explain why a reflecting telescope produces better images than a simple optical telescope.

____

______________________________ [2 marks]

b The Hubble telescope is a space telescope. Explain what is meant by a **space telescope**.

____

______________________________ [2 marks]

B–A*

c Describe **two advantages** of having a telescope in orbit.

____

______________________________ [2 marks]

D–C

3 Match words, **A**, **B**, **C** and **D**, with the numbers **1 – 4** in the sentences.

A gamma **B** infrared **C** ultraviolet **D** visible

We can see ____ **1** ____ light from stars using binoculars. ____ **2** ____ rays emitted by neutron stars can be detected using space-based telescopes. These are also used to detect ____ **3** ____ radiation from red giants and ____ **4** ____ radiation from quasars.

1	2	3	4

[4 marks]

Gravity

Grades

D–C

1 Match words, **A**, **B**, **C** and **D**, with the numbers **1 – 4** in the sentences.

A force **B** gravity **C** mass **D** separation

All objects are attracted to each other because of ____ **1** ____.

The attractive ____ **2** ____ is greater if the ____ **3** ____ of the objects is small or if the ____ **4** ____ of the objects is big.

1	2	3	4

[4 marks]

D–C

2 The diagram shows how one spacecraft used gravity to travel across the Solar System.

a Explain what caused the spacecraft to change direction as it passed the different planets.

____

____

[1 mark]

B–A*

b Explain why astronauts in an orbiting spacecraft feel weightless.

______________________________ [1 mark]

Birth of a star

D–C

3 The statements, **A**, **B**, **C** and **D**, describe the stages in the formation of a star. Put them in the correct order, matching the statements with the correct places **1 – 4** in the boxes.

A gas molecules are pulled together by gravity to form a cloud
B nuclear fusion starts
C the cloud attracts more particles and gets larger
D the core of the nebula heats up dramatically

1	→	2	→	3	→	4	→	the star begins to shine

[4 marks]

B–A*

4 a Explain why nuclear fusion results in heavier elements forming.

____

______________________________ [2 marks]

b Explain how we know that the Solar System must have contained pieces of a bigger star than our Sun.

____

______________________________ [2 marks]

Formation of the Solar System

D–C

1 The diagram shows one idea of how the Solar System was formed.

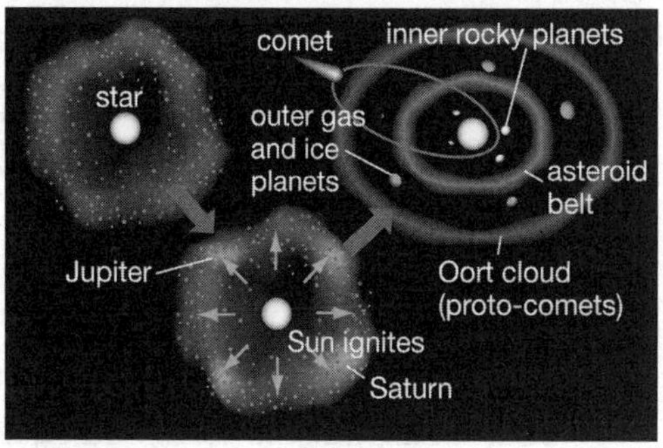

a Explain what happens to the dust cloud surrounding the newly formed star when it ignited.

____

____

[1 mark]

b Explain **one** reason why the rocky planets formed closest to the Sun and the gas and ice planets further away.

____

______________________________________ [2 marks]

B–A*

c The Sun constantly emits electromagnetic radiation. What evidence is there that it also emits electrically charged particles?

______________________________________ [1 mark]

Life and death of a star

D–C

2 Match words, **A**, **B**, **C** and **D**, with the numbers **1 – 4** in the sentences.

A black hole **B** neutron star **C** red giant **D** supernova

A star the size of our Sun will eventually expand to form a ____ **1** ____.
Stars larger than our Sun become red supergiants, which collapse and explode as a spectacular ____ **2** ____. The core remains an incredibly dense ____ **3** ____. If the star was very massive, a ____ **4** ____ remains instead. [4 marks]

1	2	3	4

D–C

3 Stars begin as a cloud of dust and gas which are crushed by gravity until nuclear fusion reactions begin.

a Explain why the Sun is constantly losing mass.

______________________________________ [1 mark]

b Write down the three stages that a star, like our Sun, will go through before it eventually dies.

____________ → ____________ → ____________ [3 marks]

B–A*

c What **two** factors are balanced when a star is in the stable phase?

____

______________________________________ [2 marks]

d How do these two forces control the life cycle of the star?

____

____

______________________________________ [3 marks]

Grades

In the beginning

D–C

1 a When is the Big Bang thought to have taken place?

____________________ [1 mark]

b Explain what is meant by the **Big Bang theory**.

____

____________________ [2 marks]

c Explain how heavier elements were created from **quarks**.

____

____________________ [3 marks]

B–A*

d Why is the presence of background microwave radiation additional evidence for the Big Bang?

____________________ [2 marks]

The expanding Universe

D–C

2 Astronomers looking at distant galaxies have noticed an effect they called the red shift.

a Explain what is meant by the **red shift**.

____

____________________ [2 marks]

b What does the red shift tell astronomers about the **motion** of galaxies?

____________________ [1 mark]

D–C

3 a Galaxies close to us have a smaller red shift than galaxies further away. This tells us that galaxies are moving

A at the same speed.
B away from us at different speeds.
C towards us at different speeds.
D slowly. [1 mark]

b A spectroscope is used to

A measure your heart.
B display the electromagnetic spectrum.
C split light into different colours.
D rejoin light that has been split up. [1 mark]

c An element emits a unique pattern of light when it is

A heated.
B looked at under high pressure.
C reacting with other elements.
D cooled. [1 mark]

d The change in the colour of light from galaxies occurs because of

A the great distance the light travels.
B the different temperatures of the stars.
C interference from our atmosphere.
D the change in wavelength of the light. [1 mark]

B–A*

4 a It is thought that the Universe is still expanding as a result of the Big Bang. Explain what is meant by the **Big Crunch** and what it depends on.

____

____________________ [3 marks]

b State **two** other possible futures for the Universe.

____

____________________ [2 marks]

P1b revision checklist

I know:

what the uses and hazards of the waves that form the electromagnetic spectrum are

- ☐ from longest to shortest wavelength: radio waves, microwaves, infrared, visible light, ultraviolet, X-rays, gamma rays
- ☐ electromagnetic radiation has many uses in communication, e.g. radio, TV, satellites, cable and mobile phone networks
- ☐ communication signals can be digital or analogue
- ☐ some forms of electromagnetic radiation can damage living cells: ionising radiation (ultraviolet, X-rays and gamma rays) can cause cancer
- ☐ electromagnetic waves obey the wave formula:

  wave speed = frequency × wavelength

what the uses and dangers of emissions from radioactive substances are

- ☐ the uses and hazards of radioactive substances (which emit alpha particles, beta particles and gamma rays) depend on the wavelength and frequency of the radiation they emit
- ☐ background radiation is all around us, e.g. granite rocks can emit gamma rays and form radioactive radon gas
- ☐ the relative ionising power, penetration through materials and range in air of alpha, beta and gamma radiations
- ☐ the activity (count) rate of a radioisotope is measured as its half-life

about the origins of the Universe and how it continues to change

- ☐ the Universe is still expanding; in the beginning, matter and space expanded violently and rapidly from a very small initial point, i.e. the Big Bang
- ☐ red shift indicates that galaxies are moving apart; the further away a galaxy, the faster it is moving away from us
- ☐ telescopes on Earth and in space give us information about the Solar System and the galaxies in the Universe

Cells

D–C

1 The diagram shows a plant cell.

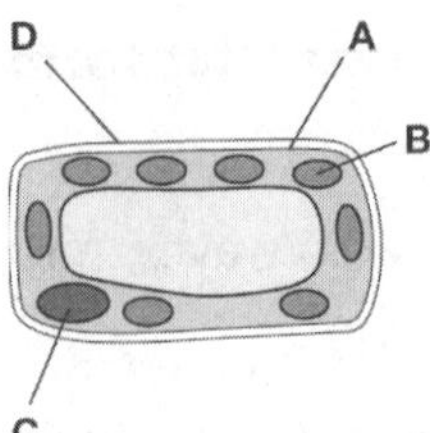

a Name parts B and C.

B ______________________ C ______________________ [2 marks]

b Describe the functions of parts A and D.

A ____

D ____ [2 marks]

c Describe the functions of each of these organelles.

Ribosomes ______________________________________

Mitochondria ____________________________________ [2 marks]

Specialised cells

D–C

2 The diagram shows a ciliated cell. Cells like this are found in the lining of the trachea and bronchi leading down to the lungs.

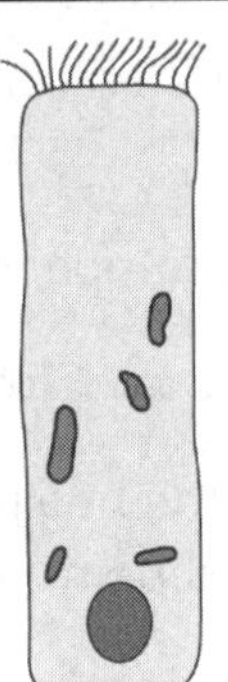

a Describe the function of ciliated cells.

____ [1 mark]

b State **two** ways in which ciliated cells are adapted for their function.

1____

2____ [2 marks]

B–A*

3 The diagram shows the tip of a plant root, magnified.

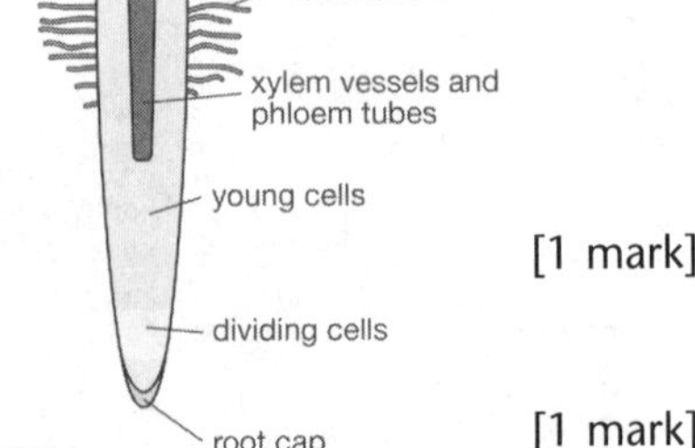

a What is the function of root hair cells?

____

____ [1 mark]

b Suggest how their shape adapts them for their function.

____ [1 mark]

c Suggest **two** ways in which the young cells in the plant root would differ from the cell shown in Question **1**.

1____

2____ [2 marks]

Grades

Diffusion 1

D–C

1 a Complete the definition of diffusion.

Diffusion is the spreading of the ______________ of a ______________ , or of any substance in solution. This results in a ______________ movement from a region where they are in a ______________ concentration. [4 marks]

b Explain why diffusion happens faster when the temperature is higher.

____

____ [2 marks]

D–C

2 The diagram shows a cell.

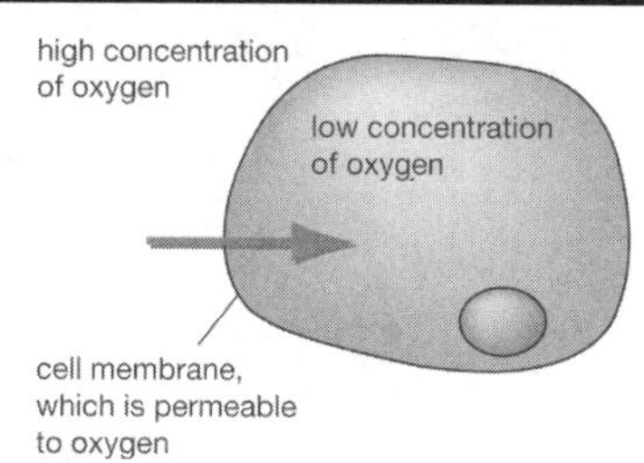

a Name the part of the cell that uses oxygen. (These parts are not shown on the diagram.) ____ [1 mark]

b Explain why oxygen diffuses into the cell more rapidly, when these parts are working faster.

____

____ [2 marks]

B–A*

c The lungs are full of tiny spaces called alveoli, which contain air that has been breathed in. Blood that has travelled around the body is pumped from the heart to the lungs. It flows through capillaries that are wrapped closely around the alveoli. Explain why oxygen diffuses into the blood, and carbon dioxide diffuses out of the blood, in the lungs.

____

____ [2 marks]

Diffusion 2

D–C

3 The cells inside a plant leaf photosynthesise during the daytime. There are tiny holes in the underside of the leaf called stomata. Gases are able to diffuse in and out of the leaf through the stomata.

a Which gas will diffuse **into** the leaf during the daytime?

____ [1 mark]

b Explain why this gas diffuses into the leaf.

____

____ [2 marks]

B–A*

c Suggest how the rate of diffusion of this gas might differ on a bright, sunny day compared with a dull day. Explain your answer.

____

____ [2 marks]

Osmosis 1

Grades

D–C

1 Complete the definition of osmosis.

Osmosis is the ______________________ of ______________________

from a ______________________ to a more concentrated solution through a

______________________ permeable membrane. [4 marks]

D–C

2 The diagram shows a piece of Visking tubing containing sugar solution. The tubing is in a beaker of water.

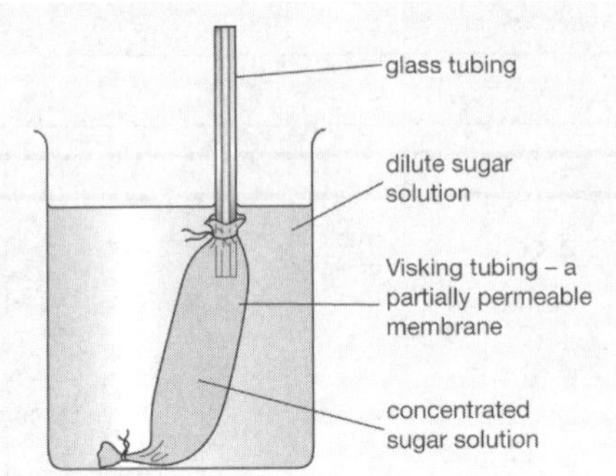

a Explain why the level of liquid in the tubing rises. Your answer should refer to water molecules, sugar molecules, and the Visking tubing membrane.

____

____

____ [3 marks]

B–A*

b Suggest **two** ways in which you could speed up the rate at which the level of the liquid rises in the tube.

1 ____

2 ____ [2 marks]

Osmosis 2

D–C

3 The diagram shows an animal cell that has been placed in water.

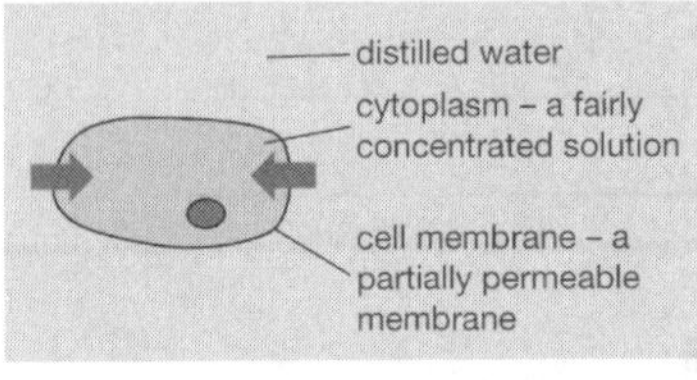

a What will eventually happen to the cell? Explain your answer.

____

____ [2 marks]

b Explain why this does **not** happen to a plant cell in pure water even though water enters the cell by osmosis.

____

____ [2 marks]

B–A*

c Look carefully at this plant cell.
What has made the cell look like this? Explain your answer.

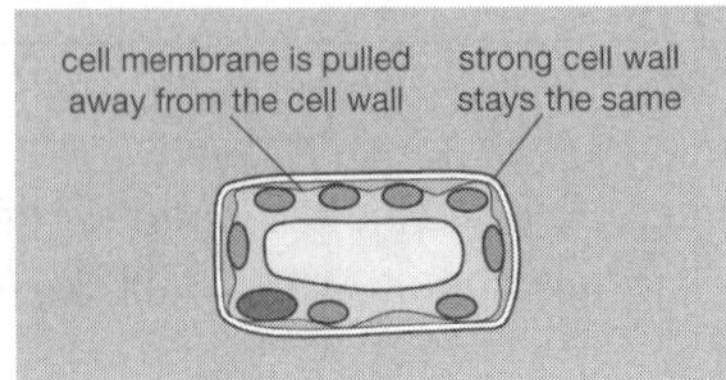

____

____

____

[3 marks]

Grades

Photosynthesis

D–C

1 a Write the word equation for photosynthesis.

______________________________ [2 marks]

b Describe the role of chlorophyll in photosynthesis.

____

______________________________ [2 marks]

B–A*

c The first substance that plants make in photosynthesis is glucose. Name **two** substances that the plant can make from the glucose, and state the function of each.

1st Substance ______________________________

Function ______________________________

2nd Substance ______________________________

Function ______________________________ [4 marks]

Leaves

D–C

2 a This diagram shows the internal structure of a leaf as it would look under a microscope.

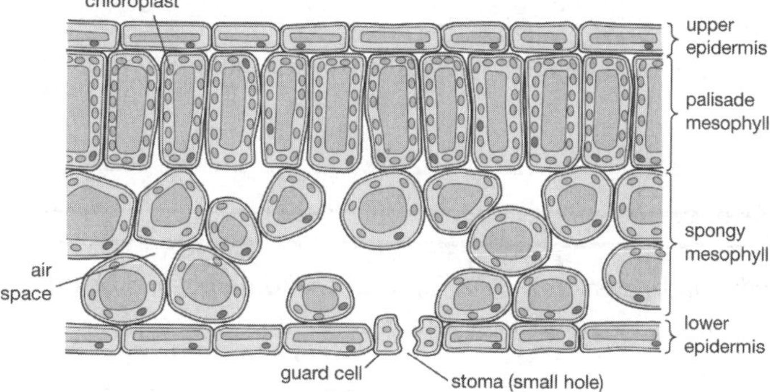

i Explain how the **position** of the palisade mesophyll cells helps them to carry out their function.

____

______________________________ [2 marks]

ii Name **one** type of cell, labelled in the diagram, that cannot photosynthesise.

______________________________ [1 mark]

B–A*

b Many of the cells in the leaf contain stores of starch. Explain why starch is better than glucose for storage purposes.

____

______________________________ [2 marks]

Limiting factors

Grades

D–C

1 Grace wants to find out if a water plant makes oxygen more rapidly when it has more light.

She put a piece of pond weed into a test tube of water. She took it into a room where the light was dim. She counted the number of bubbles that the pond weed gave off in one minute. She did this twice more at the same light intensity. She then repeated this at three other light intensities.

The table shows her results. The higher the number for light intensity, the brighter the light.

Light intensity	**Number of bubbles given off in one minute**			
	1st try	**2nd try**	**3rd try**	**Average**
1	6	9	10	8.3
2	16	18	20	18.0
3	28	27	29	28.0
4	30	32	34	

a Fill in the empty box in the table. [1 mark]

b What was the dependent variable in Grace's experiment? ____________________ [1 mark]

c Grace's teacher says that Grace's results suggest she should let the plant settle down at each new light intensity before beginning to count the bubbles. Explain how Grace's results support this suggestion.

____________________ [2 marks]

d Grace thinks that her results show that light can be a limiting factor for photosynthesis. Do you agree? Explain your answer.

____________________ [2 marks]

B–A*

e Predict what Grace's results would be if she increased the light intensity to 8 units. Explain your answer.

____________________ [2 marks]

B–A*

2 Gardeners often burn paraffin in heaters in a glasshouse where tomatoes are being grown. How might this help the tomatoes to grow faster and produce a better crop?

____________________ [2 marks]

Healthy plants

D–C

3 a Complete the sentences.

Plants obtain mineral ions from the ____________________. They need nitrate ions for producing ____________________ acids, which are then used to form ____________________.

They need ____________________ ions for making chlorophyll. [4 marks]

B–A*

b This graph shows how adding fertiliser containing nitrate to the soil affected the growth of wheat in a field.

i Explain why adding nitrate-containing fertiliser up to 150 kg per hectare increased the yield of grain.

____

[2 marks]

ii Suggest why adding more than 150 kg per hectare decreased the yield of grain.

____

[2 marks]

Grades

B2A DISCOVER MARTIAN LIVING!

Food chains

D–C

1 a Energy enters a food chain as solar energy, in light. In what form is the energy passed along the food chain?

____ [1 mark]

b Explain why green plants only capture a small part of the solar energy that reaches them.

____

____ [2 marks]

B–A*

c The efficiency of energy transfer from light energy to chemical energy in a plant is about 20%.

i Name the part of a cell where this energy transfer takes place.

____ [1 mark]

ii Sunlight containing 1000 units of light energy falls onto a leaf. Calculate the quantity of energy that will be transformed to chemical energy in the leaf.

____ [1 mark]

D–C

2 Mammals and birds keep their body temperature constant even when the temperature around them is low.

Explain why this means that energy transfer from mammals or birds to the next organism in a food chain is not very efficient.

____

____ [2 marks]

Biomass

B–A*

3 The diagram shows a pyramid of biomass.

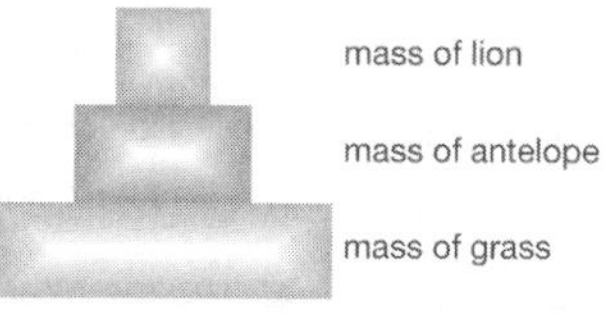

a Write the name of the trophic level next to each step in the pyramid.

____ [1 mark]

b Several hundred caterpillars live on a cabbage plant. They are eaten by small birds, which in turn are eaten by sparrowhawks.

i Sketch a pyramid of biomass for this food chain. [1 mark]

ii Explain why the pyramid is the shape that you have drawn it.

____

____ [2 marks]

Grades

Food production

1 Some farmers rear chickens indoors, in heated sheds. Explain how this can help to reduce energy losses and improve the efficiency of the production of chickens. D–C

____

____ [2 marks]

2 It has been suggested that, if we all became vegetarians, we could produce more food on each hectare of land. Discuss this suggestion. B–A*

____

____

____

____ [4 marks]

The cost of good food

3 Hens that are kept to produce eggs may be kept in three different ways: D–C

Ways of keeping hens	Freedom of hens	Cost of a box of 6 eggs
Free range	Can roam outside	99p
In barns	Can move around freely	73p
In battery cages	Cannot move freely	54p

a Explain why free range eggs are more expensive than barn eggs and battery eggs.

____

____ [2 marks]

b Suggest why many people buy free range eggs even though they are more expensive than battery eggs.

____

____ [2 marks]

4 It has been calculated that growing tomatoes in Britain in winter is more expensive and does more damage to the environment than importing tomatoes from Spain. B–A*

a Explain how transporting tomatoes from Spain could harm the environment.

____

____ [2 marks]

b Suggest why growing tomatoes in Britain in the winter might damage the environment even more than this.

____

____ [2 marks]

Death and decay

B–A*

1 Karen and Joanna did an investigation to compare the rate of decay in two different areas of soil. Area A was inside a wood. Area B was on a flower bed.

They cut some graph paper into two 10 cm x 10 cm squares. They place one piece of paper on the soil in area A, and the other in area B. They weighed the paper down with pebbles. They looked at the paper after six weeks, and counted how many of of the 1 cm x 1 cm squares on the paper had decayed.

They found that 45 of the 1 cm x 1 cm squares had disappeared from the paper in area A, and 34 of the squares had disappeared in area B.

a How many small 1 cm x 1 cm squares were there on each piece of graph paper at the start of the experiment? ____________________ [1 mark]

b State the percentage of paper that had decayed in each area after six weeks.

Area A ______________ Area B ______________ [1 mark]

c What caused the paper to decay?

____ [1 mark]

d Suggest reasons for the results that the students obtained.

____

____ [2 marks]

e i What was the independent variable in this experiment?

____ [1 mark]

ii What was the dependent variable?

____ [1 mark]

f How could the students have made their results more reliable?

____ [1 mark]

Cycles

D–C

2 a Complete the sentences.

Organisms such as earthworms, which eat dead leaves and other plant remains, are called ________________ feeders. They help to recycle the materials in the plant remains, so that they become available to other members of the ________________ of organisms in the ecosystem. For example, they release some of the carbon in the leaves back into the air, in the form of carbon dioxide, by the process of ________________. [3 marks]

b Which of the organisms in a food web may be eaten by detritus feeders?

____ [1 mark]

c Explain why detritus feeders are said to be **consumers** in a food chain.

____

____ [2 marks]

The carbon cycle 1

B–A*

1 a The Martian atmosphere is mostly carbon dioxide. If pressurised, it would become suitable for plant life. Suggest what is most likely to be the limiting factor for photosynthesis on Mars if plants were growing in a pressurised atmosphere.

______________________________ [1 mark]

b Explain why humans living on Mars would need to grow plants inside the enclosed environments where they would live.

____

______________________________ [2 marks]

The carbon cycle 2

D–C

2 The diagram shows parts of the carbon cycle. The numbers indicate how many billions of tonnes of carbon pass along each pathway in one year.

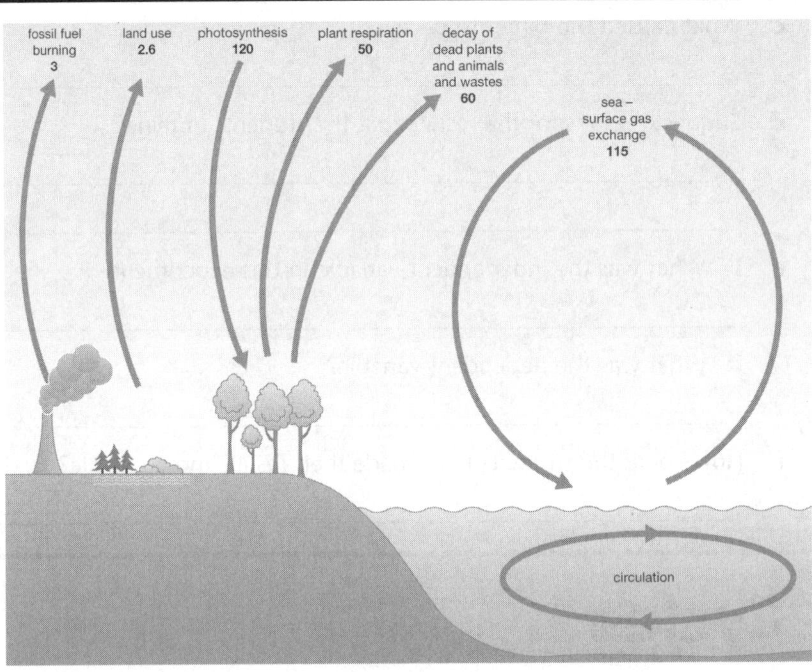

a Explain how decay of dead plants and animals releases carbon dioxide into the air.

____

______________________________ [2 marks]

b Calculate the difference between the quantity of carbon removed from the air by photosynthesis by land plants and the quantity returned to the air by plant respiration.

______________________________ [1 mark]

c Explain how this supports the idea that deforestation could contribute to global warming.

____

______________________________ [3 marks]

d 'Land use' refers to the disturbance of soil, which contains a lot of carbon locked up in the form of carbon compounds. Using what you know about decomposers, suggest how this carbon got into the soil.

____

______________________________ [2 marks]

e Suggest what is meant by **sea surface gas exchange**.

____

______________________________ [2 marks]

B2a revision checklist

I know:

what animals and plants are built from

- ☐ animal cells and plant cells have a membrane, cytoplasm and a nucleus; plant cells also have a cell wall and may have a vacuole and chloroplasts
- ☐ in multicellular organisms, different cells are specialised for different functions
- ☐ the chemical reactions inside cells are controlled by enzymes

how dissolved substances get into and out of cells

- ☐ diffusion is the net movement of particles of gas, or substances dissolved in a solution, from a region of high concentration to a region of lower concentration
- ☐ oxygen required for respiration passes through cell membranes by diffusion
- ☐ osmosis is the diffusion of water molecules through a partially permeable membrane

how plants obtain the food they need to live and grow

- ☐ green plants use chlorophyll to trap light energy from the Sun to photosynthesise
- ☐ leaves are specially adapted for photosynthesis – they can be broad, flat, thin and have lots of stomata
- ☐ the rate of photosynthesis is affected by light intensity, carbon dioxide concentration and temperature
- ☐ mineral salts in the soil are used to make proteins or chlorophyll; lack of a mineral ion results in a deficiency symptom in a plant

what happens to energy and biomass at each stage in a food chain

- ☐ energy passes along food chains but some energy is lost at every stage
- ☐ the shorter the food chain, the less energy is lost
- ☐ the mass of biomass at each stage in a food chain is less than it was at the previous stage; this can be shown in a pyramid of biomass
- ☐ reducing energy loss increases the efficiency of food production
- ☐ decomposers and detritus feeders feed on dead organisms and their waste

Enzymes – biological catalysts

D–C

1 Complete these sentences.

Enzymes are biological ____________________. They are ____________________ molecules.

Each kind of enzyme only works on a particular kind of ____________________ , which fits perfectly into a fold in the enzyme called the ____________________ site. [4 marks]

D–C

2 This graph shows how temperature affects the activity of an enzyme.

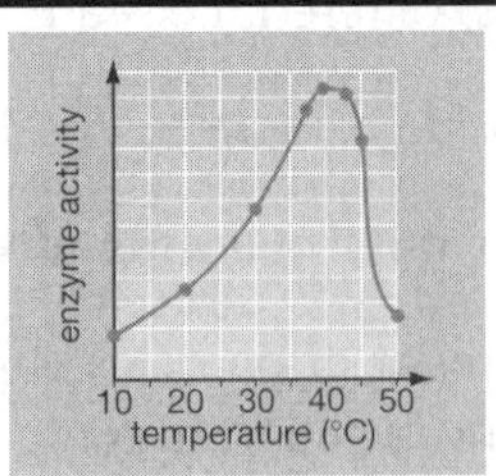

a Using what you know about rates of reaction, explain why the activity of the enzyme increases as the temperature is increased from 10 °C to 20 °C.

____

____ [3 marks]

b Lucy says that the enzyme stops working at high temperatures because it has been killed. Explain why Lucy is wrong, and give the correct explanation for the shape of the graph at temperatures above 40 °C.

____

____ [3 marks]

Enzymes and digestion

D–C

3 The diagram shows the sites of production of some of the enzymes that help with digestion.

mouth
salivary glands
oesophagus
liver
stomach
gall bladder
pancreas
small intestines
large intestine
appendix
rectum
anus

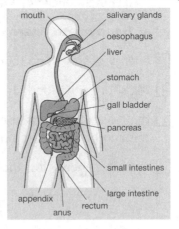

a Name **two** organs, shown on the diagram, where amylase is produced.

1 ____________________ 2 ____________________

[2 marks]

b What does amylase do?

____

____ [1 mark]

B–A*

c Protease enzymes and hydrochloric acid are produced in the stomach. Sketch a graph, with pH on the *x* axis, to show how you would expect pH to affect the activity of the protease enzyme. [3 marks]

d The enzymes that are found in the small intestine require a pH that is neutral or slightly alkaline. Explain how bile helps to provide these conditions.

____

____ [2 marks]

Grades

Enzymes at home

D–C

1 a Explain how proteases in biological washing powders can help to remove blood stains from clothes.

____

______________________________ [2 marks]

b Many of the enzymes that are used in biological washing powders have been obtained from bacteria that are adapted to live in hot springs, at temperatures up to 80 °C. Suggest why these enzymes are especially useful in washing powders.

____

______________________________ [2 marks]

c When biological washing powders were first introduced, some people found that they harmed their hands. Now, in most biological washing powders, the enzymes are inside tiny particles surrounded by a harmless covering.

Suggest how biological washing powders could harm people's hands if the enzymes were not enclosed like this.

____

______________________________ [2 marks]

Enzymes and industry

B–A*

2 Read this information about a new variety of potato.

> Potatoes contain starch. A new, genetically modified, breed of potato has been produced that contains genes for making the enzymes amylase, maltase and isomerase.
>
> The enzymes begin to work when the temperature reaches about 40 °C, so the GM potatoes produce fructose when they are cooked. The cooked potatoes can be mashed and used as a source of fructose in the food industry.

a Explain what is meant by **genetically modified**.

____

______________________________ [2 marks]

b Complete this flow diagram to show how fructose is produced in the GM potatoes, by writing the names of the missing enzymes.

starch ⟶ maltose —maltase⟶ glucose ⟶ fructose

[2 marks]

c Fructose is twice as sweet as sucrose. Explain why this makes fructose useful in the food industry.

______________________________ [2 marks]

d Suggest why the potatoes do not make fructose until they are heated to 40 °C.

______________________________ [1 mark]

e Discuss why some people might not want to eat food made from fructose made by GM potatoes.

______________________________ [2 marks]

Respiration and energy

Grades

D–C

1 a Name the waste product of aerobic respiration.

______________________________ [1 mark]

b In which cells is this waste product made?

______________________________ [1 mark]

c Sperm cells contain a large number of mitochondria. With reference to respiration, explain why this is so.

____

______________________________ [2 marks]

Removing waste: lungs

D–C

2 The table shows the approximate composition of inhaled air and exhaled air.

Name of gas	In inhaled air (approximate %)	In exhaled air (approximate %)
Carbon dioxide	0.04	4
Oxygen	21	
Nitrogen		79
Water vapour	Variable	High

a Complete the table by writing in values for the percentage of oxygen in exhaled air and the percentage of nitrogen in inhaled air. [2 marks]

b Explain why there is less oxygen in exhaled air than in inhaled air.

____

______________________________ [2 marks]

B–A*

c Using what you know about diffusion, explain why there is still quite a lot of oxygen in the air that we breathe out.

____

____

____

______________________________ [4 marks]

d Predict what would happen to the concentration of carbon dioxide in exhaled air when a person is exercising vigorously. Explain your prediction.

____

______________________________ [2 marks]

Grades

Removing waste: liver and kidneys

D–C

1 The kidneys filter the blood and allow some of the substances to pass out of the body in urine. The table shows the concentrations of six substances in the blood and in urine.

Substance	**Found in blood plasma (%)**	**Found in urine (%)**
Water	92	95
Amino acids	0.05	0
Proteins	8	0
Glucose	0.1	0
Salt	0.37	0.6
Urea	0.03	2

a Explain the difference between the concentration of urea in blood plasma and in urine.

______________________________ [1 mark]

b Suggest why the kidneys do not allow glucose to pass out of the body in urine.

____

______________________________ [2 marks]

c Salt is made up of sodium ions and chloride ions. The kidneys are able to adjust the quantity of these ions that are lost in urine. Suggest how this can help to keep the ion content of the body constant.

____

______________________________ [2 marks]

Homeostasis

D–C

2 When there is too much water in the body, the kidneys excrete large amounts of dilute urine. When there is not enough, they excrete small amounts of concentrated urine.

a State **two** ways in which the body loses water, other than in urine.

1______________________________

2______________________________ [2 marks]

b Explain why a person's kidneys produce less urine on hot days than on cold days.

____

______________________________ [2 marks]

c State **two** other factors, apart from water, that are kept constant inside the body.

1______________________________

2______________________________ [2 marks]

B–A*

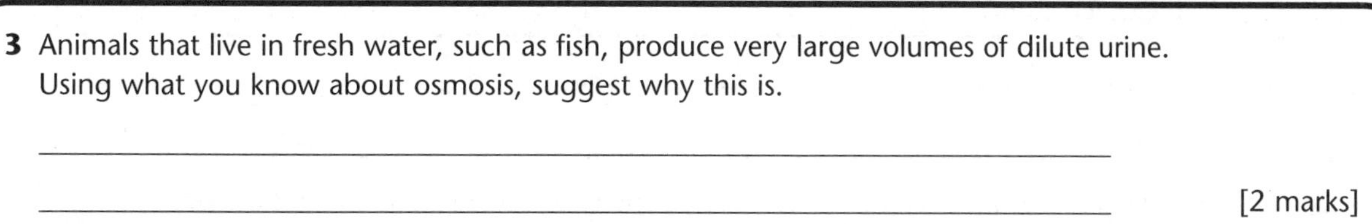

3 Animals that live in fresh water, such as fish, produce very large volumes of dilute urine. Using what you know about osmosis, suggest why this is.

____

______________________________ [2 marks]

Keeping warm, staying cool

D–C

1 a Suggest why cells cannot work properly if the core body temperature gets too high.

[2 marks]

b Describe how the body monitors temperature.

[2 marks]

B–A*

c The diagram shows the appearance of the skin when the body is too hot.

i Explain how sweating helps the body to lose heat.

[2 marks]

ii What do the arterioles supplying the skin capillaries with blood do when the body is too hot?

[1 mark]

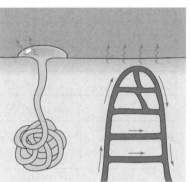

iii Explain how this helps the body to lose heat.

[2 marks]

Treating diabetes

D–C

2 Andrew has diabetes. His body does not produce insulin.

a Describe what would happen to Andrew's blood glucose concentration if he ate a meal containing a lot of sugar.

[2 marks]

b Explain your answer to part **a**.

[1 mark]

c Andrew injects himself with insulin each day. What else can he do to control the concentration of glucose in his blood?

[1 mark]

D–C

3 In 1922, F. G. Banting and C. H. Best published a research paper in which they described experiments that they had done on dogs that had had their pancreases removed. They were testing their hypothesis that the pancreas produced something that reduced blood glucose concentration. This is a summary of one of their experiments.

Time	**Treatment**	**Blood glucose concentration**	**Other results**
10.30 am	None	0.35%	
11.00 am	Injection of glucose given	0.40%	Glucose was excreted in the urine for the next four hours
3.00 pm	Injection of pancreas extract given	Rapidly fell to 0.09%	

a Explain why the dog's blood glucose concentration rose to 0.40% at 11.00 am.

[2 marks]

b Explain why glucose was excreted in the dog's urine between 11.00 am and 3.00 pm.

[2 marks]

c Did Banting's and Best's results support their hypothesis? Explain your answer.

[2 marks]

Cell division – mitosis

D–C

1 This cell contains one pair of chromosomes.

a The cell divides by mitosis. Complete the diagrams to show the chromosomes in the two daughter (new) cells produced after the cell has completed its division.

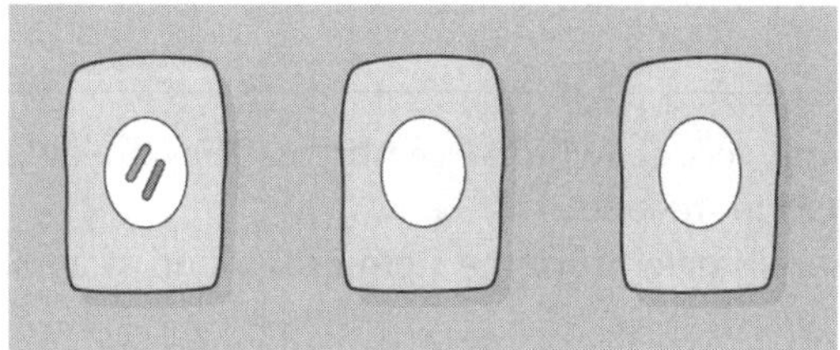

[2 marks]

b Choose the best word to complete this sentence.

The new cells that are produced by mitosis are ______________ identical to the parent cell. [1 mark]

B–A*

2 a Humans have two sets of 23 chromosomes. How many copies of each gene are there in a skin cell?

____ [1 mark]

b Name the kind of cell division that takes place in a skin cell in order to heal a cut.

____ [1 mark]

c Cell division is usually controlled by certain genes. Ultraviolet light can damage these genes in skin cells. Explain what could happen as a result.

____

____ [2 marks]

Gametes and fertilisation

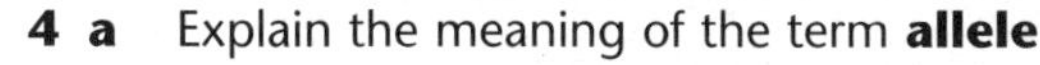

D–C

3 Sperm cells and egg cells are gametes.

a Explain why gametes must have only 23 chromosomes each.

____ [1 mark]

B–A*

b i Name the type of division that is used when a cell divides to form gametes.

____ [1 mark]

ii Outline what happens when a cell divides in this way.

____

____

____ [3 marks]

D–C

4 a Explain the meaning of the term **allele**.

____ [1 mark]

b Explain how alleles can cause variation in the characteristics of offspring that are produced by sexual reproduction.

____

____ [3 marks]

Stem cells

Grades

1 a Explain the meaning of these words. D–C

i Stem cell

____ [1 mark]

ii Differentiation

____ [1 mark]

b Stem cells can be obtained from very early human embryos. They are able to divide to form every kind of cell in the body.

i Explain how embryo stem cells differ from the stem cells found in adult bone marrow.

____

____ [2 marks]

ii Describe how the use of embryo stem cells might be able to treat conditions such as paralysis after the spinal cord has been damaged.

____

____ [2 marks]

2 Parkinson's disease is a condition in which some of the cells in the brain die. These cells are specialised to make a substance called dopamine. Lack of dopamine causes symptoms such as loss of control over movement. B–A*

Suggest how stem cells could one day be used to treat Parkinson's disease.

____

____ [2 marks]

Chromosomes, genes and DNA

3 This diagram shows a cell, and some of the contents of its nucleus. D–C

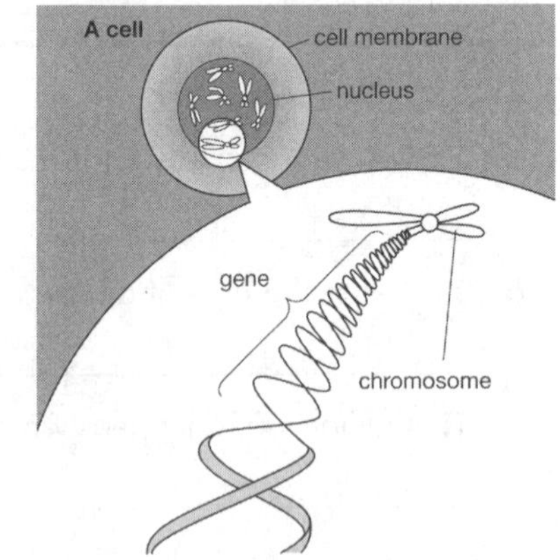

a Name the chemical that chromosomes are made of.

____ [1 mark]

b Each chromosome contains many genes. Describe how genes control the function of a cell. B–A*

____ [1 mark]

c The functions of many different human genes are now known, and it is possible to find out which varieties of these genes a person has in their cells.

i Suggest **one** advantage of this.

____ [1 mark]

ii Suggest **one** disadvantage.

____ [1 mark]

Inheritance

D–C

1 In pea plants, height is controlled by a single gene. The gene has two alleles, **T** and **t**. A plant with genes **T** and **t** is tall.

a Which allele is dominant? Explain your answer.

____

______________________________ [2 marks]

b Write down the alleles that a short plant has.

______________________________ [1 mark]

B–A*

2 In dalmatian dogs, the allele for black spots, B, is dominant to the allele b for brown spots.

a What are the alleles of a homozygous black dog?______________ [1 mark]

b What colour spots does a homozygous recessive dog have?______________ [1 mark]

c What alleles are contained in each sperm of a brown-spotted dog?______________ [1 mark]

d Construct a genetic diagram to show the puppies you would expect to get if a female, homozygous dog with black spots was crossed with a male dog with brown spots.

[3 marks]

How sex is inherited

D–C

3 Which determines a child's sex – its mother's egg, or its father's sperm? Explain your answer.

____

____

______________________________ [3 marks]

D–C

4 This diagram shows the chromosomes in a person's cells. Is the person male or female? Explain your answer.

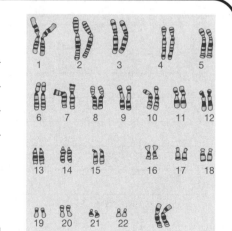

1 2 3 4 5
6 7 8 9 10 11 12
13 14 15 16 17 18
19 20 21 22

____

____

____

______________________________ [2 marks]

Inherited disorders

Grades

D–C

1 Huntington's disease is a disorder of the nervous system. A person with only one allele for Huntington's disease has this disorder.

Explain why it is not possible for a child to inherit Huntington's disease unless one of her parents has this disorder.

____

____

____ [3 marks]

B–A*

2 Angela and Sam have two children. Their son has cystic fibrosis. Neither Angela nor Sam has cystic fibrosis and their daughter does not have it either.

a Using the symbols F and f for the alleles, draw a genetic diagram to explain how Angela and Sam had a son with cystic fibrosis, even though they do not have it themselves.

[4 marks]

b When their daughter grows up, could she have a child with cystic fibrosis? Explain your answer.

____

____

____ [3 marks]

DNA fingerprinting

D–C

3 A mother wants to know which of two possible men is her child's father. DNA fingerprints are made of the mother, child and the two men. Their DNA fingerprints are shown in the diagram.

Which man is the child's father? Explain how you worked it out.

____

____

____

____

____

____

____

____

____

[4 marks]

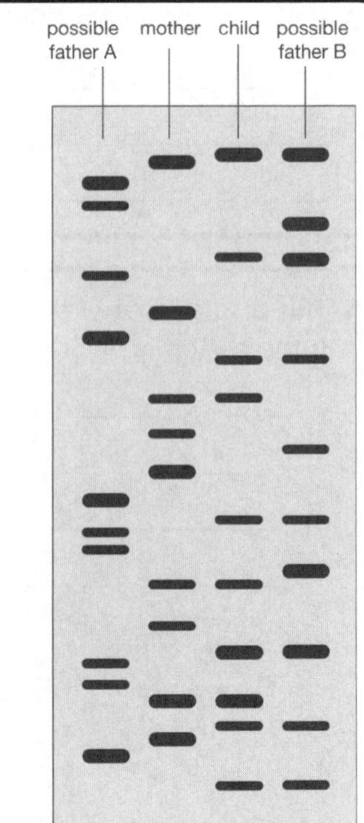

B2b revision checklist

I know:

what enzymes are and what their functions are

- ☐ enzymes are proteins that act as biological catalysts, speeding up chemical reactions
- ☐ each enzyme works at an optimum temperature and pH
- ☐ high temperatures or extremes of pH denature enzymes by affecting the shape of their active sites
- ☐ they are involved in respiration, photosynthesis, protein synthesis and digestion
- ☐ enzymes are used in washing powders and in industry

how our bodies keep internal conditions constant

- ☐ blood sugar levels are controlled by the pancreas, which makes insulin to bring down blood sugar levels
- ☐ waste products, e.g. carbon dioxide and urea, must be removed from the body
- ☐ sweating cools the body down and helps to maintain a steady body temperature
- ☐ if core body temperature is too high, blood vessels supplying the skin capillaries dilate so more blood flows through capillaries and more heat is lost
- ☐ if core body temperature is too low, blood vessels supplying the skin capillaries constrict to reduce the flow of blood through capillaries; muscles may shiver

some human characteristics show a simple pattern of inheritance

- ☐ some inherited characteristics are controlled by a single gene
- ☐ different forms of a gene are called alleles; in homozygous individuals the alleles are the same, in heterozygous individuals they are different
- ☐ how to construct/interpret a genetic diagram; and how to predict/explain the outcome of crosses between individuals for each possible combination of dominant and recessive alleles of the same gene
- ☐ in mitosis each new cell has the same number of identical chromosomes as the original
- ☐ sex chromosomes determine the sex of the offspring (male XY, female XX)
- ☐ stem cells can specialise into many types of cells

Atomic structure 2

D–C

1 Lithium is a metal in Group 1 of the periodic table.

a Describe the structure of a lithium nucleus.

____

____ [2 marks]

B–A*

b Three reactions that result in changes to the atom are shown in the table. Complete the table by filling in the missing information.

Type of reaction	Ionisation	Nuclear fission	Nuclear fusion
Change in mass of nucleus (increases/decreases/ stays the same)			
Change to the atom	Loss or gain of electrons		Two nuclei join together

[5 marks]

Electronic structure

D–C

2 The table shows some information about how electrons are arranged in a sodium atom.

a Complete the table to show the same information for magnesium and fluorine.

Periodic table element	23 Na 11 sodium	24 Mg 12 magnesium	19 F 9 fluorine
Number of electrons	11	12	
Electron arrangement	Na	Mg	F
Notation	2, 8, 1		

[5 marks]

b Sodium loses an electron to form a stable ion. What is the electronic structure of a sodium ion?

____ [1 mark]

D–C

3 The following compounds occur naturally in calcite and dolomite ore.

$CaCO_3$ **compound in calcite ore** $CaMg(CO_3)_2$ **compound in dolomite ore**

a What is the name of each compound?

Compound in calcite ____

Compound in dolomite ____ [2 marks]

b Explain how the compound in dolomite ore forms.

____

____ [2 marks]

Mass number and isotopes

D–C

1 This question is about the periodic table.

a Lithium and sodium are both in Group 1 of the periodic table. Give **one similarity** and **one difference** between the electronic configurations of atoms of lithium and sodium.

Similarity ____________________________________

Difference ____________________________________ [2 marks]

B–A*

b Chlorine has a mass number of 35.5. Explain why the mass number of chlorine is not a whole number.

____

____

____________________________________ [3 marks]

Ionic bonding

D–C

2 The diagram shows what happens when a calcium atom reacts with chlorine atoms.

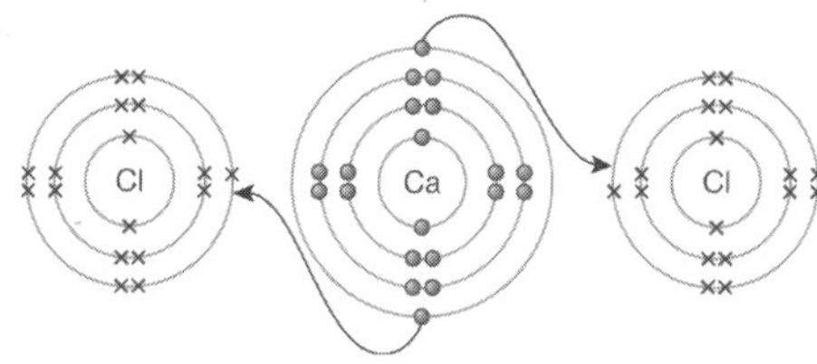

a What is the symbol for the calcium ion that forms? Draw a ring around the **correct** answer.

Ca **Ca^+** **Ca^{2+}** **Ca^-** **Ca^{2-}** [1 mark]

b What is the formula for calcium chloride?

Formula ____________________________________ [1 mark]

D–C

3 Car batteries contain a solution of an ionically bonded compound dissolved in water. The solution conducts electricity.

a Describe how ionic compounds conduct electricity when they are dissolved in water.

____

____________________________________ [2 marks]

b If a car battery was filled with pure water, the battery would not work. Explain why.

____

____________________________________ [2 marks]

Grades

Ionic compounds

D–C

1 The diagram shows how the ions in sodium chloride are arranged when it is molten.

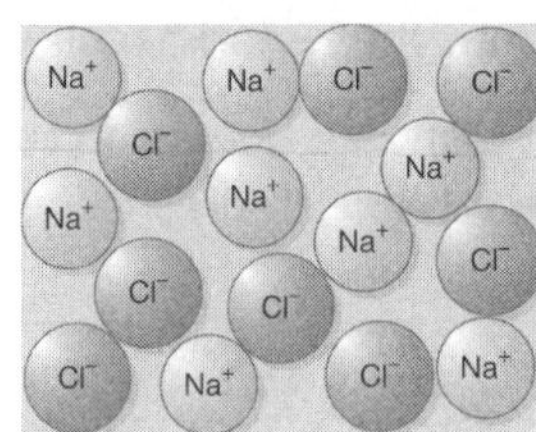

a Use the diagram to explain why sodium chloride conducts electricity when it is molten.

____

____ [2 marks]

b Give **one** reason why sodium chloride cannot conduct electricity when it is a solid.

____ [1 mark]

B–A*

2 The diagram shows how electricity can be used to plate silver onto jewellery.

a Put a cross on the diagram to show where ions **gain** electrons.

b During this process, silver atoms (**Ag**) lose electrons to form silver ions (**Ag^+**)

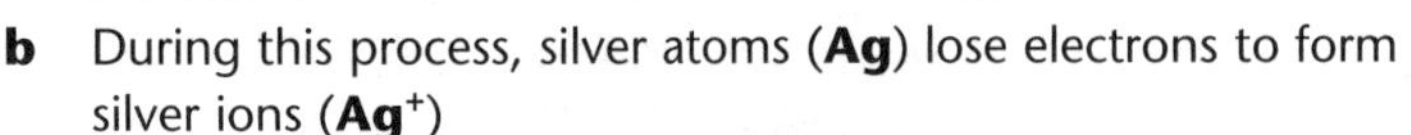

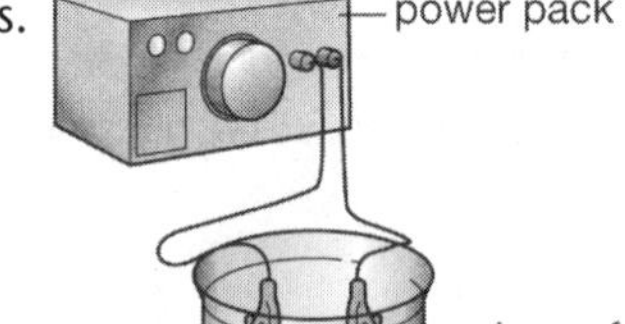

Explain what happens to a silver atom when it forms a silver ion.

____

____ [2 marks]

Covalent bonding

D–C

3 Look at the diagrams to show the bonding in chlorine and oxygen. Only the outer shell electrons are shown.

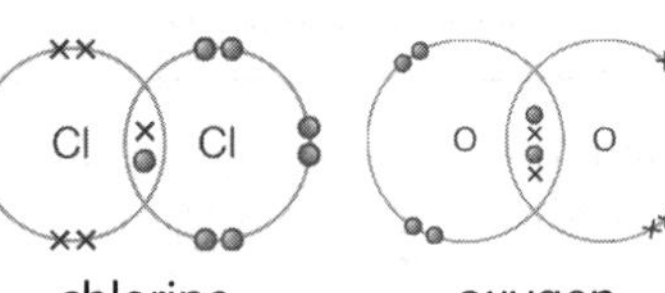

a The bond between the chlorine atoms is a **single** bond. What type of bond joins the oxygen atoms together?

____ [1 mark]

b Explain how you can tell.

____ [1 mark]

c Explain why the oxygen atoms are **not** joined together by a single bond.

____

____ [2 marks]

B–A*

d The two chlorine atoms in each chlorine molecule are very difficult to break apart yet chlorine is a gas at room temperature. Use this information to explain the difference between **bonds** and **intermolecular forces**.

____

____ [2 marks]

Simple molecules

B–A*

1 Perfumes are made by dissolving fragrant oils in a solvent. The table shows some information about possible solvents to use in a new perfume.

Name of solvent	**Methoxyethane**	**Ethanol**	**Water**	**Naphthol**
Solubility of fragrant oil in the solvent	Very soluble	Soluble	Not soluble	Very soluble
Boiling point of solvent	7 °C	78 °C	100 °C	295 °C

a Which of the solvents has the strongest intermolecular forces?

Name of solvent ____________________

Explain your reasoning.

____________________ [1 mark]

b Complete the sentences about the solvents.

i Methoxyethane is not a suitable solvent to make perfume because

____________________ [1 mark]

ii Ethanol is the best solvent for making perfume because ____________

____________________ [1 mark]

iii Naphthol is not a suitable solvent to make perfume because ____________

____________________ [1 mark]

c Suggest one **other** property that needs to be tested to check that the solvents are suitable to use to make perfumes.

____________________ [1 mark]

Giant covalent structures

D–C

2 Diamond and graphite are both macromolecules that contain carbon atoms.

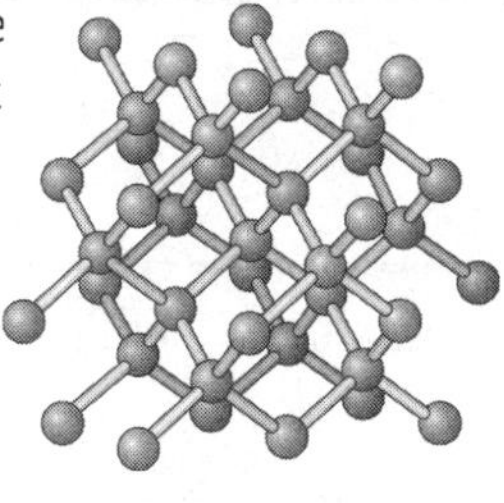

diamond

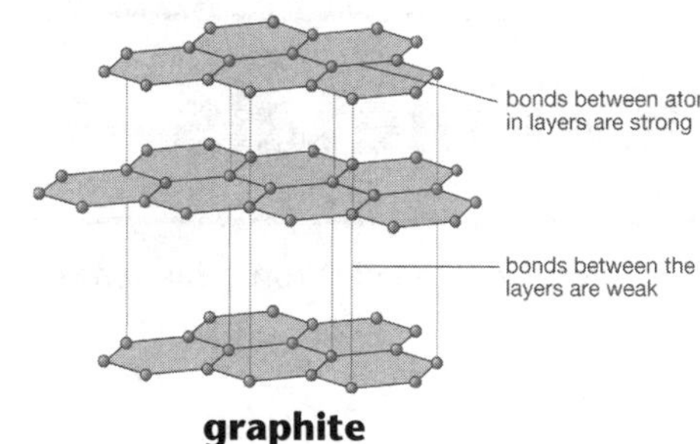
bonds between atoms in layers are strong
bonds between the layers are weak

graphite

a Compare the diagram of graphite to the diagram of diamond. Explain why graphite is soft and slippery but diamond is not.

____

____

____________________ [3 marks]

B–A*

b Graphite is used to strengthen polymers to use to build aeroplanes. One reason that graphite can be used in this way is because it is lightweight and has a low density. Use the structure of graphite to suggest why graphite has a low density.

____________________ [1 mark]

c The bonds between the layers are formed by delocalised electrons. What does the word **delocalised** mean?

____________________ [1 mark]

Grades

Metals

1 The diagram shows the structure of a metal. D–C

a Label the diagram. [2 marks]

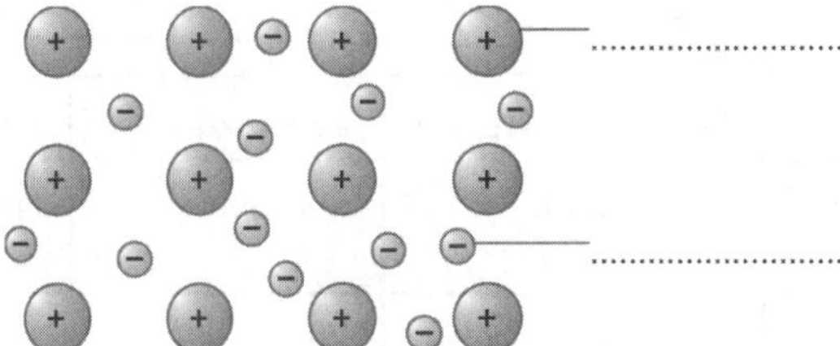

b Use the diagram to help you to explain what happens when a metal conducts electricity.

______________________________ [1 mark]

2 Read the article about mercury. B–A*

> **MERCURY**
> Mercury used to be called 'quicksilver' because it is a liquid at room temperature. Liquid mercury forms small drops that change shape and flow quickly over surfaces. Like other metals, mercury transfers both electrical and thermal energy due the delocalised electrons in its structure.

a Explain what happens to the structure of mercury when it changes shape.

______________________________ [1 mark]

b The article states that mercury 'transfers both electrical and thermal energy'.

i Explain what this statement means.

____

______________________________ [2 marks]

ii How do the delocalised electrons help the transfer of energy through metals?

______________________________ [1 mark]

Alkali metals

3 The diagrams show the electron arrangements in some atoms. B–A*

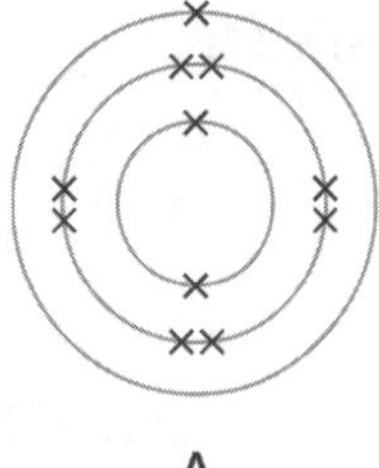

A

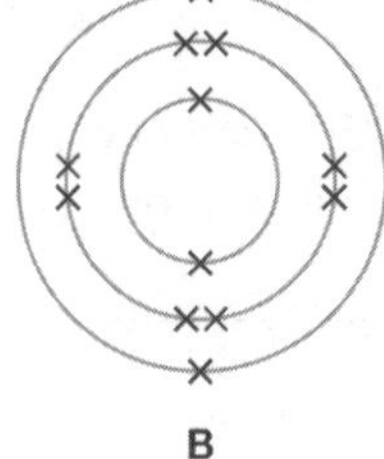

B

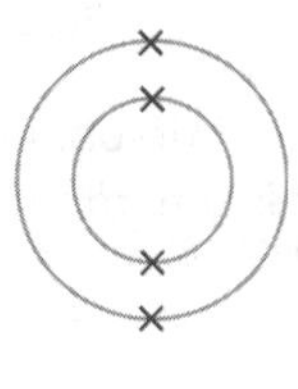

C

a All the atoms are metals. Explain how the electron arrangements show this.

______________________________ [1 mark]

b Which atom is of a metal from Group 1? ______________

Explain how you can tell. ______________ [2 marks]

c Which atom has the highest number of protons in its nucleus? ______________

Explain how you can tell. ______________ [2 marks]

Grades

Halogens

D–C

1 The diagram shows the electron arrangement in a fluorine atom.

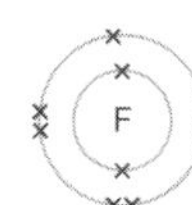

fluorine atom

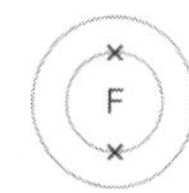

fluoride ion

a Complete the second diagram to show the electron arrangement in a fluoride **ion**. [1 mark]

b What is the charge on a fluoride ion?

______________________________ [1 mark]

c Which of the following elements will react with fluorine gas to produce fluoride ions? Draw a ring around the **correct** answer.

chlorine **oxygen** **sodium** **helium** [1 mark]

D–C

2 Candice bubbles some fluorine gas through solutions of some compounds of halogens. **Table 1** shows her results.

Solution of compound	Does it react with fluorine gas?	Colour after reaction
Sodium chloride	Yes	Very pale green
Sodium bromide	Yes	
Sodium iodide	Yes	

Table 2

Solution of compound	Does it react with bromine?
Sodium fluoride	
Sodium chloride	
Sodium iodide	

a **i** Complete **Table 1** by filling in the missing colours. [2 marks]

ii Give the names of the two products of the reaction between fluorine and sodium chloride.

______________________________ [1 mark]

b Candice carries out a similar experiment. This time she adds bromine instead of fluorine to three halogen compounds. Complete **Table 2** to show whether or not a reaction will happen in each case. [2 marks]

Nanoparticles

B–A*

3 Nanoparticles can be used to carry molecules of drugs into the human body. At the moment such treatments are expensive but in the future they may become routine.

a What does the term **nanoparticle** mean?

____

______________________________ [2 marks]

b Explain how the structure of nanoparticles enables them to carry other molecules.

____

______________________________ [2 marks]

c Companies paid millions of pounds for the research that was needed to develop nanoparticles. Explain how those companies benefit now that nanoparticles have been developed.

____

____

______________________________ [3 marks]

Grades

Smart materials

D–C

1 Smart materials have a wide range of uses.

A, **B**, **C** and **D**, are different types of smart materials:

A shape changing alloy
B photochromic polymer
C thermochromic material
D electroluminescent material

What type of smart material is used for each of the following uses?

Spectacle frame made from smart material type ______________________

Dials in a car light up when a current passes through them. They contain smart material type ______________________

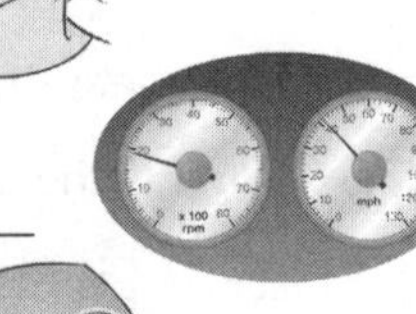

The plastic visor goes dark when it is very sunny. It is made from smart material type ______________________

The pictures on this mug change when it is filled with a hot drink. It is coated with smart material type

____

[4 marks]

Compounds

D–C

2 The table shows some information about hydrogen, oxygen and water.

Name	Hydrogen	Oxygen	Water
Formula	H_2	O_2	H_2O
Properties	Very flammable gas	Very reactive gas	Unreactive liquid

a Explain why hydrogen is an **element** but water is a **compound**.

______________________ [2 marks]

B–A*

b A mixture of hydrogen and oxygen react together explosively when they are lit. The product is water. Explain the differences between a **mixture** and a **compound** using this information.

____

____

______________________ [3 marks]

c When hydrogen reacts with oxygen, the volume of hydrogen used up is always twice the volume of oxygen that is used up. Use ideas about the formula of water to explain why.

____

______________________ [2 marks]

B–A*

3 Complete these equations.

a N_2 + ________ H_2 → ________ NH_3

b Mg + ________ HNO_3 → $Mg(NO_3)_2$ + ________

[4 marks]

Grades

Percentage composition

D–C

1 Harvey works out the percentage of elements in some compounds. He tests the following compounds:

O_2 CO CO_2 O_3 CH_4

Here are his results.

Percentage of oxygen in compound A = 57%
Percentage of oxygen in compound B = 72%

D–C

a Which of the compounds that Harvey tests are most likely to be compound A and compound B?

Compound A ____________ Compound B ____________ [2 marks]

b Explain your reasoning.

____ [1 mark]

Moles

D–C

2 Sulfur dioxide is a pollutant gas that comes from power stations. It is made when sulfur impurities in fuel burn. Some people are protesting against the amount of sulfur dioxide coming from a power station.

Sulfur in fuel produces twice the amount of sulfur dioxide when it burns.

a Laura does some calculations to check to see if the sign is true. Complete her calculations by filling in the spaces.
(atomic mass of sulfur, S: 32; atomic mass of oxygen, O: 16)

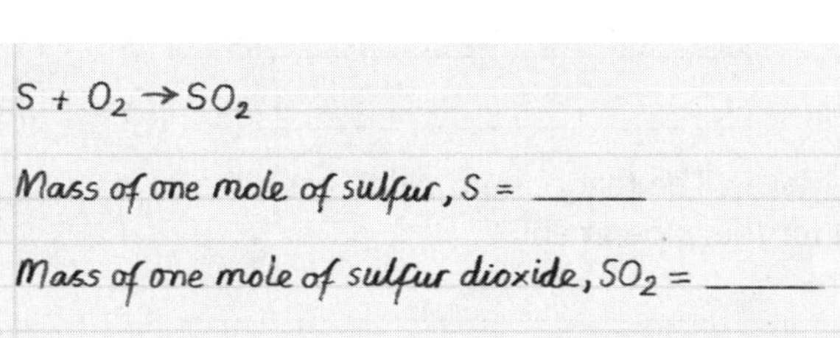
$S + O_2 \rightarrow SO_2$

Mass of one mole of sulfur, S = ______

Mass of one mole of sulfur dioxide, SO_2 = ______

[2 marks]

b Laura calculates the percentage mass of sulfur in sulfur dioxide. Complete her calculation.

Atomic mass of sulfur, S = 32

Atomic mass of oxygen, O = 16

Percentage mass of sulfur in SO_2 =

______________ %

____

____ [2 marks]

B–A*

c A power station burns 80 tonnes of coal each day. Eighty tonnes of coal contains 320 kg of sulfur. What mass of sulfur dioxide will 320 kg of sulfur produce when it burns?

____

____ [2 marks]

Yield of product

Grades

D–C

1 Harry works in a factory that makes a chemical compound that is added to painkiller tablets.

Our method for making this compound has a very high yield. This is important because it makes the process more profitable for the company. We work to make our actual yield as near as possible to our theoretical yield so that we have a high atom economy for the process.

a Give **two** reasons why having a high yield makes the process more profitable.

____

____

______________________________ [2 marks]

b What does Harry mean when he says the process has a high **atom economy**?

____

______________________________ [2 marks]

B–A*

c Explain the difference between **theoretical** and **actual** yield.

____

______________________________ [2 marks]

D–C

2 Rose makes some magnesium oxide by burning magnesium in air. She writes an equation for the reaction.

$$Mg \quad + \quad \tfrac{1}{2}O_2 \quad \rightarrow \quad MgO$$

a Eve uses 2.4 g of magnesium. Calculate the theoretical yield of magnesium oxide from 2.4 g of magnesium.
(relative atomic mass, Mg: 24; relative atomic mass, O: 16)

____

____

______________________________ [3 marks]

b Rose weighs the magnesium oxide she makes. The magnesium oxide has a mass of 3.0 g. Calculate the percentage yield for the experiment.

____

______________________________ [2 marks]

Reversible reactions

D–C

3 Ethanol (**C_2H_5OH**) is a solvent used to make perfumes. It can be made by reacting ethene gas (**C_2H_4**) with water. The reaction is reversible.

a Write the equation for the reaction. Include the sign to show that the reaction is reversible.

______________________________ [1 mark]

b Use the substances in the equation to answer these questions.

i Give the fomula of **one** substance from the equation that has the **same** empirical formula and molecular formula.

______________________________ [1 mark]

ii Give the fomula of **one** substance from the equation that has a **different** empirical formula and molecular formula.

______________________________ [1 mark]

Equilibrium 1

D–C

1 Ethanol is made on an industrial scale by reacting ethene gas with water vapour in a reactor.

ethene + water vapour ⇌ ethanol

Conditions
High temperature: 300 °C
High pressure: 70 atm
Catalyst: phosphoric acid

a Which of the reaction conditions will change the rate of reaction?

A all of them

B only temperature

C temperature and pressure

D only the catalyst [1 mark]

b When the gases come out of the reactor the ethanol is separated out. The rest of the gases are recycled back into the reactor again. Explain why this is necessary.

____

____ [2 marks]

B–A*

c Complete the table to show what effect a change in each of the reaction conditions would make to the yield.

Change to reaction condition	**Effect on yield (increases/ decreases/stays the same)**
A higher temperature	
A higher pressure	
Using less catalyst	

[3 marks]

Haber process

D–C

2 The diagram shows a plan of the Haber process for making ammonia.

$N_2 + 3H_2 \rightleftharpoons 2NH_3$

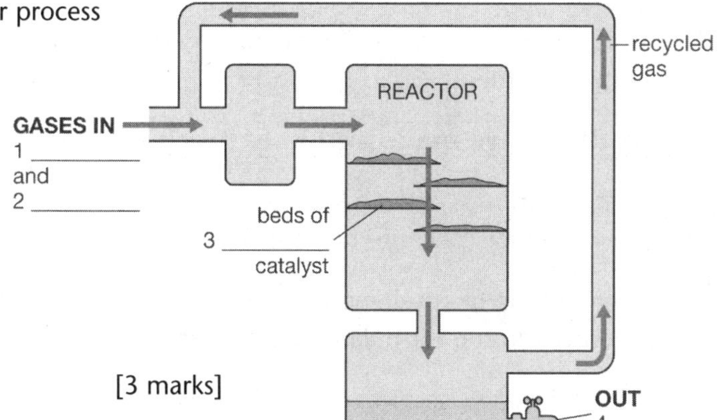

a Complete the diagram by adding the missing labels. [3 marks]

b A pressure of 200 atmospheres is used for the process.

i Give **two** advantages of running the process at a high pressure.

____

____ [2 marks]

ii Give **one** reason why it is not practical to run the process at even higher pressures.

____ [1 mark]

c The reaction to make ammonia is an equilibrium reaction. Gases from the end of the process are recycled back to the start. Explain why this is necessary.

____

____ [2 marks]

C2a revision checklist

I know:

how sub-atomic particles help us to understand the structure of substances

- ☐ an element's mass number is the number of protons plus the number of neutrons in an atom
- ☐ an element's atomic number is the number of protons in an atom
- ☐ electrons arranged in shells around the nucleus have different energy levels and this can be used to explain what happens when elements react and how atoms join together to form different types of substances
- ☐ how to write balanced chemical equations for reactions
- ☐ metals consist of giant structures of atoms arranged in a regular pattern, with delocalised electrons

how structures influence the properties and uses of substances

- ☐ ionic bonding is the attraction between oppositely charged ions
- ☐ ionic compounds are giant lattice structures with high melting points that conduct electricity when molten or dissolved
- ☐ non-metal atoms can share pairs of electrons to form covalent bonds
- ☐ giant covalent structures are macromolecules that are hard, have high melting points but do not conduct electricity
- ☐ simple molecular elements (e.g. oxygen) and compounds (water) have weak intermolecular forces
- ☐ delocalised electrons in metals and graphite enable them to conduct heat and electricity as they are free to move through the whole structure
- ☐ nanoparticles are very small structures with special properties because of their unique atom arrangement

how much can we make and how much we need to use

- ☐ the relative masses of atoms can be used to calculate how much to react and how much we can produce, because no atoms are gained or lost in chemical reactions
- ☐ the percentage of an element in a compound can be calculated from the relative masses of the element in the formula and the relative formula mass of the compound
- ☐ how to calculate chemical quantities involving empirical formulae, reacting masses and percentage yield and how to balance symbol equations
- ☐ high atom economy (atom utilisation) is important for sustainable development and economic reasons
- ☐ reversible reactions carried out in a 'closed' system will eventually reach equilibrium

Grades

Rates of reactions

D–C

1 Salma reacts magnesium ribbon with an acid. She uses a gas syringe to measure the volume of gas that is given off after 50 seconds.

a Draw a fully labelled diagram to show how Salma sets up her experiment.

[3 marks]

B–A*

b Salma carries out her experiment three times with three different concentrations of acid. Here are her results.

Acid	Volume of gas given off in 50 s	Rate of reaction (cm^3 of gas per second)
Concentration A	40	0.8
Concentration B	160	3.2
Concentration C	90	

i Calculate the rate of reaction for the acid with concentration C.

____

______________________________ [2 marks]

ii Which acid is the most concentrated? Explain your reasoning.

____

______________________________ [2 marks]

c Salma keeps a record of the concentrations of acid she uses. She knows that she needs to control all factors to make sure her results can be repeated. List **three other** factors that Salma needs to control.

1__________ 2__________ 3__________ [3 marks]

Following the rate of reaction

D–C

2 Callum is investigating the rate of reaction of marble chips with acid. He knows that the reaction produces carbon dioxide gas. He set up this experiment to find out the rate of reaction.

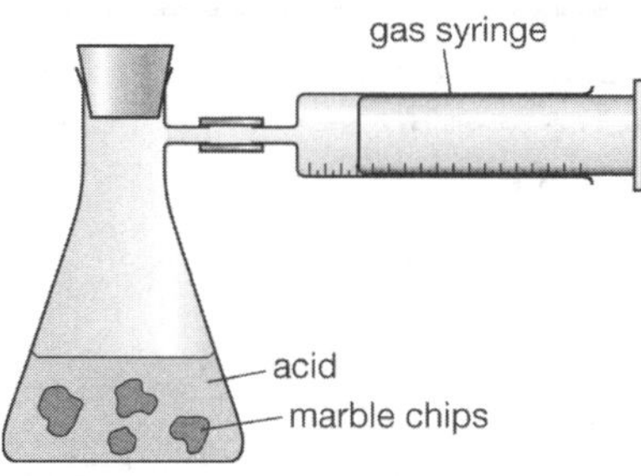

a What **two** measurements will Callum need to make to follow the rate of reaction?

1 ______________________________

2 ______________________________ [2 marks]

b How will Callum be able to tell when the reaction has stopped?

______________________________ [1 mark]

c What **other** change could Callum measure to follow the rate of reaction? Draw a ring around the **correct** answer.

colour **mass** **volume of acid** **size of chips** [1 mark]

Collision theory

D–C

1 The diagram in **box A** shows acid particles reacting with a large lump of zinc.

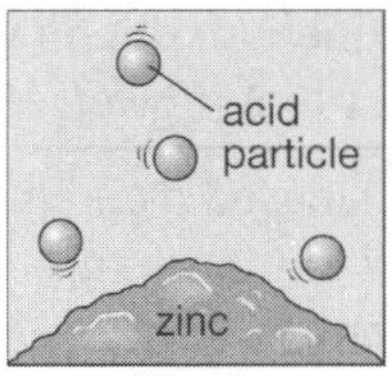

acid particle

zinc

Box A

a How would the diagram change if

i the acid was more concentrated?

____

______________________________ [1 mark]

ii the temperature was increased?

______________________________ [1 mark]

b When zinc powder is used instead of lumps of zinc the reaction is faster. Use ideas about particle collisions to explain why.

____

______________________________ [2 marks]

B–A*

2 Gases behave very differently in the ozone layer. Air at the surface is at a much higher pressure than gases in the ozone layer. Some parts of the ozone layer are very hot. These conditions alter the concentrations of gases.

a What happens to the concentration of a gas if its **temperature** is increased? Explain your reasoning.

____

______________________________ [2 marks]

b What happens to the concentration of a gas if its **pressure** is increased? Explain your reasoning.

____

______________________________ [2 marks]

Heating things up

D–C

3 Carbon dioxide is made when copper carbonate reacts with hydrochloric acid. David investigates the reaction. The diagram shows how he set up his experiment.

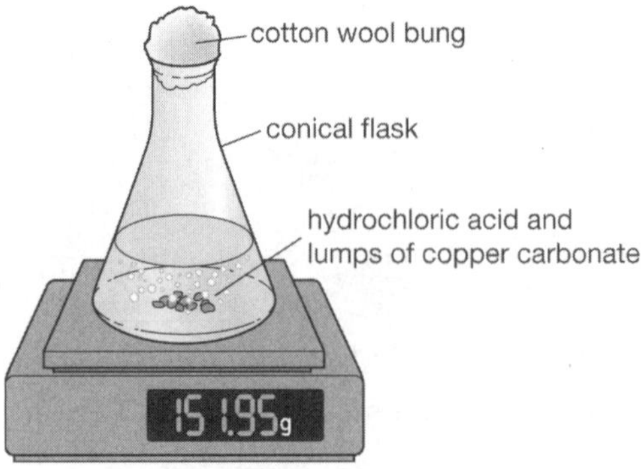

cotton wool bung

conical flask

hydrochloric acid and lumps of copper carbonate

151.95g

a What would you expect to happen to the mass of the flask during the experiment? Explain your reasoning.

____

____

____

______________________________ [3 marks]

B–A*

b David carries out his experiment at different temperatures. He calculates that the rate of reaction doubles with each 10 °C rise in temperature. Use ideas about collisions and energy to explain why this happens.

____

____

______________________________ [3 marks]

Grades

Grind it up, speed it up

D–C

1 Isabel carries out some experiments. She reacts zinc with acid in a flask. She uses different conditions. The table shows the conditions she uses.

Experiment	A	B	C	D
Size of zinc lumps	Large lumps	Small lumps	Large lumps	Small lumps
Concentration of acid	Low	High	High	Low
Temperature	20 °C	50 °C	50 °C	20 °C

a Explain why the zinc pieces dissolve fastest in experiment B.

____

______________________________ [3 marks]

B–A*

b In experiment A, hydrogen was produced at a rate of 4.5 cm^3/s.

i Suggest a value for the rate of reaction in experiment D. Explain your reasoning.

Rate ______________________________

Reason ______________________________ [3 marks]

ii Suggest a value for the rate of reaction in experiment B. Explain your reasoning.

Rate ______________________________

Reason ______________________________ [3 marks]

Concentrate now

D–C

2 Leo does an experiment (experiment 1). He reacts some acid with some marble chips. He measures the change in mass as the reaction happens. He stops measuring when no more carbon dioxide is given off. Leo repeats the experiment. He uses the same amount of acid and marble chips, but this time he adds some water to the flask (experiment 2).

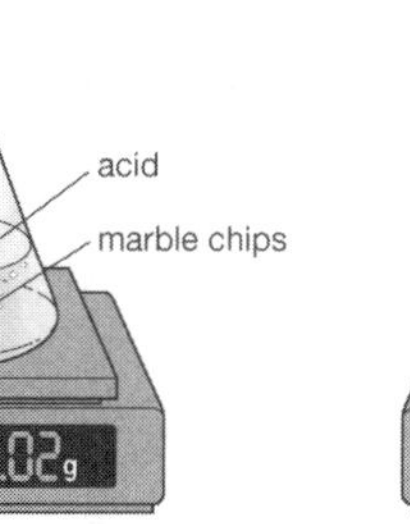

at the start

151.95g

at the end

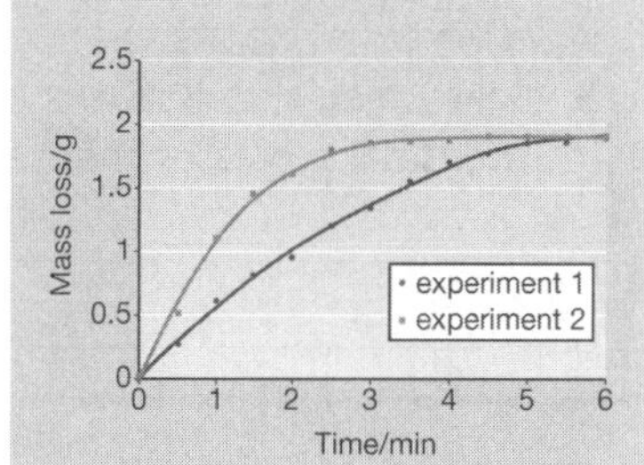

a What has been used up when the reaction stops? ______________________________ [2 marks]

b The reactions for both experiments were fast at the beginning and then slowed down. Explain why.

____

______________________________ [3 marks]

c Explain why the graphs for experiment 1 and experiment 2 are different shapes.

______________________________ [2 marks]

d In each experiment, Leo uses 25 cm^3 of 2 mol/dm^3 hydrochloric acid.

i What volume of water would Leo need to add to the acid so that the reaction rate was exactly half as fast? ______________________________

ii Explain your reasoning. ______________________________

______________________________ [2 marks]

Catalysts

Grades

1 Hydrogen peroxide (H_2O_2) is used to bleach hair. It slowly decomposes to form water and oxygen gas. D–C

a Complete the equation for this reaction.

$2H_2O_2 \rightarrow$ ______________________ [2 marks]

b Sophie did some experiments using hydrogen peroxide. B–A*
In experiment 1 she used hydrogen peroxide with no catalyst.
In experiment 2 she added 0.2 g of manganese dioxide catalyst.

The diagram shows the activation energy for experiment 1. Draw and label a second curve on the diagram to show the activation energy for experiment 2.

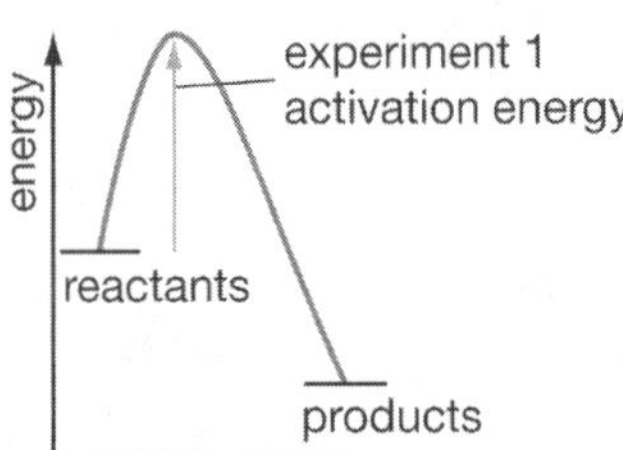

[2 marks]

c Sophie filtered off the manganese dioxide at the end of experiment 3 and dried it thoroughly before weighing it. What mass of manganese dioxide should Sophie have? Explain your reasoning.

Mass ____

Reason ____ [2 marks]

2 The Haber process is used to make ammonia for making fertilisers. Iron is used as catalyst in the reaction. D–C

a The iron catalyst is packed into the reaction tower. It is only replaced about every 15 years. Why does the iron catalyst have such a long lifetime?

____ [1 mark]

b Many metal catalysts are very expensive. Give **two** reasons why they make industrial processes cheaper in the long run.

1 ____

2 ____ [2 marks]

Energy changes

3 **a** The table shows some information about temperature and energy changes during three reactions. Complete the table by filling in the boxes. D–C

Reaction	Temperature change	Exothermic or endothermic?
Dissolving ammonium nitrate in water	Decreases	
Adding zinc powder to copper		Exothermic
Adding magnesium ribbon to an acid	Increases	

[2 marks]

b Which of these reactions is likely to be exothermic? Draw a ring around the **two correct** answers.

oxidation **thermal decomposition** **neutralisation** **evaporation** [2 marks]

Equilibrium 2

B–A*

1 Hydrogen for making ammonia can be made by reacting methane gas with steam. The reaction is reversible.

$$CH_4 + H_2O \underset{\text{exothermic}}{\overset{\text{endothermic}}{\rightleftharpoons}} CO + 3H_2$$

a What is the name of the other product of the reaction? ____________________ [1 mark]

b Would a low or a high temperature give the highest yield of hydrogen? Explain your reasoning.

____

____________________ [2 marks]

c In practice, higher temperatures are almost always used for industrial processes. Explain why.

____________________ [1 mark]

Industrial processes

D–C

2 The flow chart shows how ammonia is made in the Haber process.

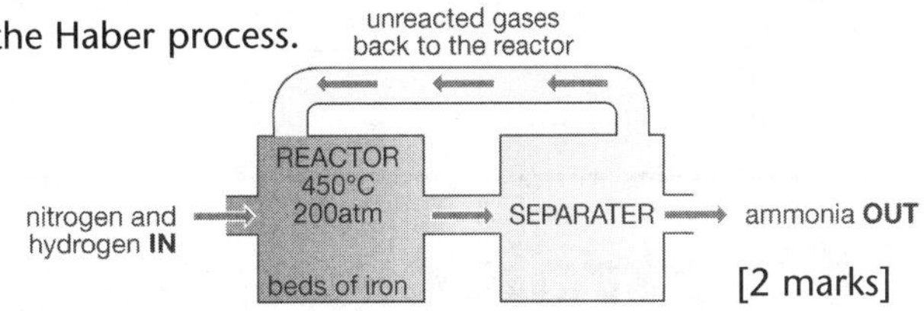

a Write a word equation for the reaction that happens in the reactor. Include the sign to show that the reaction is reversible.

____________________ [2 marks]

b **i** Why does the reactor contain iron?

____________________ [2 marks]

ii The iron is in the form of very small pieces rather than large lumps. Explain why.

____________________ [2 marks]

c The unreacted gases travel back to the reactor through a pipe.

i Name the **two** gases that travel through the pipe.

____________________ and ____________________ [1 mark]

ii Give a reason why this reaction never gives 100% yield.

____________________ [1 mark]

D–C

3 The table shows some data about the yield of ammonia from the Haber process under different conditions.

Temperature (°C)	Pressure (atm)	Yield
400	200	40%
400	100	25%
500	100	15%

a Predict the yield of ammonia at 400 °C and 50 atm pressure.

____________________ % [1 mark]

B–A*

b In industry, the optimum conditions for the Haber process are 450 °C and 200 atm pressure.

i Explain why this temperature is better to use than either a lower or a higher temperature.

____________________ [2 marks]

ii Explain why a relatively high pressure is chosen.

____________________ [2 marks]

iii Use your answers to explain what the term 'optimum conditions' mean.

____________________ [2 marks]

Grades

Free ions

D–C

1 The diagram shows the arrangement of ions in solid lithium chloride.

a Explain how the structure of lithium chloride is typical of an ionic solid.

Li+ Cl− Li+ Cl−
Cl− Li+ Cl− Li+
Li+ Cl− Li+ Cl−
Cl− Li+ Cl− Li+

[2 marks]

b Solid lithium chloride does not conduct electricity but a solution of lithium chloride dissolved in water is a good conductor. Explain why.

[2 marks]

c When an electric current passes through a solution of lithium chloride, a greenish gas forms.

B–A*

i What is the name of this gas?

[1 mark]

ii At which electrode does this gas form? Explain your reasoning.

[2 marks]

Electrolysis equations

B–A*

2 Read the article about the electrolysis of molten sodium chloride.

> **ELECTROLYIS OF MOLTEN SODIUM CHLORIDE**
> Sodium chloride is usually electrolysed by passing an electric current through a solution of the salt in water. However, this does not make sodium metal – a gas forms at the negative electrode instead. Electrolysing **molten** sodium chloride is very difficult because it has to be carried out at about 1000 °C. At this temperature, the electricity splits the salt to form sodium metal and chlorine gas.

a **i** Name the gas that forms at the negative electrode during the electrolysis of aqueous sodium chloride solution.

[1 mark]

ii Explain why this gas forms instead of sodium metal.

[1 mark]

b Why must a high temperature be used for the electrolysis of molten sodium chloride?

[2 marks]

c **i** Complete the half equation to show what happens to a sodium ion during the electrolysis of molten sodium chloride.

$Na^{+} \quad \longrightarrow$

[2 marks]

ii Explain why this reaction is an example of a **reduction** reaction.

[1 mark]

Grades

Uses for electrolysis

B–A*

1 The diagram shows what happens when sodium chloride is electrolysed. Complete the labels on the diagram to show the names of each product of the electrolysis and give a use of each.

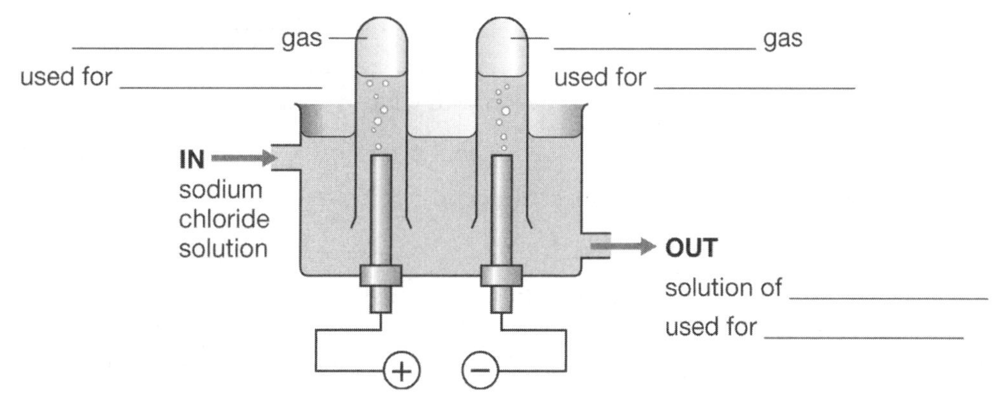

[6 marks]

D–C

2 Copper is purified by electrolysis.

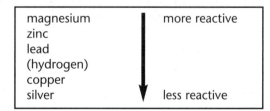

electrode made of pure copper
electrode made of impure copper
solution containing copper ions

a Complete the sentences to explain what happens during the electrolysis.

When copper is purified, both electrodes are made from ____________________ .

A block of impure copper is used as the ____________________ electrode.

Copper ions from the solution form pure copper by ____________________ two electrons.

[3 marks]

B–A*

b Complete the equations to show what happens at each electrode.

Positive electrode	**Negative electrode**
Cu(s) g	Cu^{2+}(aq)

[2 marks]

Acids and metals

D–C

3 The table shows the reactivity of some metals.

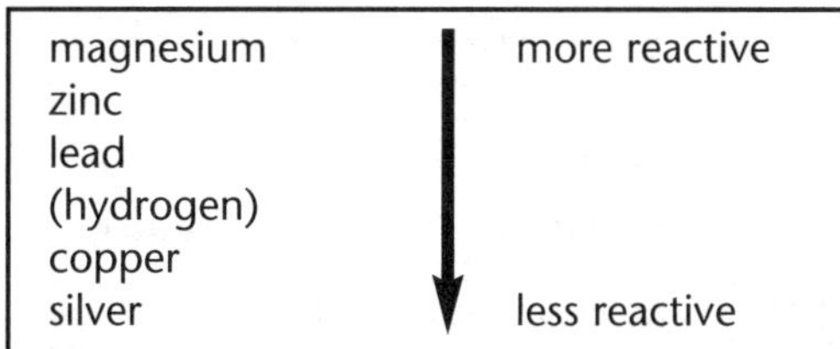

magnesium	more reactive
zinc	
lead	
(hydrogen)	
copper	
silver	less reactive

a Holly has samples of zinc and copper metals and some dilute hydrochloric acid. Explain how she uses these substances to show that zinc is more reactive than hydrogen in the reactivity series but copper is less reactive.

____

____

____ [3 marks]

Making salts from bases

D–C

1 Karim makes some salts. The table shows what substances he uses.

Substances used	**Salt formed**
Zinc and dilute hydrochloric acid	
Copper oxide and dilute sulfuric acid	Copper sulfate

a What is the name of the salt formed when zinc reacts with dilute hydrochloric acid?

______________________________ [1 mark]

B–A*

b **i** Explain why Karim uses copper oxide rather than copper metal to make copper sulfate.

______________________________ [1 mark]

ii Give the name of another copper compound that Karim could add to the acid to make copper sulfate. ______________________________ [1 mark]

Acids and alkalis

D–C

2 Max makes sodium chloride. He puts some sodium hydroxide solution (an alkali) in a flask and adds hydrochloric acid from a burette. He adds litmus solution to the sodium hydroxide to act as an indicator. He adds hydrochloric acid to the sodium hydroxide until he reaches the end point when all the sodium hydroxide has been neutralised. The table shows the litmus at different pHs.

pH below 7	**pH above 7**
red	blue

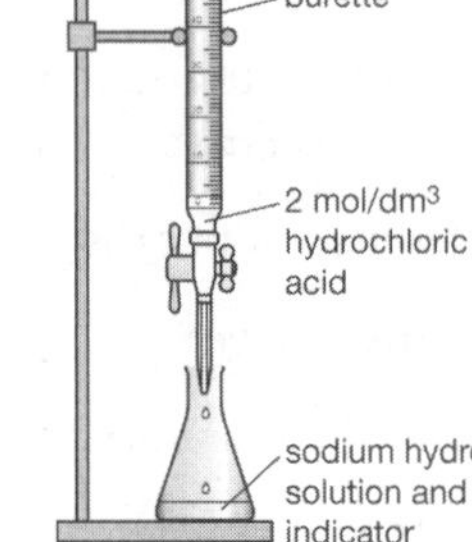

a What colour change will the litmus indictor show at the end point of the experiment?

From ______________________ to ______________________ [1 mark]

b Why could litmus indicator **not** be used to measure the exact pH of the sodium hydroxide?

______________________________ [1 mark]

c Max wants to make a pure solution of sodium chloride without an indicator.

i Describe how he can do this.

____

______________________________ [3 marks]

ii How can Max make sodium chloride crystals from his solution?

______________________________ [1 mark]

B–A*

3 Compounds A, B and C are used in fertilisers.

NH_4NO_3	$(NH_4)_2SO_4$	$K(NH_4)_2PO_4$
compound A	**compound B**	**compound C**

a Explain why all three compounds are suitable to use as fertilisers.

______________________________ [1 mark]

b Name compound A. ______________________________ [1 mark]

c Give the name and fomula of the **acid** that would be needed to make compound B.

______________________________ [2 marks]

d The three most important elements for plant growth are potassium, nitrogen and phosphorus. Which compound contains all three elements?

______________________________ [1 mark]

Grades D–C, B–A*

Neutralisation

D–C

1 The table shows some information about the ions in some solutions.

a Complete the table by filling in the empty boxes. [2 marks]

Solution	Type of positive ion	Type of negative ion
NaOH	Na^+	OH^-
H_2SO_4	H^+	SO_4^{2-}
	Na^+	SO_4^{2-}
HBr	H^+	

b **i** Which solution in the table is an alkali?

______________________________ [1 mark]

ii Explain how you can tell.

______________________________ [1 mark]

B–A*

2 This question is about how ions react during neutralisation reactions. This equation shows the ions produced by lithium hydroxide when it dissolves in water.

$$LiOH(aq) \rightarrow Li^+(aq) + OH^-(aq)$$

a Write a similar equation to show the ions produced by hydrochloric acid, HCl(aq), in water.

______________________________ [2 marks]

b When lithium hydroxide and hydrochloric acid react together, water is formed by the reaction between hydroxide ions and hydrogen ions.

i Write an ionic equation to show the reaction that happens when water forms.

______________________________ [1 mark]

ii Explain why this equation is the same for all neutralisation reactions.

____

______________________________ [3 marks]

iii Give the name of the salt that forms during the reaction.

______________________________ [1 mark]

Precipitation

B–A*

3 Read the article about removing pollutants from water.

> **REMOVING POLLUTANTS FROM WATER**
> Most pollutants can be removed from water by precipitation reactions. Positively charged metal ions are usually removed by reacting them with sodium carbonate or sodium hydroxide. Negative ions containing non-metals can be removed by adding calcium compounds. Once the precipitates are formed, they are easy to remove from the water.

a Give the name of **two** compounds that form when lead ions are removed from the water.

______________________________ [1 mark]

b Give the name of the compound that forms when phosphate ions are removed from the water.

______________________________ [1 mark]

c Explain why the compounds that form are easily removed from the water.

____

______________________________ [2 marks]

C2b revision checklist

I know:

how we can control the rates of chemical reactions

- ☐ the rate of a reaction can be found by measuring the amount of a reactant used or the amount of product formed over time
- ☐ reactions can be speeded up by increasing the: temperature; concentration of a solution; pressure of a gas; surface area of a solid; and by using a catalyst
- ☐ particles must collide with sufficient energy in order to react; the minimum energy required is the activation energy
- ☐ concentrations of solutions are given in moles per cubic decimetre (mol/dm^3); equal volumes of solutions of the same molar concentration contain the same number of particles of solute
- ☐ equal volumes of gases contain the same number of molecules

whether chemical reactions always release energy

- ☐ chemical reactions involve energy transfers
- ☐ exothermic reactions give OUT energy; endothermic reactions take IN energy
- ☐ in reversible reactions, equilibrium is reached at a point when the rate of the reverse reaction balances the rate of the forward reaction
- ☐ the relative amounts of all the reacting substances at equilibrium depend on the conditions of the reaction; this principle is used to determine the optimum conditions for the Haber process

how can we use ions in solutions

- ☐ when molten or dissolved in water, ions in ionic compounds are free to move
- ☐ passing an electric current through an ionic compound breaks it down into its elements: this is called electrolysis
- ☐ at the negative electrode, positively charged ions gain electrons (reduction) and at the positive electrode, negatively charged ions lose electrons (oxidation)
- ☐ electrolysis of sodium chloride solution makes hydrogen, chlorine and sodium hydroxide
- ☐ how to complete and balance supplied half equations for the reactions occurring at the electrodes during electrolysis.
- ☐ metal oxides and hydroxides are bases and react with acids to form salts
- ☐ soluble salts can be made from reacting an acid with a metal or a base and insoluble salts can be made by mixing solutions of ions
- ☐ in neutralisation reactions, H^+ ions from acids react with OH^- ions to produce water

See how it moves!

D–C

1 The distance-time graph shows the motion of a bus.

a Describe the motion of the bus at

i section A ______________________________

____

[2 marks]

ii section B ______________________________

[1 mark]

b In which section did the bus travel fastest? ____________________ [1 mark]

B–A*

c Describe the appearance of the graph if the bus is accelerating.

____

______________________________ [2 marks]

D–C

2 a A coach driver travels 90 km in 2 hours. Calculate his average speed.

__________ km/h [3 marks]

b Explain why the coach does not always travel at this average speed.

____

______________________________ [2 marks]

Speed isn't everything

D–C

3 a What is velocity?

______________________________ [2 marks]

B–A*

b Describe the velocity of a bung on a piece of string as Imran swings it around his head.

____

______________________________ [2 marks]

c What force stops the bung flying off in a straight line?

______________________________ [1 mark]

D–C

4 The diagram shows Simon firing an arrow.

a What is the speed of the arrow just before it leaves the bow?

______________________________ m/s

[1 mark]

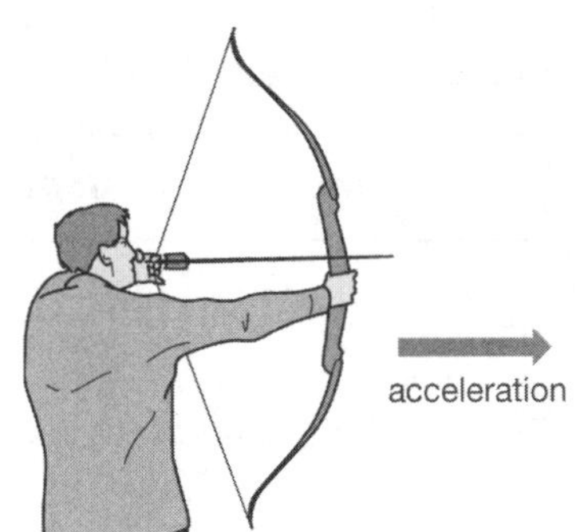

b After 0.5 s, the arrow reaches its top speed of 150 m/s. Calculate the acceleration of the arrow.

______________________________ m/s^2

[3 marks]

Velocity-time graphs

D–C

1 The velocity-time graph shows the motion of Alice on her bicycle.

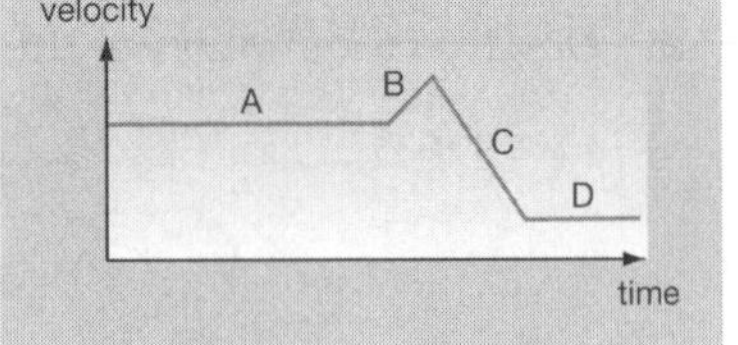

a Describe Alice's motion at

i section A ______________________________ [1 mark]

ii section B ______________________________ [1 mark]

b In which section is Alice travelling slowest? ______________________ [1 mark]

c How could you tell if Alice stopped moving?

______________________________ [1 mark]

B–A*

2 The diagram shows the distance-time graph for a cyclist.

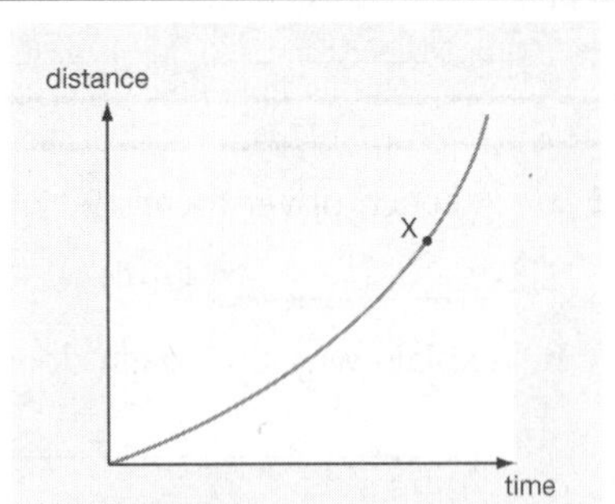

a Describe how you could use the graph to find his speed at any time.

____

____

______________________________ [2 marks]

b Explain how you could find the total distance travelled at any time from the graph.

______________________________ [1 mark]

Let's force it!

D–C

3 a The diagram shows the forces acting on a car moving forwards.

backward force = 400N

forward force = 800N

i What is the size of the resultant force?

______________________________ [1 mark]

ii Describe the motion of the car including its direction.

____

______________________________ [2 marks]

b The driver reduces the force from the engine until is the same size as air resistance. Complete the sentence by choosing the correct word from the box.

balanced	**unbalanced**	**constant**	**changing**

The forces are ______________________________ . [1 mark]

c Describe the motion of the car now. ______________________________

______________________________ [2 marks]

B–A*

d Describe how you could find the resultant force if the forces act in different directions.

____

______________________________ [3 marks]

Force and acceleration

Grades

D–C

1 a Pete is pushing a shopping trolley. He puts more shopping in.
What happens to the force needed to make the trolley move faster?

____ [1 mark]

b The mass of the trolley is 20 kg. What force is needed to make it accelerate by 4 m/s^2? Include the units in your answer.

____

____

____ [3 marks]

B–A*

c Describe how the mass of the trolley could be found using a forcemeter and an accelerometer.

____

____

____ [3 marks]

D–C

2 Kamal is investigating the acceleration caused by different forces. Here is a table of his results.

Force	Acceleration
1 N	2 m/s^2
2 N	4 m/s^2
3 N	6 m/s^2
4 N	7 m/s^2
5 N	10 m/s^2

a Explain whether his results prove that acceleration is proportional to the force used.

____

____

____

[2 marks]

b Explain which one of his results should be repeated.

____

____ [2 marks]

Balanced forces

D–C

3 A van is trying to pull another car out of a ditch. They are not moving.

a What can you say about the size of the force from the van and the size of the force from the car?

____ [1 mark]

b A tractor is used instead of the van. It pulls the car out. How does the force from the tractor compare with the force from the car?

____ [1 mark]

B–A*

c As the tractor accelerates, so does the car. Explain why this makes the car feel heavier for the tractor to pull.

____

____ [2 marks]

Terminal velocity

Grades

D–C

1 The cyclist is travelling at a steady speed. Friction is one force that slows him down.

a Describe how the size of this force changes as he travels faster.

____

[1 mark]

b What is **terminal velocity**?

____

[1 mark]

c The cyclist starts to pedal harder and accelerates until he reaches a faster terminal velocity. Describe how the forces on him change as his speed changes.

____

____

____ [3 marks]

D–C

2 Skydivers open parachutes to help them land safely.

a What force increases when the parachute is open?

____ [1 mark]

b Explain why this helps the parachutist to slow down.

____

____ [2 marks]

B–A*

c Explain how the shape of the parachute can affect the terminal velocity of a skydiver.

____

____ [2 marks]

Stop!

D–C

3 a Stopping distance can be split into two parts. What are these called?

____

____ [2 marks]

b Nicole is taking her first driving lesson. Why will her thinking distance be longer than her driving instructor's?

____ [1 mark]

c On her second lesson, Nicole was very tired. How will this affect her stopping distance?

____

____ [2 marks]

B–A*

d Nicole's car has been fitted with new tyres. Explain how these will help to reduce her stopping distance.

____

____ [3 marks]

Grades

P2A DISCOVER FORCES!

Moving through fluids

D–C

1 Charlotte is comparing the speed that a marble falls through different liquids.

a First she drops the marble in a cylinder of water. Explain why it reaches its terminal velocity.

____

____ [2 marks]

b Next, she drops the marble through thick honey in the same sized cylinder. Explain why the terminal velocity will change.

____

____ [2 marks]

B–A*

c Explain why the density of the honey affects the terminal velocity of the marble.

____ [1 mark]

Energy to move

D–C

2 a Complete the sentences.

A hairdryer is designed to transfer electrical energy as ________________

energy to ________________ energy. [2 marks]

b What is the main energy change in a light bulb?

________________ to ________________ [2 marks]

D–C

3 Greg is driving his electric car. The batteries ran out and the car stopped.

a Explain what has happened to the energy from the battery.

____

____ [2 marks]

b Why did the batteries run out more quickly when he drove the car on a rough surface?

____

____ [2 marks]

B–A*

c Some electric vehicles could use flywheels to increase the efficiency of their batteries. Explain why a flywheel could help reduce energy losses caused when a vehicle brakes.

____

____ [2 marks]

Working hard

Grades

1 Muhammed did 12 J of work when he lifted an apple.

D–C

a How much energy was transferred to the apple?

_______________ J [1 mark]

b What force was he working against when he lifted up the apple?

_______________ [1 mark]

c Muhammed dragged a box of apples 2 metres along the ground. He measured the force needed as 25 N. How much **work** did he do?

_______________ J [3 marks]

B–A*

d Explain why he did more work on the box when he started dragging the box up a slope.

____

_______________ [2 marks]

How much energy?

D–C

2 Complete the spaces in the sentences.

Andrea has a large elastic harness attached to her when she does a bungee jump from the top of a bridge. As she jumps, the _______________ energy she has on the bridge changes into _______________ energy as she falls. At the bottom of the jump, the rope is fully extended and it gains _______________ energy. Eventually she stops moving because all the energy has spread as _______________ energy to the surroundings.

[4 marks]

D–C

3 Ryan is playing rugby. He runs fast holding the ball, giving the ball kinetic energy.

a Explain what is meant by **kinetic energy**.

_______________ [1 mark]

b The ball is kicked upwards. When the ball is at its highest point, what forms of energy does it have?

_______________ [2 marks]

B–A*

c Calculate the kinetic energy the ball has when it travels at 10 m/s. The ball's mass is 0.3 kg.

_______________ [3 marks]

Momentum

D–C

1 a Complete the sentences:

i One example of a scalar quantity is ______________________ . [1 mark]

ii One example of a vector quantity is ______________________ . [1 mark]

b What is the **momentum** of a ball? Its mass is 0.5 kg and it travels at 8 m/s.

____________ kg m/s [3 marks]

D–C

2 The diagram shows two trolleys colliding.

Inelastic collision

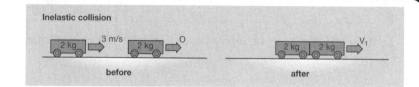

2 kg → 3 m/s 2 kg → O

before

2 kg 2 kg → V_1

after

a Calculate the total momentum before the collision.

____________ kg m/s [3 marks]

b Write down the total momentum after the collision.

______________________________ kg m/s [1 mark]

c Calculate the speed that the joined-up trolley moves off with.

____

______________________________ [3 marks]

Off with a bang!

D–C

3 a Jake blows up a balloon but does not tie a knot in it. Describe how the balloon and air move when he lets it go.

____

____

______________________________ [2 marks]

b What is the total momentum of the balloon before he lets go?

______________________________ [1 mark]

c Describe what happens to the total momentum after he lets go.

______________________________ [1 mark]

d Use the idea of momentum to explain

i why a swimmer moves forward in the water.

____

______________________________ [3 marks]

B–A*

ii why a swimmer spins in the water if they push in the same direction with both arms.

____

______________________________ [3 marks]

Keep it safe

Grades

1 a Explain why a force affects the momentum of an object. D–C

[1 mark]

b A car seat belt is slightly elastic. If the car is in a crash, the belt stretches slightly and the passenger is less likely to be hurt.

i What is momentum of a car that is stopped?

[1 mark]

ii The elastic in the seat belt lets the momentum of the car change over a longer time. Why does this help passengers avoid getting hurt?

[2 marks]

c Describe how a car's crumple zone affects the momentum change of passengers in a head-on collision.

[2 marks]

2 Adam drops his mobile phone onto a concrete path and it breaks. Sarah drops her mobile phone onto a carpet and it does not break. D–C

a What is the momentum of each mobile phone after it lands?

[1 mark]

b Each mobile phone has the same change in momentum.

i Explain why Sarah's phone did not break.

[3 marks]

ii If the momentum of Adam's phone immediately before it landed was 1 kg m/s, and it took 0.02 s to stop, calculate the force it felt. B–A*

[3 marks]

Static electricity

3 Owen rubs a balloon on some cloth. The balloon becomes charged. D–C

a What particles are negatively charged?

[1 mark]

b Explain how rubbing the balloon makes it negatively charged.

[3 marks]

c The negatively charged balloon is held near another negatively charged balloon. Describe what happens.

[2 marks]

Grades

P2A DISCOVER FORCES!

Charge

D–C

1 a Explain how electrical charge behaves differently in electrical conductors and in electrical insulators.

____

____ [2 marks]

b Damien is rubbing a balloon with a piece of woollen cloth.

i Explain how a static charge can build up on the balloon.

____

____

____ [3 marks]

B–A*

ii Explain why the charged up balloon can stick on an uncharged wall.

____

____

____ [3 marks]

D–C

2 a What is a gold leaf electroscope used to detect?

____ [1 mark]

b Explain you can use a gold leaf electroscope to compare a charged up comb and an uncharged comb.

____

____

____ [3 marks]

Van de Graaff generator

D–C

3

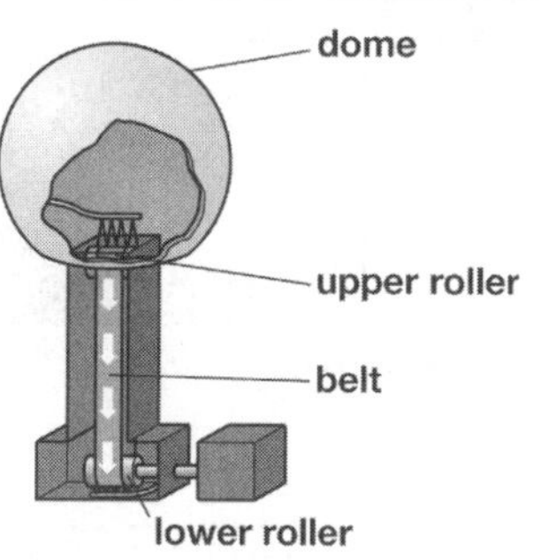

a What type of material are the belt and rollers made of?

____ [1 mark]

b Explain how the belt and rollers create a charge.

____

____ [2 marks]

c Explain why the equipment must stay on an insulated mat.

____ [1 mark]

Sparks will fly!

D–C

1 a During a thunderstorm, clouds lose their electric charge as lightning. Many buildings have lightning conductors. Explain how a lightning conductor can protect the building during a thunderstorm.

____

____

____

____

[3 marks]

b How can you tell if a charged object loses its charge suddenly?

____ [1 mark]

c Why are structures in an oil refinery earthed?

____

____ [2 marks]

D–C

2 A photocopier uses static electricity to transfer negatively charged toner to a sheet of paper. First, the toner passes over a positively charged drum.

a Why does the toner stick to the drum?

____ [1 mark]

b Paper passes over the drum and the toner sticks to the paper. What charge does the paper need to have?

____ [1 mark]

D–C

3 When paint is sprayed from a can it becomes electrically charged.

a Explain why the paint becomes positively charged.

____

____

____

[2 marks]

b The particles spread thinly and evenly on the car's surface. Explain why the particles spread out.

____

____ [2 marks]

B–A*

4 Describe how a smoke precipitator uses electrostatics to reduce pollution.

____

____

____ [3 marks]

P2a revision checklist

I know:

how we can describe the way things move

- ☐ how to calculate the speed of a body from the slope of a distance-time graph
- ☐ how to calculate the acceleration of a body from the slope of a velocity-time graph
- ☐ how to calculate the distance travelled by a body from a velocity-time graph

how we make things speed up or slow down

- ☐ to change the speed of an object, an unbalanced force must act on it
- ☐ forces can add up or cancel out to give a resultant force; when the resultant force is *not* zero, an object accelerates:

  resultant force (N) = mass × acceleration: $F = ma$

- ☐ an object falling through a fluid accelerates until it reaches a terminal velocity, when the resultant force is zero
- ☐ the stopping distance of a car is the thinking distance plus the braking distance; this increases as the speed increases

what happens to the movement energy when things speed up or slow down

- ☐ when a force causes an object to move, energy is transferred and work is done
- ☐ every moving object has kinetic energy that can be transformed into other forms
- ☐ the kinetic energy of a body depends on its mass and its speed:

  kinetic energy = $\frac{1}{2}$ × mass × speed^2

what momentum is

- ☐ every moving object has momentum that depends on its mass and its velocity:

  momentum = mass × velocity

- ☐ momentum has size and direction and is conserved in collisions and explosions
- ☐ how to use the equation:

  $$\textbf{force} = \frac{\textbf{change in momentum}}{\textbf{time taken for the change}}$$

what static electricity is, how it can be used and the connection between static electricity and electric currents

- ☐ rubbing electrical insulators together builds up static electricity because electrons are transferred
- ☐ if an object gains electrons it has a negative charge; if it loses electrons it has a positive charge
- ☐ electrostatic charges can be used in photocopiers, smoke precipitators and paint sprayers
- ☐ when electrical charges move we get an electrical current
- ☐ if potential difference becomes high enough, a spark may jump across the gap between a body and any earthed conductor which is brought near it

Circuit diagrams

Grades

1

1 2 3 4

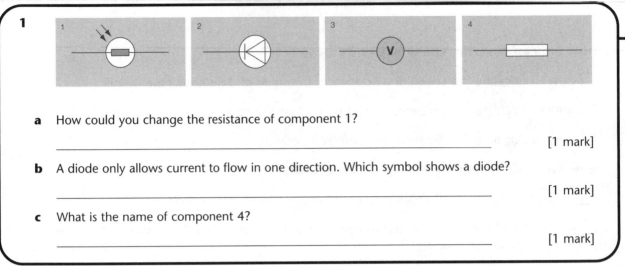

D–C

a How could you change the resistance of component 1?

[1 mark]

b A diode only allows current to flow in one direction. Which symbol shows a diode?

[1 mark]

c What is the name of component 4?

[1 mark]

2 The circuit shows the wiring of a circuit in a reading lamp.

D–C

a What is the name of component X?

[1 mark]

X

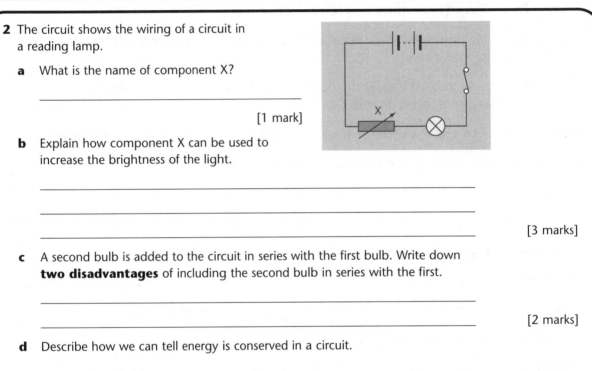

b Explain how component X can be used to increase the brightness of the light.

[3 marks]

c A second bulb is added to the circuit in series with the first bulb. Write down **two disadvantages** of including the second bulb in series with the first.

[2 marks]

d Describe how we can tell energy is conserved in a circuit.

B–A*

[2 marks]

Resistance 1

3 a What is meant by **electrical current**?

D–C

[2 marks]

b How is the resistance in a conductor different from the resistance in an insulator?

[2 marks]

c Give **one** reason why metals are better conductors of electricity than plastics.

[1 mark]

Grades

Resistance 2

D–C

1 a The resistance of a wire depends on collisions inside the wire. Explain what the collision theory of resistance is.

____

____

____ [3 marks]

b Use this idea to explain why different materials have a different resistance.

____ [1 mark]

Ohm's Law

D–C

2 The graph shows the results of an experiment to find out how current changes when the voltage changes in a wire.

a What pattern does the graph show?

____

____

____

____ [2 marks]

b Use the graph to write down the current when the voltage is 6 V.

____ [1 mark]

c Use your answer to part **b** to calculate the resistance of the wire at 6 V.

____

____

____ [3 marks]

d The length and thickness of the wire stay the same. Write down **two** other factors to keep constant for the resistance of the wire to stay the same.

____

____ [2 marks]

e The graph shows the experiment repeated using a filament bulb. Write down **one** reason why this graph is a different shape from the graph in part **a**.

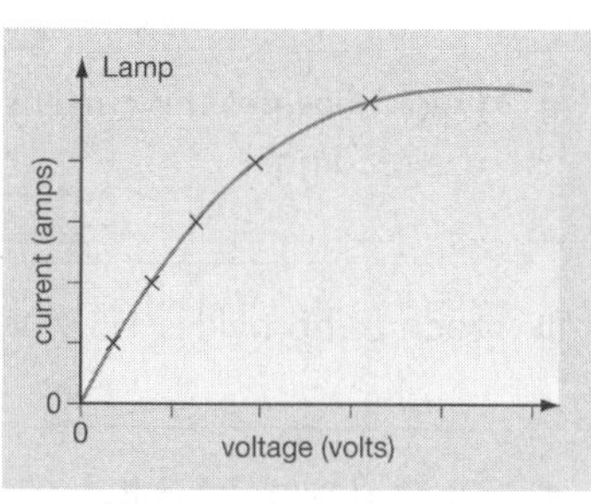

____

____

____

____ [2 marks]

B–A*

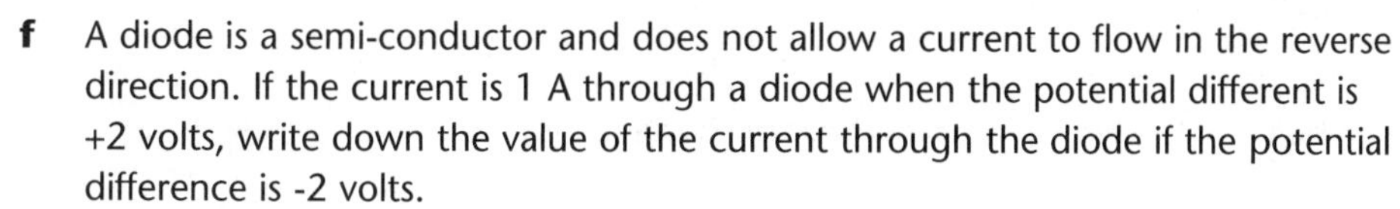

f A diode is a semi-conductor and does not allow a current to flow in the reverse direction. If the current is 1 A through a diode when the potential different is +2 volts, write down the value of the current through the diode if the potential difference is -2 volts.

____ [1 mark]

More components

Grades

D–C

1 a How could you change the resistance of the thermistor?

____

____

[1 mark]

B–A*

b What happens inside the thermistor when it is heated up?

____

______________________________ [2 marks]

c Explain how a digital thermometer takes temperature readings.

____

______________________________ [2 marks]

D–C

d How does the resistance of the light dependent resistor change if a light shines onto it?

____

____

[1 mark]

B–A*

e Explain how automatic lighting systems are designed to switch off at night.

____

______________________________ [2 marks]

Components in series

D–C

2 The circuit shows three bulbs wired in series.

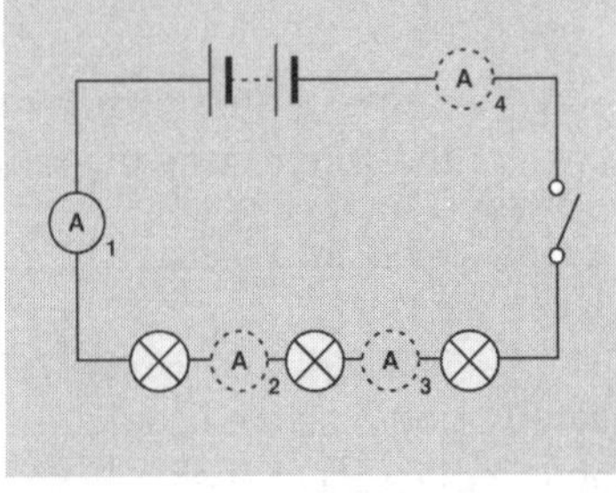

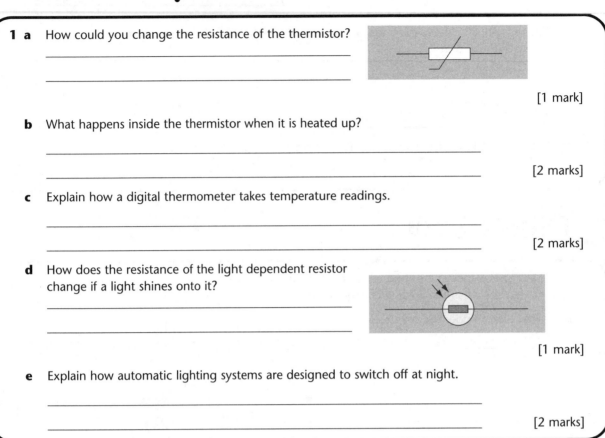

a Fred measures the current in four places in the circuit. What can you say about the readings?

______________________________ [1 mark]

b Each cell provides 1.5 V. What is the size of the voltage supplied to the circuit?

______________________________ [1 mark]

c Write down the voltage across each of the three bulbs.

1__________ 2__________ 3__________ [1 mark]

d If the number of bulbs in the circuit doubles, how does the resistance of the circuit change?

______________________________ [1 mark]

Components in parallel

D–C

1 The circuit shows three bulbs wired in parallel. The circuit is switched on.

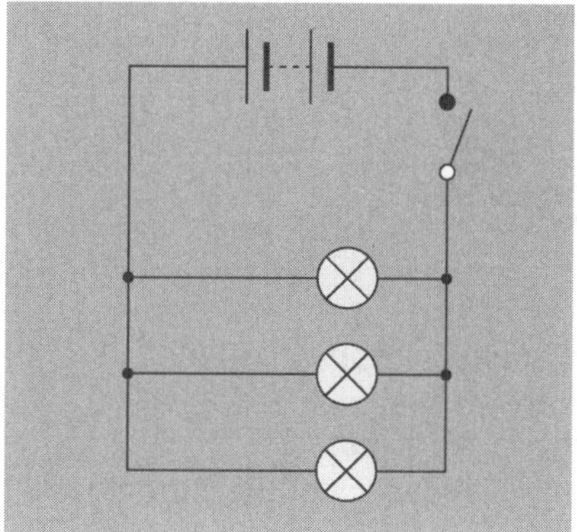

a If the current through each bulb is 0.5 A, how big is the current through the battery?

______________________________ [1 mark]

b Each cell provides 1.5 V. What is the size of the voltage supplied to the circuit?

______________________________ [1 mark]

c Write down the voltage across each of the three bulbs.

______________________________ [1 mark]

d Write down **one** reason to wire the bulbs in parallel rather than in series.

______________________________ [1 mark]

The three-pin plug

D–C

2 The diagram shows a plug connecting the lead from an iron.

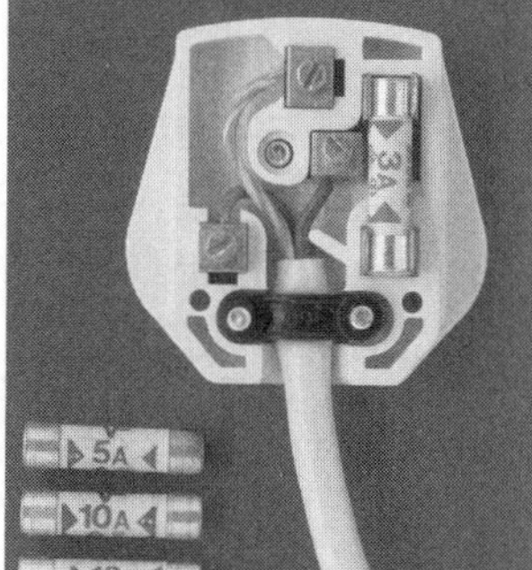

a Why are the cable and casing made of plastic?

____

______________________________ [1 mark]

b Explain why the fuse must be connected in series to the live wire.

____

____

____

______________________________ [2 marks]

c Describe the structure of the cable that connects the plug to the equipment.

____

____

______________________________ [3 marks]

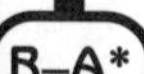

B–A*

d Explain why a radio does not need to be earthed but is still safe to use.

____

______________________________ [2 marks]

Domestic electricity

Grades

D–C

1 X Y

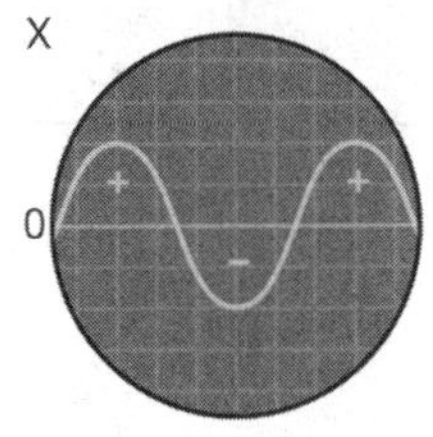

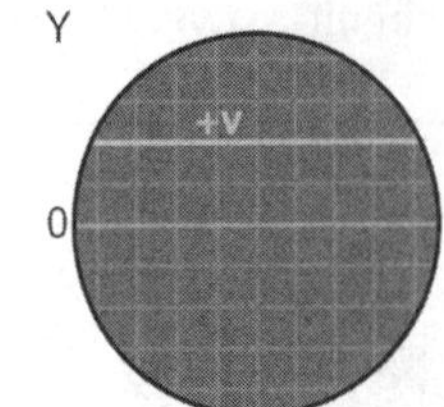

a Which diagram shows the trace you would get from a battery?

______________________________ [1 mark]

b Explain what is meant by the phrase 'the frequency of the current is 50 Hz'.

____

______________________________ [2 marks]

c What equipment is used to show the picture of changing voltage in a circuit?

______________________________ [1 mark]

B–A*

d Describe how the potential difference of mains supply changes through one cycle.

____

____

______________________________ [3 marks]

Safety at home

D–C

2 Brad's radio stopped working until his dad changed the fuse for him.

a How does the fuse protect the radio from damage?

____

____

______________________________ [2 marks]

b Why is a different fuse needed for Brad's electric heater?

______________________________ [1 mark]

c Explain why Brad's electric heater needs to be earthed but his radio does not.

______________________________ [1 mark]

d Explain how earthing the heater protects the user.

____

____

______________________________ [3 marks]

e Explain why a circuit breaker is safer to use than a fuse.

____

______________________________ [2 marks]

Grades

P2B DISCOVER NUCLEAR FUSION!

Which fuse?

D–C

1 a Nathan has to put a fuse into a plug for a radio. He has a choice of fuses rated at 3 A, 5 A or 13 A. The current flowing through the radio is 0.4 A.

i Which fuse should he use?

______________________________ [1 mark]

ii One fuse has '13 A' printed on it. What does this tell you about the fuse?

____

______________________________ [3 marks]

b i Nathan also has a heater. Its power rating is 2000 W. Calculate the size of the current flowing through the heater, if mains electricity is supplied at 230 V.

__________ A [3 marks]

ii What would happen if Nathan put a fuse in with too low a rating?

____

______________________________ [2 marks]

B–A*

iii Calculate the charge flowing through the heater each second. The voltage of mains supply is 230 V. Include units in your answer.

____

______________________________ [3 marks]

Radioactivity

D–C

2 a i In an atom, which particle has no charge?

______________________________ [1 mark]

ii In an atom, which particles are found in the nucleus?

______________________________ [2 marks]

b What is an ion?

______________________________ [1 mark]

c This table gives information about different atoms.

Element	Atomic No.	Notation
Hydrogen	1	${}^{1}_{1}H$
Helium	2	${}^{4}_{2}He$
Lithium	3	${}^{7}_{3}Li$
Beryllium	4	${}^{9}_{4}Be$
Boron	5	${}^{11}_{5}B$
Carbon	6	${}^{12}_{6}C$

i Write down the mass number for beryllium.

______________________________ [1 mark]

ii How many protons has boron got in its nucleus?

______________________________ [1 mark]

iii How many neutrons has helium got in its nucleus?

______________________________ [1 mark]

d Carbon has several isotopes.

i What is different for different isotopes of carbon?

______________________________ [1 mark]

ii What is the same for all isotopes of carbon?

______________________________ [1 mark]

Alpha, beta and gamma rays 2

Grades

D–C

1 a What is meant by the term

i unstable ______________________ [1 mark]

ii radioisotope ______________________ [1 mark]

b Alpha radiation is the most **ionising** type of radiation. Explain what ionising means.

____

______________________ [2 marks]

c Explain why gamma rays can travel further in air than alpha particles.

____

______________________ [2 marks]

B–A*

d Americium-241 decays by losing an alpha particle, forming a new element. Complete the values of X and Y in this equation.

$^{241}_{X}Am \rightarrow ^{Y}_{93}Np + ^{4}_{2}\alpha$

X = __________

Y = __________ [2 marks]

Background radiation 2

D–C

2 The diagram shows the main sources of background radiation.

Fallout from nuclear weapons testing 0.5%
Work-related 0.4%
Total discharge from all nuclear power industry 0.1%
Medical, mainly from X-rays 11.5%
Other – mainly air travel 0.5%
Cosmic rays from outer space 14%
All living things give off radiation from their bodies and from the food they eat 17%
Gamma rays from rocks and soil 19%
Radon and thoron gas is released from soil, rocks and building materials 37%

a What is meant by **background radiation**?

____

____

____

______________________ [2 marks]

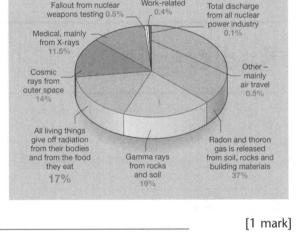

b Write down **one** source of background radiation from outer space.

______________________ [1 mark]

c Write down **one** source of man-made radiation.

______________________ [1 mark]

d Is most background radiation from natural sources or from man-made sources?

______________________ [1 mark]

D–C

3 Radon gas is radioactive and comes from rocks like granite.

a Why do some parts of the United Kingdom have a higher background radioactivity than others?

______________________ [2 marks]

b Explain whether we should allow people to live in places where there is granite rock.

____

____

______________________ [3 marks]

Grades

P2B DISCOVER NUCLEAR FUSION!

Inside the atom

D–C

1 The diagram shows one of the first models of the atom.

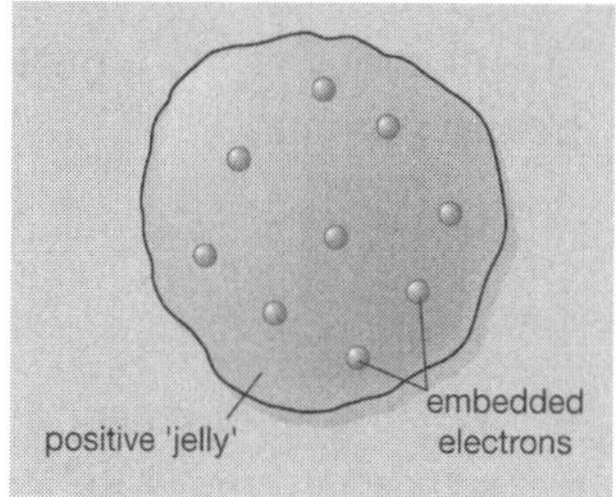

a Write down the three main features of the nuclear model of the atom.

____

____

______________________________ [3 marks]

b What evidence is there that inside an atom

i there is a positively charged nucleus?

______________________________ [1 mark]

ii the nucleus is very small?

______________________________ [1 mark]

iii electrons orbit outside the nucleus?

______________________________ [1 mark]

B–A*

c Describe how different forces interact in the nucleus.

____

______________________________ [3 marks]

Nuclear fission

D–C

2 a What is **nuclear fission**?

____

______________________________ [2 marks]

b Write down **one** use for nuclear fission.

______________________________ [1 mark]

c The picture shows a chain reaction.

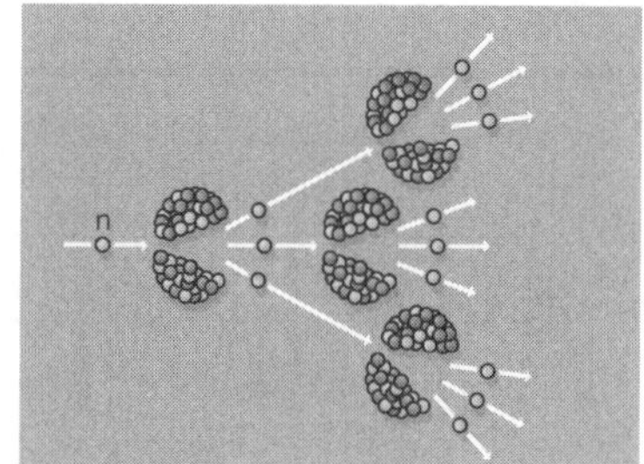

i How many neutrons are produced from each reaction?

____

______________________________ [1 mark]

ii Why does a chain reaction make the size of the reaction increase?

____

______________________________ [2 marks]

d Explain why only certain elements can undergo fission.

____

______________________________ [2 marks]

B–A*

e This equation shows one stage in a chain reaction.

$^{235}_{92}U + 10n \rightarrow {}^{144}_{56}Ba + {}^{90}_{36}Kr + {}^{1}_{0}n + {}^{1}_{0}n +$ **ENERGY**

Explain why this chain reaction will involve more atoms at each stage.

______________________________ [1 mark]

Nuclear power station

Grades

1 The diagram shows a nuclear power station.

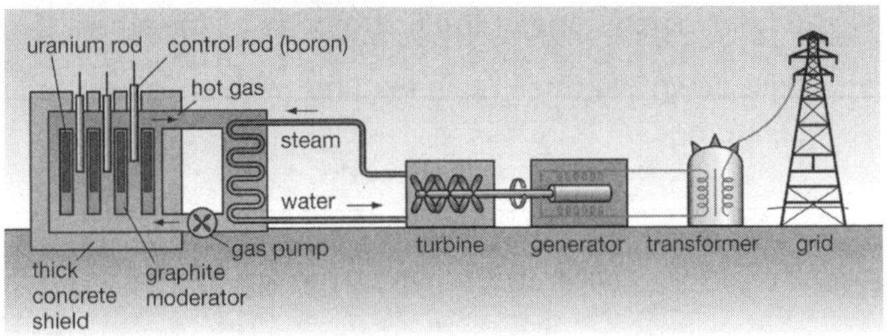

D–C

a What type of reaction takes place in the uranium rod?

_______________________________________ [1 mark]

b How are the nuclear reactions controlled?

____

_______________________________________ [2 marks]

c What energy change takes place in the generator?

_______________________________________ [1 mark]

d How does the nuclear reactor

i create steam?

____

____

_______________________________________ [3 marks]

ii generate electricity?

____

_______________________________________ [2 marks]

B–A*

e Describe what happens to the spent nuclear fuel.

____

_______________________________________ [2 marks]

Nuclear fusion

D–C

2 a What is **nuclear fusion**?

____

_______________________________________ [2 marks]

b Write down **one** place where nuclear fusion takes place.

_______________________________________ [1 mark]

c One nuclear fusion reaction involves hydrogen nuclei.
What is created when hydrogen nuclei fuse?

____

_______________________________________ [2 marks]

d Why are large amounts of heat and pressure needed for nuclear fusion to take place?

____

_______________________________________ [2 marks]

B–A*

e Describe **two** difficulties when trying to recreate nuclear fusion to generate electricity.

____

_______________________________________ [2 marks]

P2b revision checklist

I know:

what the current through an electrical current depends on

- ☐ the symbols for components shown in circuit diagrams
- ☐ resistance is increased in long, thin, heated wires
- ☐ the current through a component depends on its resistance; the greater the resistance the smaller the current for a given p.d. across the component

potential difference = current × resistance

- ☐ in a series circuit: the total resistance is the sum of the resistance of each component; the current is the same through each component; the total p.d. of the supply is shared between the components
- ☐ in a parallel circuit: the total current through the whole circuit is the sum of the currents through the separate components; the p.d. across each component is the same

what mains electricity is and how it can be used safely

- ☐ mains electricity is an a.c. supply of 230 V and has a frequency of 50 Hz; it is very dangerous
- ☐ fuses and earth wires protect appliances from damage and people from harm
- ☐ three-pin plugs must be wired correctly and hold the correct fuse
- ☐ how to interpret diagrams of oscilloscope traces

why we need to know the power of electrical appliances

- ☐ the power of an electrical appliance is the rate at which it transforms energy:

$$\textbf{power} = \frac{\textbf{energy transformed}}{\textbf{time taken}}$$

- ☐ **energy transformed = potential difference × charge**

  charge = current × time
- ☐ most appliances have their power and the p.d. of the supply they need printed on them so we can calculate the current and fuse required

what happens to radioactive substances when they decay

- ☐ isotopes (elements with the same number of protons but a different number of neutrons) with unstable nuclei emit energy as radiation
- ☐ how the Rutherford and Marsden scattering experiment revealed the structure of the atom
- ☐ background radiation comes from rocks, soil, cosmic rays, living things and medical X-rays

what nuclear fission and nuclear fusion are

- ☐ nuclear fission is the splitting of an atomic nucleus; it is used in nuclear reactors
- ☐ nuclear fusion is the joining of two smaller nuclei to form a larger one; stars release energy by nuclear fusion

Periodic table

Key

relative atomic mass
atomic symbol
name
atomic (proton) number

		1 H hydrogen 1															
1	2											3	4	5	6	7	8
																	4 He helium 2
7 Li lithium 3	9 Be beryllium 4											11 B boron 5	12 C carbon 6	14 N nitrogen 7	16 O oxygen 8	19 F fluorine 9	20 Ne neon 10
23 Na sodium 11	24 Mg magnesium 12											27 Al aluminium 13	28 Si silicon 14	31 P phosphorus 15	32 S sulfur 16	35.5 Cl chlorine 17	40 Ar argon 18
39 K potassium 19	40 Ca calcium 20	45 Sc scandium 21	48 Ti titanium 22	51 V vanadium 23	52 Cr chromium 24	55 Mn manganese 25	56 Fe iron 26	59 Co cobalt 27	59 Ni nickel 28	63.5 Cu copper 29	65 Zn zinc 30	70 Ga gallium 31	73 Ge germanium 32	75 As arsenic 33	79 Se selenium 34	80 Br bromine 35	84 Kr krypton 36
85 Rb rubidium 37	88 Sr strontium 38	89 Y yttrium 39	91 Zr zirconium 40	93 Nb niobium 41	96 Mo molybdenum 42	[98] Tc technetium 43	101 Ru ruthenium 44	103 Rh rhodium 45	106 Pd palladium 46	108 Ag silver 47	112 Cd cadmium 48	115 In indium 49	119 Sn tin 50	122 Sb antimony 51	128 Te tellurium 52	127 I iodine 53	131 Xe xenon 54
133 Cs caesium 55	137 Ba barium 56	139 La* lanthanum 57	178 Hf hafnium 72	181 Ta tantalum 73	184 W tungsten 74	186 Re rhenium 75	190 Os osmium 76	192 Ir iridium 77	195 Pt platinum 78	197 Au gold 79	201 Hg mercury 80	204 Tl thallium 81	207 Pb lead 82	209 Bi bismuth 83	[209] Po polonium 84	[210] At astatine 85	[222] Rn radon 86
[223] Fr francium 87	[226] Ra radium 88	[227] Ac* actinium 89	[261] Rf rutherfordium 104	[262] Db dubnium 105	[266] Sg seaborgium 106	[264] Bh bohrium 107	[277] Hs hassium 108	[268] Mt meitnerium 109	[271] Ds darmstadtium 110	[272] Rg roentgenium 111	Elements with atomic numbers 112–116 have been reported but not fully authenticated.						

* The Lanthanides (atomic numbers 58–71) and the Actinides (atomic numbers 90–103) have been omitted.
Cu and Cl have not been rounded to the nearest whole number.

Notes

Notes

Notes

Notes

Answers

Remember:
Check which grade you are working at.

B1a Human biology

Page 148 Coordination

1 a i Motor *(1)*

ii

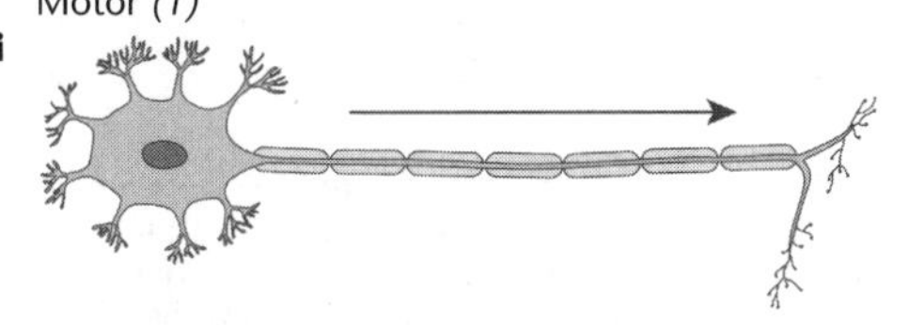

(1)

b As an electrical impulse *(1)*

Page 148 Receptors

2 a Temperature receptor in skin *(1)*; muscle in arm / hand *(1)*

b Heat *(1)*; electrical *(1)*

3 a i Heart *(1)*

ii Provide more oxygen to muscles (so you can run faster) *(1)*

b *(Any 2:)* Information travels more slowly using hormones; travels as chemical using hormones but electrical impulse in neurone; travels in blood using hormones but along cells using neurones; lasts longer using hormones than using neurones.

Page 149 Reflex actions

1 a i Transmitter substance /acetylcholine *(1)*

ii Transmitter substance / contents of X; released into cleft; diffuses across cleft; sets up impulse in the next neurone *(2)*

b i One *(1)*

ii (Any 1:) Each neurone can have many synapses with other neurones; the impulse can also pass on to other neurones when it gets to the end of the sensory neurone, for example to a neurone that carries the impulse up to the brain so that you can be aware of what is happening.

Page 149 In control

2 a To replace (sodium and chloride ions) lost in sweat *(1)*; to maintain their concentration in the blood / so that cells can work effectively *(1)*

b Makes the liver take glucose from the blood *(1)*; storing it as glycogen *(1)*

c i If there is no insulin, then there is nothing to make her liver store glucose *(1)*; she has no stores from which her body can take glucose between meals *(1)*

ii *(Any 2:)* Glucose is needed for respiration, to provide energy to cells; if there is no glucose, cells can run out of energy; she will feel very tired, and may even go into a coma (because her brain cells have no energy supply)

Page 150 Reproductive hormones

1 a i Pituitary *(1)* **ii** Ovary *(1)*

b i Arrow at day 14 *(1)* **ii** Days 1 to 5 *(1)*

iii A is oestrogen *(1)*; B is LH *(1)*

Page 150 Controlling fertility

2 a The older the woman, the lower the success rate *(1)*

b *(Any 2:)* The success rate is too low; uses up money better spent on other health treatments; refer to possible problems for children of (very) elderly parents

3 *(Any 2:)* FSH stimulates eggs to mature; it is difficult to find the correct dose, as women don't all respond in the same way to this hormone; if the dose is too high, then more than one egg may mature at the same time; there could therefore be two or more eggs in the oviduct at the same time; if more than one is fertilised, then more than one zygote is produced, each of which may develop into an embryo

Page 151 Diet and energy

1 a i The rate of chemical reactions in cells / the body *(1)*

ii She uses up energy faster *(1)*; so more of her food is used for releasing energy rather than being stored as fat *(1)*

iii She could do more exercise *(1)*

b i Fat *(1)*

ii The carbohydrate (in the form of glucose) is broken down in the process of respiration *(1)*; glucose + oxygen carbon dioxide + water, which releases energy for the cell to use *(1)*

Page 151 Obesity

2 a In Russia, there are more obese women than obese men, in Canada, the percentages obese men and obese women are the same *(1)*

b If obesity was caused mostly by genes, you would expect approximately the same proportion of women and men to be obese, because they would have similar genes *(1)*; the difference could be explained if men live a different lifestyle from women, for example doing more exercise and eating less *(1)*

c Obese people put more, stress / weight, on their joints *(1)*; cause damage to the joint surface / wear away cartilage *(1)*

Page 152 Not enough food

1 a i (Any 1:) Energy taken in depends on the total amount eaten; what matters is the balance between energy taken in and energy used; weight is lost only if energy taken in **is less than energy used**

ii *(Any 2:)* They will not be able to maintain this weight loss / weight will easily go back on again; they may feel ill / weak (after losing so much weight so quickly); they may have lost a lot of water

iii High fat levels in the diet can raise the level of cholesterol in the blood *(1)*; which can increase the risk of developing heart disease *(1)*

Page 152 Cholesterol and salt

2 a C *(1)*

b *(Any 2:)* The sample of people should have included the same age range; the sample should have been selected in the same way in both cities (for example at random, or the people visiting a doctor's surgery); there should have been a large number of people in each sample

c *(Any 2:)* If fat is reduced in the diet, this may lead to a lower blood cholesterol level; this will cause the liver to produce more cholesterol, so you won't be able to get the levels down much further; however, if the enzyme that makes cholesterol is not working, then the liver will not be able to respond to the low blood cholesterol levels, and they will stay low

Page 153 Drugs

1 a You cannot manage without it *(1)*; if you stop taking the drug you experience withdrawal symptoms *(1)*

b i More in 16-39 age group than in the 30-59 age group *(1)*; 7.3 % of frequent nightclub visitors in 16-39 age group took cocaine, but only 1.4% in the 30-59 age group *(1)*

ii In both age groups, those who often visited night clubs regularly were more likely to take cocaine than those who only visited night clubs rarely *(2)*

Page 153 Trialling drugs

2 a A *(1)*

b B, C, D and E *(1)*

c B *(1)*

d *(Any 2:)* The trials involve only a limited number of people; once many more people are taking the drug, rare side effects have more chance of showing up; the trials only go on for a limited time; the side effects may not show up until people have been taking the drug for a long time

Page 154 Illegal drugs

1 a Syringe may be contaminated *(1)*; with viruses *(1)*

b *(Any 2:)* May become addicted; because it changes the behaviour of synapses in the brain; suffer withdrawal symptoms if they stop taking it

Page 154 Alcohol

2 a A *(1)*

b No, we only know how many were admitted to hospital *(1)*; perhaps women take less of the drug than men / are less likely to be made ill by taking the drug *(1)*

c Liver cells are damaged as they try to break down alcohol *(1)*

d *(Any 2:)* Alcohol is a legal drug; the others are all illegal; alcohol alone caused more hospital admissions than all the others put together; 390 admissions compared with 230 admissions, i.e. 160 more

Page 155 Tobacco

1 a *(Any 2:)* There is a time lag between smoking and getting; cancer early increase in deaths from lung cancer may not have been noticed; took a while for people to make a link between the two

b Some other factor could be causing both increases *(1)*
c Carbon monoxide combines with haemoglobin *(1)*; this reduces the oxygen-carrying capacity of the blood the baby does not get so much oxygen *(1)*; so its cells may not be able to respire as they should and release energy for growth *(1)*

Page 155 Pathogens

2 a B *(1)*
b He realised that the ward with most deaths was visited by doctors; he suggested they were passing disease-causing substances between patients *(1)*; he asked doctors to wash their hands (before and after touching patients) *(1)*
c *(Any 2:)* People are living closely together in a hospital, so it is easier for pathogens to get from one to another; people may go into hospital because they have an infectious disease, which could spread to other people; people in hospital may have weakened immune systems because they are ill, so they may not be likely to suffer from infections

Page 156 Body defences

1 a Phagocytosis *(1)*
b It is destroyed / killed *(1)*; digested by enzymes *(1)*
2 a It was caused by a pathogen (a virus) which was carried in a person's body *(1)*; who travelled from China / Vietnam / Singapore, to Canada / Switzerland *(1)*
b *(Any 2:)* An epidemic is an outbreak of an infectious disease in one area; a pandemic is a world-wide outbreak of a disease; almost all SARS cases were in a few countries close together, not widely spread around the world

Page 156 Drugs against disease

3 a C *(1)*
b i 47.9 – 38.1 = 9.8 % fewer *(1)*
ii To reduce the risk of bacteria developing resistance *(1)*; the more they are exposed to antibiotics, the more likely it is that a resistant strain will arise *(1)*

Page 157 Arms race

1 D *(1)*
2 a *(Any 2:)* They have increased; slowly at first and then more rapidly; almost doubling every two years between 2000 and 2004
b *(Any 2:)* Mutations in bacteria happen randomly (they do not happen because of the antibiotic); causing some bacteria to have resistance to the antibiotic; these bacteria survive and reproduce even when the antibiotic is used; while the non-resistant ones die

Page 157 Vaccination

3 a *(Any 2:)* No, because we cannot be sure the drop in cases was caused by the vaccination; some other factor might have caused them to fall; but there is a correlation between the use of the vaccine and the fall in the number of cases
b *(Any 2:)* Not everyone was vaccinated straight away; only 10–14 year olds vaccinated; as time went on a bigger proportion of people had immunity
c *(Any 4:)* Reference to antigens; white blood cells respond lymphocytes; secrete antibodies specific to, antigen / TB / bacterium; multiply; form memory cells; on next infection bacteria killed immediately on a bigger proportion of people had immunity (accept other well-argued suggestions).

B1b Evolution and environment

Page 159 Hot and cold

1 a 38 °C *(1)*
b It loses heat to the cold, air / ground *(1)*
c The cold blood from the feet is warmed by the warm blood in the artery as it flows upwards *(1)*
2 a Small body and large ears provides a larger surface area than large body and small ears less *(1)*; surface area means less heat loss, so good in cold conditions / more surface area means more heat loss, so good in hot conditions *(1)*
b Reference to camouflage *(1)*

Page 159 Adapt or die

3 Plants and animals are **adapted** for survival in their habitat. Some plants have **poisons** in their leaves, which harm insects that eat the leaves. Some plants have warning colours to deter **predators**. *(3)*

4 Competition for light *(1)*; for water *(1)*

Page 160 Two ways to reproduce

1

1	2	3	4
D	A	C	B

(4)

2 *(Any 2:)* In asexual reproduction, there in only one parent; the offspring get all of their genes from this parent; in sexual reproduction, two gametes fuse together; the gametes may have different genes (even if they both come from the same parent) so the offspring can each have a different mixture of genes

Page 160 Genes and what they do

3 a In the nucleus of the cell, making up the chromosomes *(1)*
b Each DNA molecule is made up of many different genes *(1)*; the genes determine many of the characteristics of the cell, and of the organism *(1)*
4 a The male dog *(1)*; because all of his chromosomes came from there *(1)*
b i 39 *(1)*
ii 78 *(1)*
iii 78 *(1)*

Page 161 Cuttings

1 a *(Any 2:)* They might have different genes; they might be growing in different kinds of soil; one might have more water than the other; one might have more light than the other
b Acorns are produced by sexual reproduction, so they are genetically different *(1)*; cuttings only have genes from one parent, so they are genetically identical *(1)*
2 a B *(1)*
b *(Any 2:)* All banana plants are genetically identical; as they cannot reproduce sexually, new combinations of genes cannot be produced; so all the plants will remain susceptible to the disease

Page 161 Clones

3 a They are all formed from one zygote *(1)*; so they all have the same genes / they are genetically identical *(1)*
b Many eggs could be obtained from one female of the rare species; and fertilised; each embryo could be split into many embryos; which could be put back into different females; or females of a different (but closely related) species; so you could get many more offspring from just one female parent

Page 162 Genetic engineering

1 a i (Cheaper because:) He does not have to buy pesticides / he does not have to spend time or fuel spraying pesticides
ii (Cheaper because:) The farmer's costs in growing the maize are less / there will be no pesticides contaminating the maize
b They may think that eating GM foods is harmful to health *(1)*; they may think that the GM maize might harm other organisms in the environment *(1)*
2 The bacteria contain the human gene for making insulin *(1)*; the gene carries the, code / instructions, for making human insulin *(1)*

Page 162 Theories of evolution

3 a C *(1)*
b *(Any 2:)* Only the ones with long necks reproduce; they pass on their genes to their offspring; all offspring contain the genes for long necks

Page 163 Natural selection

1 a D *(1)*
b A *(1)*
c *(Any 3:)* Pale moths now better camouflaged / dark moths less well camouflaged; so more pale moths survive / fewer dark moths survive; pale moths more likely to breed / dark moths less likely to breed; more genes for pale wings passed on to offspring; so most moths in the population now pale
2 *(Any 2:)* A population on an island is isolated from the populations on the mainland; there may be different selection pressures on them / natural selection gives different features an advantages; so individuals with different characteristics are more likely to breed on the island and on the mainland; over time, different characteristics will evolve in the two places

Page 163 Fossils and evolution

3 a From fossils *(1)*

b *(Any 2:)* Idea that they were adapted to their environment; more toes better for running on wet ground; climate became dryer; horses with fewer toes more likely to survive; and pass on their genes for fewer toes to their offspring; over time / many generations, all horses had only one toe

c *(Any 2:)* There may be other species that we have not yet found fossils of; which may have been the ancestors of the third type of horse; fossils do not tell us about ancestry

Page 164 Extinction

1 *(Any 3:)* Competition with the introduced crayfish / introduction of new competitor; compete for limited resources / not enough food for all; new disease may kill them; not enough new crayfish born to replace the ones that die

2 *(Any 2:)* Killed by humans; easy to kill because they could not fly; and because they were not afraid; other animals (rats, cats, dogs) introduced by humans killed the dodos; or were competitors of dodos (needing same food, same nesting sites); humans may have destroyed the dodo's habitat; so they could not find food / could not nest

Page 164 More people, more problems

3 a i The right hand part beyond the current year

ii About 2 billion *(1)*

iii B *(1)*

b *(Any 3:)* More land is used for building houses; more land is used for agriculture or food production; more land is used for roads; more land is used for quarries or mines; more land is polluted

c *(Any 2:)* Improvements in diet; improvements in living conditions: clean water, good sewage disposal; improvements in health care: antibiotics, other medicines, pain-free surgery

Page 165 Land use

1 a C *(1)*

b i No set-aside land in 1986 *(1)*; increased to maximum in 2001 fell slightly to 2005 *(1)*

ii Either grazing land or crop land both of which have reduced in area over this time period *(1)*

iii *(Any 2:)* Biodiversity is the number of different species present – would increase; no herbicides or pesticides used on set-aside land; a variety of plants would grow there rather than just one if a crop was growing; provide more niches for different species of animals

Page 165 Pollution

2 a Concentration of DDT increases up the food chain *(1)*; greatest concentration in the animals at the end of the chain *(1)*

b *(Any 2:)* DDT persists for a long time; in the environment / in the bodies of animals; perhaps DDT is still being used in other countries

Page 166 Air pollution

1 a Coal-burning power stations*(1)*

b Because bronchitis *(1)*; trigger asthma attacks (but not actually cause asthma) *(1)*

Page 166 What causes acid rain?

2 a C *(1)*

b They damage the lungs cause breathing problems / coughing / asthma attacks *(1)*

c *(Any 2:)* The gases dissolve in water droplets; react to form acids; rain droplets containing acid fall to the ground when it rains

d Kills trees by damaging their leaves / roots / washing away nutrients *(1)*; kills fish by making water acidic / toxic chemicals washed in from soil *(1)*

e *(Any 2:)* The limestone neutralises the acid; by reacting with it (to form salts and water); pH is raised

f *(Any 4:)* Rises between 1970 and 1990; because there were more vehicles on the roads; falls sharply between 1990 and 2000; significant number of cars now fitted with catalytic converters; which change NOx to nitrogen; continues to fall until 2005; more cars being fitted with catalytic converters; new designs for catalytic converters that work better

Page 167 Pollution indicators

1 a *(Any 2:)* Use the animals as indicators; different animals can live in different oxygen concentrations; find out which animals are living in the water; match them to the oxygen concentration in which they are known to live

b *(Any 2:)* Rat-tailed maggots; they can survive in the lowest oxygen concentration; raw sewage reduces the oxygen concentration in the water

c i Haemoglobin picks up oxygen so they can get oxygen *(1)*; out of the water even when the concentration is low *(1)*

ii Many other species are able to live in well-oxygenated water *(1)*; bloodworms cannot compete with them for, space / food *(1)*

Page 167 Deforestation

2 a Marbled cats and oriental short-clawed otters *(1)*; they were present before logging, but there were none in the forest even two years after logging *(1)*

b *(Any 2:)* Their numbers fell immediately after logging but then increased; although not back to their former level; logging destroyed their habitat / they prefer to live amongst large trees; some returned and were able to live in the regenerating forest

c *(Any 2:)* These were more common after logging than before; their preferred habitat may be smaller trees; their competitors were not present in the forest after logging

d *(Any 2:)* We may find useful drugs in the rainforest plants; deforestation increases carbon dioxide in the air; because there is less photosynthesis; this could add to global warming

Page 168 The greenhouse effect – good or bad?

1 a A *(1)*

b *(Any 2:)* Increase in world population; increase in energy use; more fossil fuels burned

Page 168 Sustainability – the way forward?

2 a i *(Any 2:)* Making sure we do not use resources faster than they are being replaced; using the environment in such a way that we do not spoil it for the future; improving the quality of people's lives without damaging the chances of our survival in the future

ii *(Any 2:)* Less petrol used so less oil needs to be taken from the ground; less carbon dioxide released to the atmosphere (by the combustion of fuels); so less contribution to global warming

b Heat loss without insulation = 2 x 10 = 20 W; heat loss with insulation = 0.3 x 10 = 3 W; so 7 less W of heat energy are lost *(2)*

C1a Products from rocks

Page 170 Elements and the periodic table

1 a Potassium *(1)* **b** Magnesium *(1)* **c** Boron *(1)*

d Iodine is now in same group as bromine / chlorine / named halogen / other halogens *(1)*; with other elements with similar properties *(1)*

Page 170 Atomic structure 1

2 a 35 *(1)*; 35 *(1)*; 45 *(1)*

b Electrons have no / negligible mass *(1)*

c i Kr *(1)* **ii** F Cl *(1)*

3 a i A *(1)*

ii 2 electrons in the outer shell (atom B is helium – a Noble Gas) *(1)*

b i B / D *(1)* **ii** Full outer shell *(1)*

c 2,1 *(1)*

Page 171 Bonding

1

1	2	3	4
D	B	C	A

(4)

2 B *(1)*

Page 171 Extraction of limestone

3 a *(Any 2:)* Making concrete; making cement; *(allow one mark for 'building materials')*; making glass; making iron; used by farmers to treat soil

b *(Any 3:)* Eyesore / scarring to landscape; damage to habitats / loss of wildlife; dust / noise problems; traffic problems

c Employment / jobs/ income for local shops / better road system / money for local health care / closed quarries provide leisure facities *(1)*

Page 172 Thermal decomposition of limestone

1 a Calcium oxide *(1)*; carbon dioxide *(1)*; CO_2 *(1)*
b B *(1)*
2 a H_2O *(1)*
b Due to $CaCO_3$ being formed *(1)*; which is a solid *(1)*

Page 172 Uses of limestone

3 a Concrete contains sand and gravel *(1)*
b *(Any 2:)* Concrete is stronger; cement is smoother to use; due to the added sand and gravel
4 C *(1)*

Page 173 The blast furnace

1 Carbon monoxide *(1)*; oxygen *(1)*; reduction *(1)*
2 D *(1)*

Page 173 Using Iron

3 a i C *(1)*
ii Steel contains different elements *(1)*; atoms are different sizes *(1)*
b A *(1)*

Page 174 Using steel

1 a *(Any 3:)* Wrought iron contains the same type of atoms; atoms in wrought iron are in regular rows; atoms in steel are of different elements; sizes of atoms are different; rows are not regular in steel
b *(Any 2:)* steel is harder; does not rust; steel is stronger; steel has a shinier finish

Page 174 Transition metals

2 a Positive ions / cations *(1)*; electrons *(1)*
b Electrons are involved in conduction *(1)*; they move *(1)*
3 a Iron *(1)*; cobalt *(1)* **b** D *(1)*

Page 175 Aluminium

1 a Reactive *(1)*; negative *(1)*; gain *(1)*
b i 4 *(1)* **ii** Oxidation *(1)*; electrons are lost *(1)*
2 D *(1)*

Page 175 Aluminium recycling

3 a *(Any 2:)* Spoiling of habitats / tree felling / ecosystem damage; burning trees leads to climate change; loss of living area for local people; health problems due to mining e.g. dust; use of old mines as dumps
b *(Any 2:)* Thrown away with other waste; difficulty of sorting / collecting; attitude of people to recycling

Page 176 Titanium

1 a *(Any 3:)* Lower density; this means that the bikes will be lighter; corrosion resistant; this means that the frame will not rust / corrode
b i Aluminium oxide is an ionic compound and titanium oxide is a covalent compound *(1)*; ionic compounds conduct electricity when they are molten, covalent compounds do not *(1)*
ii Titanium has a melting point above 900 °C / much higher melting point *(1)*

Page 176 Copper

2 a B *(1)* **b** C *(1)*
3 C *(1)*

Page 177 Smart alloys

1 a (Similar because:) Contain mainly metal atoms *(1)*; contain mixture of elements *(1)*; can be bent and stretched into different shapes *(1)*
(Different because:) Save a shape memory *(1)*
b Heating *(1)*; then cooling *(1)*

Page 177 Fuels of the future

2 a C *(1)* **b** D *(1)*
3 a *(Any 2:)* Buses burn petrol / fossil fuels; which produces pollutant gases; for example, carbon monoxide / sulphur dioxide / other named pollutant
b Electricity is produced by burning fossil fuels *(1)*; named effect on the environment e.g. produces climate change / acid rain / produces named pollutant gas *(1)*

Page 178 Crude oil

1 a

1	2	3	4
C	B	D	A

(4)

b B *(1)*

Page 178 Alkanes

2 a A *(1)* **b** B; D *Both correct = (1)*
c C *(1)* **d** C *(1)*

Page 179 Pollution problems

1 a hydrocarbon + **oxygen → carbon dioxide** + **water**
sulfur + **oxygen → sulfur dioxide** *(4)*
b *(Any 2:)* Smaller fish die due the acid rain; older ones survive with less competition; additional information: older fish are often deformed / fish stocks will die out
2 a C *(1)* **b** C only *(2)*

Page 179 Reducing sulfur problems

3 B and D *(2)*
4 a More carbon dioxide formed *(1)*; leading to increased climate change *(1)*
b Acid rain *(1)*; caused by sulphur in ship fuel *(1)*

C1b Oils, Earth and atmosphere

Page 181 Cracking

1 C *(1)*
2 D *(1)*
3 a i

```
H H
| |
C=C
| |
H H
```

(1)

ii Ethane *(1)*

Page 181 Alkenes

4 a C_5H_{10} *(1)* **b** C *(1)*
5 a Ethene speeds up the ripening of fruit *(1)*
b Petals and leaves fall off flowers *(1)*

Page 182 Making ethanol

1 a Fermentation *(1)*; using yeast *(1)*
b Renewable means that the supply will not run out *(1)*; more sugar beet can be grown to produce more fuel *(1)*; it is called a biofuel because it is made from living things *(1)*
c i $C_2H_4 + H_2O$ g C_2H_5OH *Left hand side correct = (1) right hand side correct = (1)*
ii Ethanol contains oxygen atoms (as well as carbon and hydrogen) *(1)*

Page 182 Plastics from alkenes

2 a

```
 [F  F]            F  F  F  F  F  F
 [|  |]            |  |  |  |  |  |
-[C--C]-    OR   -C--C--C--C--C--C-
 [|  |]            |  |  |  |  |  |
 [F  F]n           F  F  F  F  F  F
```

All single bonds shown with bonds coming out of each end of polymer = (1) at least three repeating units shown, or 'n' after brackets to imply 'many' = (1)

b Does not have a double bond *(1)*
3 A *(1)*

Page 183 Polymers are useful

1 a B *(1)*
b D *(1)*
c *(Any 2:)* A has shorter chains; and so is more runny; molecules can move over each other more easily
2 C *(1)*

Page 183 Disposing of polymers

3 (Burning:) saves using fossil fuels to produce electricity *(1)*; (Recycling:) Saves the use of crude oil as a raw material to produce new polymers *(1)*
4 *(Any 2:)* Polymers are non-toxic / metals might rust; (some) polymers can be made biodegradable; so that the polymer rots away naturally / the metal is permanent

Page 184 Oil from plants

1

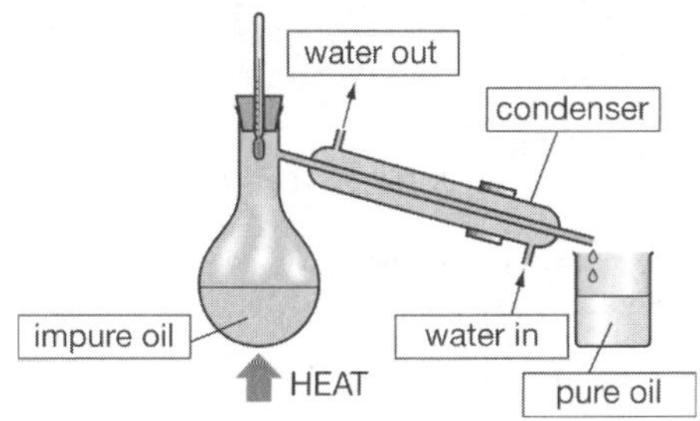

Water connections labelled correctly = (1) condenser label = (1)
impure and pure oil labelled correctly = (1)

2 D *(1)*

Page 184 Green energy

3 **a** (Any 1:) Mixed with other oils / fuels; fuels that are mixed in are usually petrol / fuels from crude oil

b *(Any 2:)* Saves fossil fuels; idea that fossil fuels are not renewable / will run out; biofuels are renewable; they are 'greenhouse neutral'

c Petrol is non-renewable / will run out *(1)*; when it becomes scarce it will be more expensive *(1)*

d *(Any 2:)* Produces less greenhouse gases / less CO_2; non-toxic; biodegradable

Page 185 Emulsions

1 **a**

Type of emulsion	Dispersed phase	Continuous phase
Shaving foam	**Air/gas**	**Liquid**
After shave cream	**Water/liquid**	**Oil/liquid**

Correct words in correct lines, in the correct order = (2)

b Do not mix together

2 C *(1)*

Page 185 Polyunsaturates

3 **a** Contains all single bonds / contains no double bonds *(1)*

b Contains lots of / many double bonds *(1)*

c

Type	Olive oil	Saturated fat	Polyunsaturated oil
Observations	Orange to colourless	**No change**	**Orange to colourless**

(2)

4 A *(1)*

Page 186 Making margarine

1 **a**

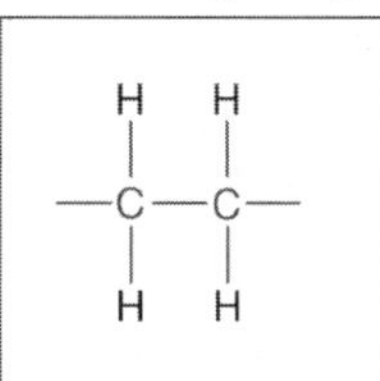

Important points: two more hydrogens shown; molecule is now straight not V shaped

b Oil is heated (to 60 °C) *(1)*; nickel catalyst used *(1)*

2

1	2	3	4
D	C	B	A

(4)

Page 186 Food additives

3 **a** *(Any 3:)* E220 is a preservative; E340 is an antioxidant (need both); preservatives help foods to keep for longer/longer shelf life; antioxidants stop loss of quality due to reacting with oxygen in the air

b Links to illness *(1)*; (specific example:) cancer/hyperactivity/allergies/asthma *(1)*

Page 187 Analysing chemicals

1 **a** A contains two ingredients *(1)*; the one that is also in D is not harmful *(1)*; the second ingredient is also in the sweet *(1)*

b Measure the distance travelled by the solvent front *(1)*; measure the distance travelled by the spot *(1)*; use the formula:

$$\text{Retention factor} = \frac{\text{distance moved by spot}}{\text{distance moved by solvent}} \quad (1)$$

c 3 cm *(1)*

Page 187 The Earth

2

1	2	3	4
B	D	A	C

(4)

3 C *(1)*

Page 188 Earth's surface

1 **a**

1	2	3	4
A	C	B	D

(4)

b A *(1)*

Page 188 Earthquakes and volcanoes

2 **a**

1	2	3	4
D	C	B	A

(4)

b Continental is less dense because it rises on top of the oceanic *(1)*

c Rock is being pushed down as one plate moves under the other *(1)*

Page 189 The air

1 D *(1)*

2 **a** 3; 2 *Both correct = (1)*

b Nitrogen; sodium **Both correct** = *(1)*

c Due to the nitrogen *(1)*; which is a gas *(1)*

Page 189 Evolution of the air

3 **a** **i** The volcano is at over 1000 °C / at a very high temperature *(1)*

ii The Earth cooled *(1)*; water vapour condensed *(1)*; to form the seas *(1)*

b **i** Carbon dioxide *(1)*

ii Dissolving in sea water *(1)*; photosynthesis *(1)*

c *(Any 2:)* Does not contain oxygen; does not contain carbon dioxide; contains large amounts of sulphur dioxide / hydrogen / carbon monoxide

4 Earth has an atmosphere and moon does not *(1)*; greenhouse effect warms earth's surface *(1)*; due to carbon dioxide in the air *(1)*

Page 190 Atmospheric change

1

Activity	Decreases amount of carbon dioxide in air	Increases amount of carbon dioxide in air	Does not affect amount of carbon dioxide in air
Using CFCs			✓
Planting more trees	✓		
Burning biofuels			✓
Burning fossil fuels		✓	

(4)

2 a Crude oil *(1)*; limestone *(1)*

b Short lived plants decay and return carbon dioxide to the air *(1)*; trees live a long time and the carbon dioxide is 'locked up' in the wood for a long time while it rots away slowly *(1)*

c If the seas warm up, not as much carbon dioxide will dissolve in them *(1)*; therefore, more carbon dioxide would be released into the air *(1)*

3 C *(1)*

P1a Energy and electricity

Page 192 Heat energy

1 a Travel at speed of light *(1)*; travel through vacuum *(1)*; travel in straight lines *(1)*

b i Drawing the curtains stops heat radiation coming into the house through the glass *(2)*

ii Paint the house white *(1)*; white reflects heat *(1)*; reduces heat transfers into the house *(1)*

2 a The biscuits are cooler *(1)*

b The object's temperature measures *(1)*; the average internal energy of all the particles *(1)*

Page 192 Thermal radiation

3

1	2	3	4
C	B	A	D

(4)

4 a The temperature *(1)*; number of particles *(1)*

b The ice cube is colder *(1)*; it has less particles *(1)*

c The camera detects thermal radiation emitted by an object *(1)*; the drink emits more thermal energy before the ice cube is added because it is warmer *(1)*

Page 193 Conduction and convection

1 a Heat conduction

b As each particle becomes warmer it vibrates more *(1)*; the vibration is passed from particle to particle *(1)*; and this transmits the energy *(1)*

c Free electrons carry heat energy quickly from one part of the substance to another *(1)*; metals contain free electrons so they conduct heat well / glass has no free electrons so cannot conduct heat well *(1)*

2

1	2	3	4
C	B	A	D

(4)

Page 193 Heat transfer

3 a D **b** A **c** B **d** C

Page 194 Types of energy

1 a B **b** A **c** C **d** D

Page 194 Energy changes

2 a Heat *(1)*; sound *(1)*

b It spreads to the surroundings making them warm up a bit *(1)*

c Chemical energy from fuel changes into (gravitational) potential energy *(1)*; and kinetic energy *(1)*; and then into heat (in the brakes) *(1)*; also wasted forms of energy: sound/heat *(1)*

3

1	2	3	4
D	C	A	B

(4)

Page 195 Energy diagrams

1 a Output *(1)*; wasted *(1)*

b As heat in moving energy parts *(1)*

c Sankey diagram with one input which splits into two output arrows *(1)*; equal width *(1)*; input labelled 1000J, each output labelled 500J *(1)*; input labelled chemical energy; output labelled kinetic energy and heat *(1)*

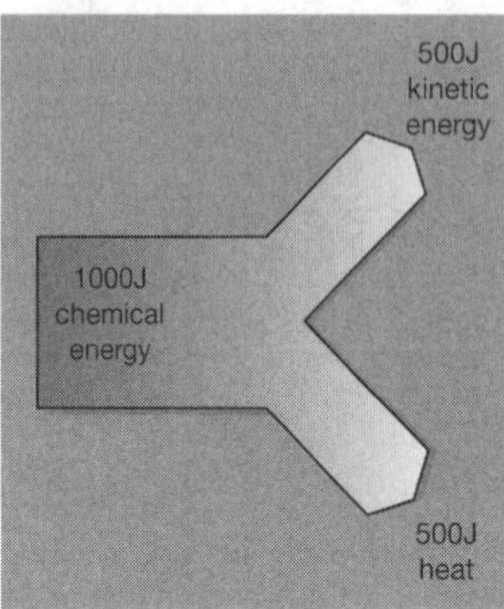

2

1	2	3	4
A	C	B	D

(4)

Page 195 Energy and heat

3 a C *(1)* **b** A *(1)* **c** B *(1)* **d** D *(1)*

Page 196 Energy, work and power

1 a Work done = force x distance *(1)*
= 500 x 60 *(1)*
= 30 000 J *(1)*

b Power = $\frac{\text{work done}}{\text{time taken}}$ *(1)*
= $\frac{30\ 000}{75}$ *(1)*
= 400 W *(1)*

c *(Any 2:)* Efficiency measures the proportion of energy usefully transferred *(1)*; The new lift wastes less energy doing the same work *(1)*; more of the input energy is usefully transformed *(1)*

d The work done (Fxd) is the same in each case, but moving vertically lifts her through a smaller distance so the force must increase to compensate *(2)*

Page 196 Efficiency

2 a B *(1)* **b** A *(1)* **c** C *(1)* **d** D *(1)*

Page 197 Using energy effectively

1 a Turn off lights *(1)*; use energy saver light bulbs *(1)*

b It reduces heat losses through doors / windows / walls / floors *(1)*

2 a Time taken for the savings *(1)*; to match the costs *(1)*

b *(Any 3:)* Annual savings are smallest *(1)*; over several years in the same home, savings using other methods will be greater *(1)*; over 10 years, draught proofing saves 150 - 45 = £105 but loft insulation saves more *(1)* £600 – 240 = £360 *(1)*

c Expensive to install *(1)*; only work if its sunny *(1)*

Page 197 Why use electricity?

3 a B *(1)* **b** C *(1)* **c** A *(1)* **d** D *(1)*

Page 198 Electricity and heat

1 a *(Any 2:)* Type of metal; thickness; length

b A smaller current flows through the radio *(1)*

c A higher rated fuse wire is thicker *(1)*; so it can carry larger current before melting *(1)*

2 a It drops to a very low value at low temperatures *(1)*

b *(Any 2:)* Large resistance causes wires to heat up; but super conductors have such low resistance; the wires don't heat up so the energy losses are very low

c Only effective at extremely low temperatures (also, cost) *(1)*

Page 198 The cost of electricity

3 a The energy transferred is 1200 J per second *(2)* Energy used = power x time *(1)*; the energy transferred is 1200J *(1)*

b 1.5 x 1/4 *(1)* = 0.375 kWh *(1)*

4 a A *(1)* **b** D *(1)* **c** C *(1)* **d** B *(1)*

Page 199 The National Grid

1

1	2	3	4
B	C	D	A

2

1	2	3	4
D	A	C	B

3 a Direct current *(1)*

b Power stations supply alternating current *(1)*; which changes direction 50 times a second *(1)*

c High voltage means small current is needed *(1)*; less energy is wasted as heat *(1)*

Page 199 Generating electricity

4 a i Wheels have kinetic energy which changes into electrical energy in the dynamo *(1)*

ii Kinetic energy *(1)* → electrical energy *(1)*

b i The moving magnet creates a current in the wire *(1)*

ii There is more kinetic energy that can be changed into electrical energy *(1)*

Page 200 Power stations

1

1	2	3	4
B	D	C	A

2 **a** **i** Chemical energy → heat energy → kinetic energy *(2)*
ii Kinetic energy → electrical energy *(1)*
b Combined cycle gas power stations use heat left in hot gases *(1)*; to produce steam for a second cycle *(1)*
c Waste energy is used to heat local buildings *(1)*
d Methane gas is harvested from domestic waste and used as a fuel for gas power stations *(2)*

Page 200 Renewable energy

3 a A barrage/dam is built across an estuary/river mouth *(1)*; as tidal water moves in or out through pipes in the wall *(1)*; it forces turbines in the pipes to spin *(1)*
b Only a few suitable sites/floods local area/effect on local ecosystems / only available at certain times of day *(1)*
c *(Any 3:)* Mirrors focus the Sun's rays onto a tower; this heats air which heats steam to run a turbine; mirrors focus the Sun's rays onto a tower; this heats pipes containing concentrated saline solution; and is used to generate steam to run a turbine

Page 201 Electricity and the environment

1 **a** Release of greenhouse gases *(1)*; causes acid rain *(1)*
b *(Any 2:)* Wind turbines can be stand-alone; power lines from the National Grid are not needed; produce electricity directly from the wind

Page 201 Making comparisons

2 **a** **i** Coal *(1)*
ii Coal *(1)*; nuclear *(1)*
b Locate near consumers *(1)*; near good transport network for fuel *(1)*; near river or coast for cooling water *(1)*
c If as it floods, vegetation dies and rots *(1)*; rotting vegetations creates methane which is a greenhouse gas *(1)*

3

1	2	3	4
B	A	D	C

(4)

4 **a** Cost of fuel *(1)*; cost of building power stations *(1)*
b New power stations will need to use fossil fuels more efficiently *(1)*; so the new design will probably be more complicated *(1)*; but they will use less fuel when generating electricity *(1)*
c *(Any 1:)* Bacteria in sewage can be oxidised; this process releases electrons and protons; the bacteria and protons cluster round different electrodes which sets up a voltage

P1b Radiation and the Universe

Page 203 Uses of electromagnetic radiation

1 **a** All travel at same speed *(1)*; can travel through vacuum *(1)*; travel in straight lines *(1)*
b 1 Microwave *(1)*; 2 visible *(1)*; 3 X-rays *(1)*
c

1	2	3	4
D	B	A	C

(4)

2

1	2	3	4
D	B	A	C

(4)

Page 203 Electromagnetic spectrum 1

3 **a**

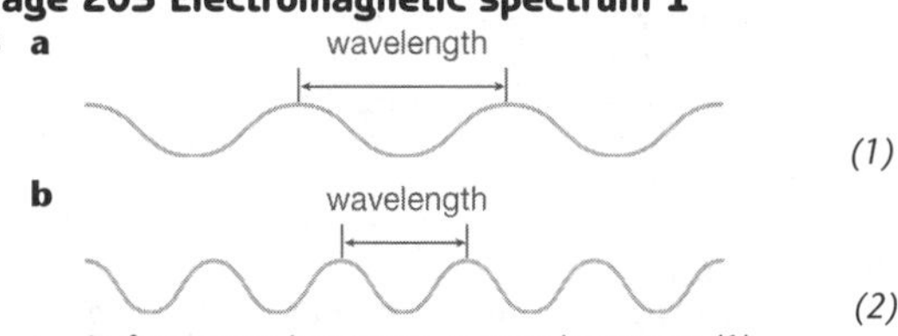

(1)

b *(2)*

c As frequency increases, energy increases *(1)*
d They are the same *(1)*

Page 204 Electromagnetic spectrum 2

1

1	2	3	4
B	C	D	A

(4)

2 **a** Transmitted *(1)*
b Absorbed *(1)*
c Reflected *(1)*

Page 204 Waves and matter

3 **a** X-rays *(1)*
b X-rays pass through soft tissue *(1)*; they are absorbed by bones *(1)*; and detected / absorbed by photographic plate under the person *(1)*
c X-rays can ionise / damage cells *(1)*

4 **a** 300 000 000 m/s *(1)*
b Speed = frequency x wavelength *(1)*
c **i**

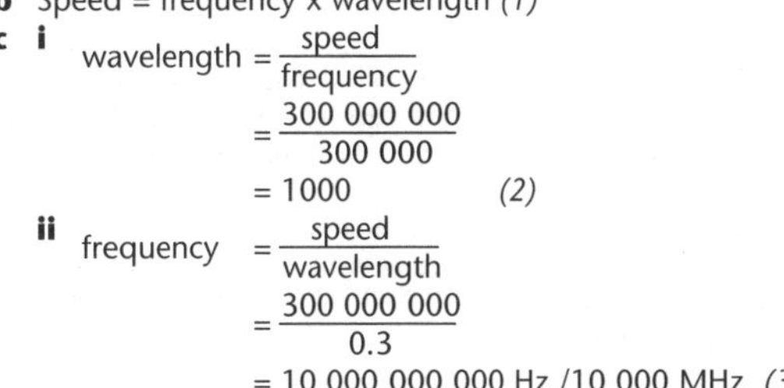

$$\text{wavelength} = \frac{\text{speed}}{\text{frequency}} = \frac{300\,000\,000}{300\,000} = 1000 \qquad (2)$$

ii

$$\text{frequency} = \frac{\text{speed}}{\text{wavelength}} = \frac{300\,000\,000}{0.3} = 10\,000\,000\,000 \text{ Hz} / 10\,000 \text{ MHz} \qquad (3)$$

Page 205 Dangers of radiation

1

1	2	3	4
A	D	B	C

(4)

2 **a** Ultraviolet *(1)*
b Sun burn *(1)*; premature skin aging / wrinkling *(1)*; skin cancer *(1)*
c Dark skin cells on the surface absorb the UV *(1)*; so it does not penetrate*(1)*; to more vulnerable cells under the surface *(1)*
d *(Any 1:)* Ultraviolet radiation, X-rays and gamma radiation all ionise atoms and molecules in living cells; damage to DNA may cause a mutation (change) in a gene; cells may mutate and grow into a cancerous tumour

Page 205 Telecommunications

3 **a** Analogue *(1)*
b Infrared / visible pulses *(1)*; carrying the information are sent down optic fibres to home *(1)*
c Disruption digging up roads / better alternatives e.g. satellite links *(1)*
d *(Any 2:)* Links possible to countries further away; information travels very rapidly; can reach remote areas; very clear signals possible

Page 206 Fibre optics: digital signals

1 **a** Pulses of light *(1)*; repeatedly reflect from the inside surface of the fibre *(1)*
b **i**

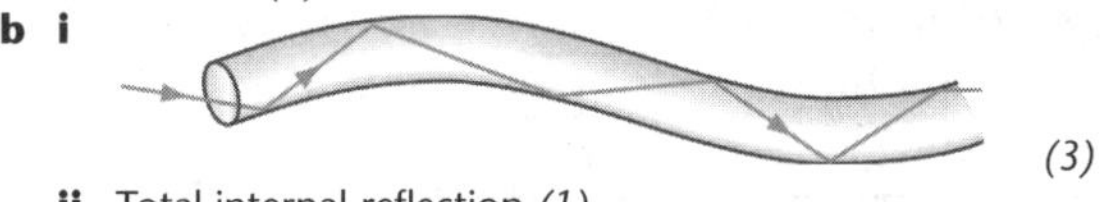

(3)

ii Total internal reflection *(1)*
c *(Any 2:)* More information can be stored; licences can be more compact; possibly easier to update

Page 206 Radioactivity

2

1	2	3	4
A	D	B	C

(4)

3 **a** A, B *(1)*
b 6 *(1)*
c 14 *(1)*
d This is when particles or bursts of radiation are emitted from the nucleus *(2)*

Page 207 Alpha, beta and gamma rays 1

1 **a** Helium nucleus / 2 protons and 2 neutrons *(1)*; emitted from nuclei *(1)*
b They cannot penetrate through skin *(1)*
c Smoke detector *(1)*; localised destruction of cancer cells *(1)*
d *(Any 2:)* Gamma rays have no charge *(1)*; and no mass *(1)*; do not interact with atoms easily *(1)*

2 **a** Beta *(1)*; alpha *(1)*
b Alpha and beta particles have opposite charges *(1)*

Page 207 Background radiation 1

3 a C *(1)* b B *(1)* c A *(1)* d D *(1)*

4 a It is detecting background radiation *(1)*

b Decay from a radioactive source is random *(1)*

Page 208 Half-life

1 a D *(1)*

Note: the multiple choice options to C and D should be 'three quarters' instead of 'half'. This means the correct answer is D: about three quarters of the atoms will decay.

b A *(1)*

c B *(1)*

d C *(1)*

2 a The half-life of a radioisotope is the average time it takes *(1)*; for half *(1)*; of its atoms to decay *(1)* / its count rate to halve

b Organic materials / materials containing carbon *(1)*

c The radioactivity of the old arrow will be less *(1)*; because more of the carbon atoms have decayed *(1)*

d Count rate has fallen to a quarter of its original value *(1)*; the arrow is 2 half lives old *(1)* i.e. it is 2 x 5700 years old = 11400 years old *(1)*

Page 208 Uses of nuclear radiation

3 a Maintains quality of product stops wastage *(1)*

b Can't be detected through paper *(1)*

c Less radiation means the thickness is too big *(1)*; so the rollers should comes closer *(1)*

Page 209 Safety first

1 a D *(1)* b B *(1)* c A *(1)* d C *(1)*

Page 209 Searching space

2 a Large diameter *(1)*; to collect more light *(1)*

b Telescope is in orbit / in space *(1)*; and transmits images back to earth *(1)*

c *(Any 2:)* Less distortion due to pollution *(1)*; better quality images; very large telescope so can collect very faint images *(1)*

3

1	2	3	4
D	A	B	C

(4)

Page 210 Gravity

1

1	2	3	4
B	A	D	C

(4)

2 a Effect of gravity *(1)*

b The force of the earth's gravity is balanced by the forces due to the spacecraft's motion *(1)*

Page 210 Birth of a star

3

1	2	3	4
C	A	D	B

(4)

4 a During fusion nuclei join together *(1)*; which means nuclei get heavier, not lighter *(1)*

b The earth contains many heavier elements *(1)*; that cannot be formed by fusion processes in a star the size of our Sun *(1)*

Page 211 Formation of the Solar System

1 a The dust was flung outwards *(1)*

b Rocks are heavier than dust particles *(1)*; so are attracted more strongly to the Sun *(1)*

c The Northern Lights / Aurora borealis *(1)*

Page 211 Life and death of a star

2

1	2	3	4
C	D	B	A

(4)

3 a The mass is lost as heat and light energy given out by the Sun *(1)*

b Red giant *(1)* → white dwarf *(1)* → black dwarf *(1)*

c Gravitational forces *(1)*; radiation pressure *(1)*

d While they are balanced, the star is stable *(1)*; if the core cools, radiation pressure falls and gravity is stronger *(1)*; causing the star to collapse (or vice versa) *(1)*

Page 212 In the beginning

1 a Billions of years ago *(1)*

b Universe began from a very small initial point that exploded apart in an instant *(2)*

c Quarks join to form atomic particles *(1)*; these join to form hydrogen atoms *(1)*; hydrogen fuses to form heavier atoms *(1)*

d It is the echo from the explosion *(1)*; caused by the Big Bang *(1)*

Page 212 The expanding Universe

2 a Light from a source moving away *(1)* is redder than expected / shifted to the red end of the spectrum *(1)*

b If light from a galaxy is red shifted, the galaxy is moving away *(1)*

3 a B *(1)*

b C *(1)*

c A *(1)*

d D *(1)*

4 a In future the Universe may start contracting *(1)* if its mass is above a critical limit *(1)*; and eventually it will collapse in a Big Crunch *(1)*

b The Universe may continue expanding *(1)*; stops expanding and stays at a constant size *(1)*

B2a Discover Martian living!

Page 214 Cells

1 a B chloroplast; C nucleus *(1)*

b A controls what enters and leaves the cell; D holds the cell in shape *(1)*

c Ribosomes – where proteins are made *(1)*; mitochondria – where (aerobic) respiration takes place *(1)*

Page 214 Specialised cells

2 a Sweep mucus (containing trapped bacteria and dirt) up from the lungs *(1)*

b They have cilia which can move in unison *(1)*; they have many mitochondria to provide energy for the movement of the cilia *(1)*

3 a To absorb water and mineral ions from the soil *(1)*

b They have a large surface area *(1)*

c *(Any 2:)* Smaller; no large vacuole; no chloroplasts

Page 215 Diffusion 1

1 a Diffusion is the spreading of the **particles** / **molecules** of a **gas**, or of any substance in solution. This results in a **net** movement from a region where they are in a **higher** concentration. *(4)*

b Particles are moving faster; have more kinetic energy so they spread out faster *(1)*; bump into each other more often *(1)*

2 a Mitochondria *(1)*

b Respiration happens faster / oxygen used up more rapidly resulting in a lower concentration of oxygen in the cell *(1)*; so there is a greater concentration gradient (between the cell and outside) *(1)*

c Blood coming to lungs is low in oxygen and high in carbon dioxide so oxygen diffuses from high concentration in alveoli to low concentration in blood carbon dioxide diffuses from high concentration in blood to low concentration in alveoli *(2)*

Page 215 Diffusion 2

3 a Carbon dioxide *(1)*

b Carbon dioxide is used in photosynthesis *(1)*; so there is a lower concentration inside the leaf than in the air / there is a concentration gradient from the air to the leaf *(1)*

c *(Any 2:)* Photosynthesis will take place faster; so concentration of carbon dioxide in the leaf will be lower; so diffusion into the leaf will be faster

Page 216 Osmosis 1

1 Osmosis is the **diffusion** of **water** from a **dilute** to a more concentrated solution, through a **partially** permeable membrane. *(4)*

2 a *(Any 3:)* Water molecules can go through the membrane; sugar molecules are too large to go through; water molecules diffuse / move by osmosis from the more dilute solution to the concentrated one; idea of random movement

b Heat the liquids *(1)*; increase the concentration of the sugar solution *(1)*

Page 216 Osmosis 2

3 **a** It will burst *(2)*
  b The strong cell wall stops it bursting *(2)*
  c *(Any 3:)* It has been put in concentrated sugar solution; it has lost water; by osmosis; because the cytoplasm is a less concentrated solution than the sugar solution; contents have shrink and pulled away from the cell wall

Page 217 Photosynthesis

1 **a** Carbon dioxide + water → glucose + oxygen *(2)*
  b *(Any 2:)* Absorbs energy from light; the energy is used to make glucose; from carbon dioxide and water
  c Starch, energy store *(1)*; cellulose, to make cell walls sucrose *(1)*; to transport to other parts of the plant *(1)*; proteins, for growth fats, as food stores in seeds *(1)*

Page 217 Leaves

2 **a** **i** Near the upper surface so light can easily reach them / so they get a lot of light *(2)*
  ii Cell in epidermis (either upper or lower) *(1)*
  b *(Any 2:)* Starch is insoluble; it cannot get out of the cell; it will not interfere with reactions in the cell; it will not affect the concentration of the cell and so will not affect osmosis

Page 218 Limiting factors

1 **a** 32.0 *(1)*
  b The number of bubbles given off per minute *(1)*
  c *(Any 2:)* The number of bubbles increases with each reading; suggesting the plant was still responding to the increase in light intensity; she should have waited until the plant was photosynthesising steadily
  d *(Any 2:)* Yes; the more light the plant is given, the faster it photosynthesises; suggesting that lack of light was stopping it from photosynthesising faster; however, at the highest light intensities the rate of photosynthesis seems to be levelling off; suggesting some other factor is beginning to be the limiting factor
  e *(Any 2:)* The number of bubbles might rise (but not so much as before) or it might stay the same; light will no longer be a limiting factor; some other factor will limit the rate of photosynthesis; for example carbon dioxide concentration

2 Carbon dioxide produced which increases rate of photosynthesis *(1)*; heat produced which raises temperature and increases rate of photosynthesis *(1)*

Page 218 Healthy plants

3 **a** Plants obtain mineral ions from the **soil**. They need nitrate ions for producing amino acids, which are then used to form **proteins**. They need **magnesium** ions for making chlorophyll *(4)*
  b **i** *(Any 2:)* The soil was short of nitrate ions; so giving the plant more allowed it to make more amino acids; which could then be used to make proteins; some of which would be used to help the grain to grow
  ii *(Any 2:)* This could have increased the concentration in the soil; so it became greater than in the plants' roots; so that they lost water by osmosis; the plants became short of water and could not grow well

Page 219 Food chains

1 **a** As chemical potential energy (in food) *(1)*
  b *(Any 2:)* Some light does not hit their leaves; some light passes straight through the leaves; some light is reflected from the leaves; chlorophyll does not absorb all wavelengths (colours) of light
  c **i** Chloroplast *(1)*
  ii 200 units *(1)*

2 *(Any 2:)* Glucose broken down by respiration; to provide heat energy (to keep warm); this is lost to the environment; so there is less energy for the next organism in the chain

Page 219 Biomass

3 **a** From bottom up: producers; primary consumers / herbivores; secondary consumers / carnivores; tertiary consumers / top carnivores *(1)*
  b **i** *(see diagram) (1)*

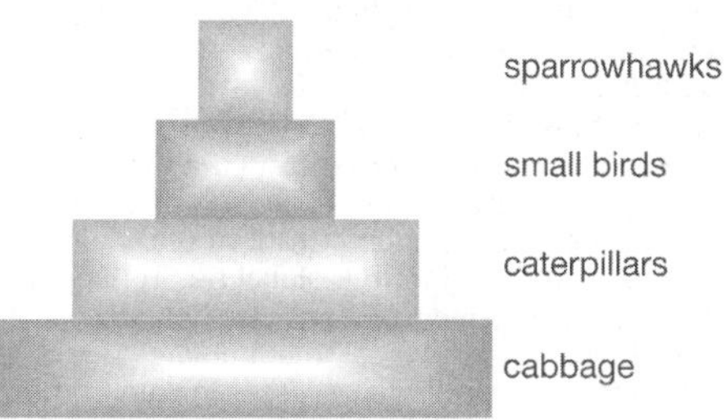

  ii Energy is lost between each trophic level as heat / in respiration / other way in which it is lost *(1)*; so there is less energy available for animals at successive trophic levels so fewer animals can be supported at successive levels *(1)*

Page 220 Food production

1 *(Any 2:)* The chickens do not need to produce heat in their bodies; which involves respiration; and uses up food so more of their food becomes chicken meat

2 *(Any 4:)* The shorter the food chain the less energy is lost; being vegetarian means eating at the end of a short food chain; however, not all land is suitable for growing crops; only suitable for animals to graze / only produces grass which we cannot eat; many subsistence farmers in developing countries need animals to provide transport / leather / other products, as well as to provide food; animals can be used for food in winter when plant crops may be in short supply

Page 220 The cost of good food

3 **a** *(Any 2:)* There are greater energy losses; because the chickens use energy moving around; so less eggs produced per, chicken / square metre / quantity of food provided
  b *(Any 2:)* They may think they taste better; they may think they are better for health; they may dislike the idea of hens being reared in battery cages

4 **a** Transport vehicles emit pollutants *(1)*; especially carbon dioxide, which may cause global warming *(1)*
  b *(Any 2:)* Glasshouses will need heating and lighting in winter; which may use electricity; generated from fossil fuels; which produces carbon dioxide; no need for this in Spain where it is warmer and light levels are higher / winter days are longer

Page 221 Death and decay

1 **a** 100 *(1)*
  b Area A 43%; Area B 34% *(1)*
  c Detritivores / microorganisms / bacteria fungi; feeding on paper; which contained cellulose *(1)*
  d *(Any 2:)* More decay organisms in the wood than on the flower bed; because of dead leaves / leaf litter; more moisture on the woodland floor than on the flower bed; because evaporation from soil reduced by dead leaves; warmer in the wood
  e **i** The type of site (wood, flower bed) *(1)*
  ii The number of squares that had decayed *(1)*
  f Use several pieces of paper in each area, count each one and calculate the average *(1)*

Page 221 Cycles

2 **a** Organisms such as earthworms, which eat dead leaves and other plant remains, are called **detritus** feeders. They help to recycle the materials in the plant remains, so that they become available to other members of the **community** of organisms in the ecosystem. For example, they release some of the carbon in the leaves back into the air, in the form of carbon dioxide, by the process of **respiration**. *(3)*
  b All of them *(1)*
  c *(Any 2:)* They eat food that has been produced by other organisms; they need organic nutrients / they need carbohydrates, fats and proteins; they do not photosynthesise

Page 222 The carbon cycle 1

1 **a** Light intensity *(1)*
  b *(Any 2:)* To remove carbon dioxide from the air (that humans would breathe out); to add oxygen to the air (that humans need to breathe in); to provide food

Page 222 The carbon cycle 2

2 a *(Any 2:)* Microorganisms; break down carbon compounds in dead bodies / wastes; use them in respiration (which produces carbon dioxide)
  b 70 billion tons *(1)*
  c *(Any 3:)* Plants take in more carbon than they give out; so they help to remove carbon dioxide from the atmosphere; reducing the concentration of carbon dioxide; which traps heat and is contributing to a global rise in temperature
  d *(Any 2:)* Carbon came from, animals / plants / waste materials; decomposer break them down; parts of the decomposers / dead bodies / faeces remain in the soil
  e Tiny plants / phytoplankton, remove carbon dioxide for photosynthesis *(1)*; they and sea animals release carbon dioxide by respiration *(1)*

B2b Discover DNA!

Page 224 Enzymes – biological catalysts

1 Enzymes are biological **catalysts**. They are **protein** molecules. Each kind of enzyme only works on a particular kind of **substrate**, which fits perfectly into a fold in the enzyme called the **active** site. *(4)*

2 a *(Any 3:)* Particles moving faster / have more kinetic energy (as temperature increases); more frequent collisions; more energetic collisions; between enzyme and substrate
  b *(Any 3:)* Enzymes are not alive so cannot be killed; the enzyme is denatured (at high temperatures); the enzyme molecule loses its shape; so its substrate cannot fit into its active site

Page 224 Enzymes and digestion

3 a Mouth *(1)*; pancreas *(1)*
  b It digests starch to maltose *(1)*
  c

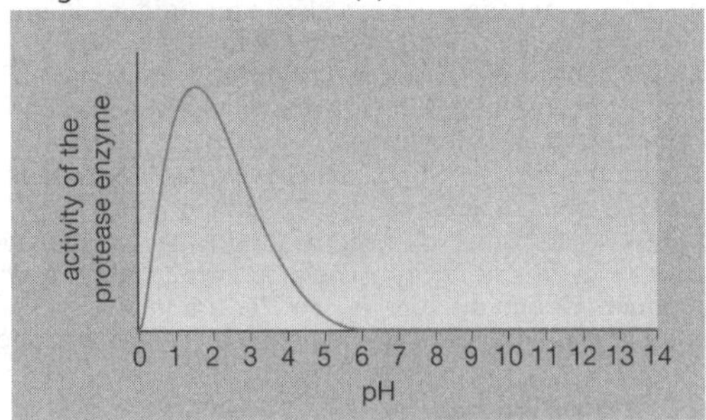

(3)

  d Bile neutralises the acid from the stomach; it is produced in the liver / stored in the gall bladder / flows in down the bile duct; when there is food in the stomach / food moving out of the stomach; contains a base / contains sodium hydrogencarbonate

Page 225 Enzymes at home

1 a *(Any 2:)* Haemoglobin (from blood) is a protein; proteases break down proteins; turns them into amino acids; that can be washed away
  b This is the optimum temperature for the enzymes in them *(1)*; enzymes are denatured if they get too hot *(1)*
  c Hands / cells, contain proteins *(1)*; the enzymes could start to digest these proteins *(1)*

Page 25 Enzymes and industry

2 a DNA / genes, have been altered *(1)*; addition of genes from a different (species of) organism *(1)*
  b

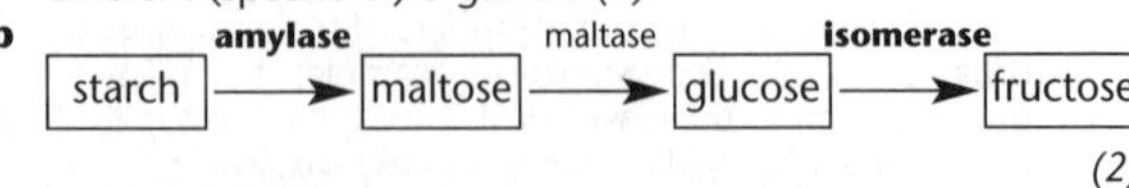

(2)

  c Less needs to be used so cheaper for manufacturers *(1)*; and fewer calories / kilojoules for dieters *(1)*
  d Reactions work too slowly at lower temperatures / optimum temperature for enzymes is 40 °C or above *(1)*
  e *(Any 2:)* Anti-GM feelings among some people much of this is not founded on any scientific understanding; it is very unlikely; there would be health risks from the GM potatoes / discussion of health risks given in answer unlikely; that the GM potatoes would cause any environmental damage / discussion of environmental damage given in answer

Page 226 Respiration and energy

1 a Carbon dioxide *(1)*
  b Every cell in the body *(1)*
  c *(Any 2:)* Aerobic respiration takes place in mitochondria so many mitochondria; supply a lot of energy; sperm need a lot of energy for swimming

Page 226 Removing waste: lungs

2 a Oxygen in exhaled air around 16% *(1)*; nitrogen in inhaled air around 79% *(1)*
  b Body cells use oxygen in respiration *(1)*; so blood arriving at the lungs is low in oxygen *(1)*
  c Oxygen moves between the alveoli and the blood *(1)*; by diffusion *(1)*; down a concentration gradient *(1)*; diffusion only occurs until concentrations are equal (another reason is that not all of the air that is breathed into and out of the lungs ever reaches the alveoli) *(1)*
  d Might have a greater concentration of carbon dioxide *(1)*; because cells are producing carbon dioxide more rapidly so concentration of carbon dioxide in the blood arriving at the lungs is greater *(1)*

Page 227 Removing waste: liver and kidneys

1 a There is more urea in the urine than in the blood because the kidneys remove urea from the blood *(1)*
  b Glucose is useful to the body *(1)*; it provides energy *(1)*
  c If there are too few sodium or chloride ions, the kidneys can keep most of them in the blood *(1)*;
  if there are too many, they can excrete more of them in the urine *(1)*

Page 227 Homeostasis

2 a *(Any 2:)* In sweat, in breath, in faeces
  b *(Any 2:)* More sweat is lost on hot days; so there is less water in the body; so the kidneys conserve water
  c Temperature *(1)*; glucose concentration *(1)*

3 *(Any 2:)* Their cells contain a more concentrated solution than water; so water enters their bodies by osmosis; through the gills; they need to excrete dilute urine to reduce the amount of water in the body

Page 228 Keeping warm, staying cool

1 a Enzymes; are denatured by high temperatures; so cannot catalyse metabolic reactions *(2)*
  b The brain measures the temperature of the blood flowing through *(1)*; it temperature sensors in the skin measure the temperature of the environment *(1)*
  c i Water in sweat evaporates *(1)*; taking heat from the skin *(1)*
    ii Dilate / get wider *(1)*
    iii More blood flows near to the surface of the skin *(1)*; loses heat by radiation *(1)*

Page 228 Treating diabetes

2 a It would rise *(1)*; and stay high for some time *(1)*
  b *(Any 1:)* No insulin will be produced to lower the blood glucose level; it will only fall as the body cells use it up (in respiration)
  c Take care about when he eats foods containing carbohydrates and how much carbohydrate he eats *(1)*

3 a *(Any 2:)* The dog had no pancreas; so did not make insulin; it was given an injection of glucose
  b *(Any 2:)* The blood glucose concentration was very high; because the dog had no pancreas / there was no insulin; that would normally lower the blood glucose concentration (in normal circumstances there is no glucose in the urine)
  c *(Any 2:)* Yes; the hypothesis is supported; in a dog with no pancreas the blood glucose concentration became too high; injecting pancreas extract lowered the blood glucose level

Page 229 Cell division – mitosis

1 a

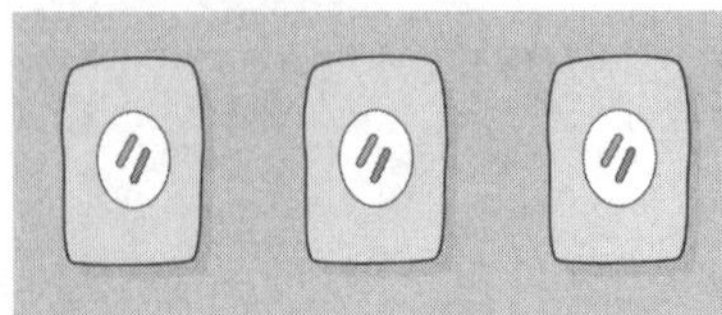

(2)

  b Genetically *(1)*

2 a 2 (one in each set of chromosomes) *(1)*
b Mitosis *(1)*
c *(Any 2:)* The cell could divide in an uncontrollable way; forming a tumour; cancer could develop

Page 229 Gametes and fertilisation

3 a So that after fertilisation the new cell has the correct number of 46 *(1)*
b i Meiosis *(1)*
ii *(Any 3:)* Copies of the chromosomes are made; chromosomes from each set pair up; then separate as the cell divides; it divides twice; forming four cells each with one set of chromosomes
4 a A variety of a gene *(1)*
b *(Any 3:)* Some characteristics are determined by genes; the parents may have different mixtures of alleles; so the gametes also have different mixtures of alleles; so the offspring have different mixtures of alleles (and therefore characteristics)

Page 230 Stem cells

1 a i A cell that is not yet specialised / a cell that can divide to form other specialised cells *(1)*
ii The development of an unspecialised cell into one that is specialised for a particular function *(1)*
b i Embryo stem cells can form every different kind of cell *(1)*; bone marrow stem cells can only form blood cells *(1)*
ii Embryo stem cells could be made to produce new nerve cells *(1)*; which could help to mend the spinal cord *(1)*
2 *(Any 2:)* Take stem cells from an embryo; place them in the brain where the cells have died; the stem cells may be able to form specialised nerve cells; that secrete dopamine

Page 230 Chromosomes, genes and DNA

3 a DNA *(1)*
b The sequence of bases in the DNA determines the sequence of amino acids in the proteins that are made *(1)*
c i Can diagnose inherited illness before a person has symptoms / check for faulty genes in an embryo during IVF so that one with correct genes can be implanted *(1)*
ii If a person is going to develop a genetic illness, they might prefer not to know / insurance companies might want to test a person to find out if they have genes that might make them ill or shorten their life *(1)*

Page 231 How sex is inherited

1 a T *(1)*; its characteristic appears even when the other allele is present *(1)*
b t and t *(1)*
2 a BB *(1)*
b bb *(1)*
c b (in each sperm) *(1)*
d parents BB x bb

gametes all (B) all (b)

offspring all Bb *(3)*

Page 231 How is sex inherited?

3 The father's sperm *(1)*; which could carry either an X chromosome *(1)*; or a Y chromosome an egg always carries an X chromosome *(1)*
4 Female *(1)*; they have two X chromosomes *(1)*

Page 232 Inherited disorders

1 The allele for Huntington's disease must be dominant *(1)*; for the child to get this allele at least one of her parent's must have it *(1)*; and will therefore have the disorder *(1)*

2 a Angela and Sam's alleles: Ff Ff

gametes (F) and (f) (F) and (f)

offspring

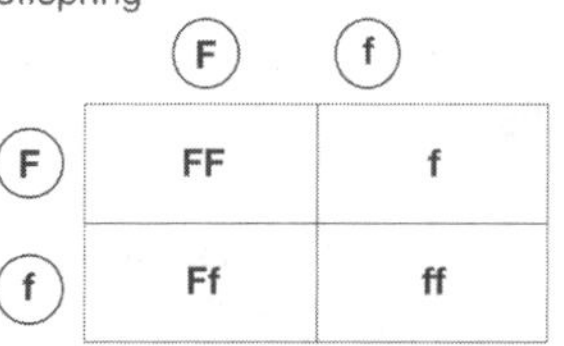

	(F)	(f)
(F)	FF	f
(f)	Ff	ff

There is therefore a 1 in 4 chance that a child will have the alleles ff and therefore have cystic fibrosis *(4)*
b She could have a child with cystic fibrosis *(1)*; if she is a carrier (Ff) and her partner is also a carrier *(1)*; but she may not be a carrier, in which case she cannot have a child with the disorder *(1)*

Page 232 DNA fingerprinting

3 *(Any 4:)* Possible father B is the actual father; the child has some bands that have not come from its mother; these bands must have come from its father;
not all of these bands are present in possible father A; they are all present in possible father B

C2a Discover Buckminsterfullerene!

Page 234 Atomic structure 2

1 a Contains 3 protons *(1)*; and 4 neutrons *(1)*
b

Type of reaction	Ionisation	Nuclear fission	Nuclear fusion
Change in mass of nucleus (increases/decreases/ stays the same)	**Stays the same**	**Decreases**	**Increases**
Change to the atom	Loss or gain of electrons	**Nucleus splits; into two**	Two nuclei join together

Page 234 Electronic structure

2 a

Periodic table element	23 Na 11 sodium	24 Mg 12 magnesium	19 F 9 fluorine
Number of electrons	11	12	**9**
Electron arrangement	Na	Mg	F
Notation	2, 8, 1	**2, 8, 2**	**2, 8**

b 2, 8
3 a Calcium carbonate *(1)*; calcium magnesium carbonate *(1)*
b Mention of exchange of ions *(1)*; magnesium exchanges for calcium *(1)*

Page 235 Mass number and isotopes

1 a (Similarity:) Both have 1 electron in the outer shell *(1)*
(Difference:) Different numbers of electrons / sodium has more electrons / has an extra shell / has 8 more electron *(1)*
b *(Any 3:)* Different atoms of chlorine have different masses; chlorine exists as isotopes; the relative atomic mass is an average value; takes into account differences in abundancy as well as mass

Page 235 Ionic bonding

2 a Ca^{2+}
b $CaCl_2$
3 a Ions are charged *(1)*; ions can move *(1)*
b Pure water does not conduct electricity *(1)*; because it does not contain any ions *(1)*

Page 236 Ionic compounds

1 a Ions are charged / have positive and negative charges *(1)*; and are free to move *(1)*
  b Ions in the solid cannot move *(1)*
2 a Cross on the jewellery at the negative electrode *(1)*
  b Ions gain electrons *(1)*; one electron is gained *(1)*

Page 236 Covalent bonding

3 a Double *(1)*
  b 4 shared electrons *(1)*
  c Oxygen atoms have 6 electrons in their outer shell *(1)*; so need to gain two electrons to form a stable arrangement *(1)*
  d Atoms in the molecule are held together by strong bonds *(1)*; forces between molecules (intermolecular forces) are very weak *(1)*

Page 237 Simple molecules

1 a Water; it has a much higher melting and boiling point; due to stronger forces between the molecules *(1)*
  b i Low boiling point so will evaporate too easily / perfume would not last in the bottle idea *(1)*
    ii Ethanol dissolves the fragrant oil and its boiling point is not far above body temperature *(1)*
    iii Boiling point too high so would not evaporate off the persons' body *(1)*
  c Toxicity/ whether harmful to people / reactivity / if it breaks down easily etc. *(1)*

Page 237 Giant covalent structures

2 a Graphite has weak bonds between layers *(1)*; layers can break off *(1)*; all bonds in diamond are very strong *(1)*
  b There are spaces between layers *(1)*
  c Electrons are free to move *(1)*

Page 238 Metals

1 a

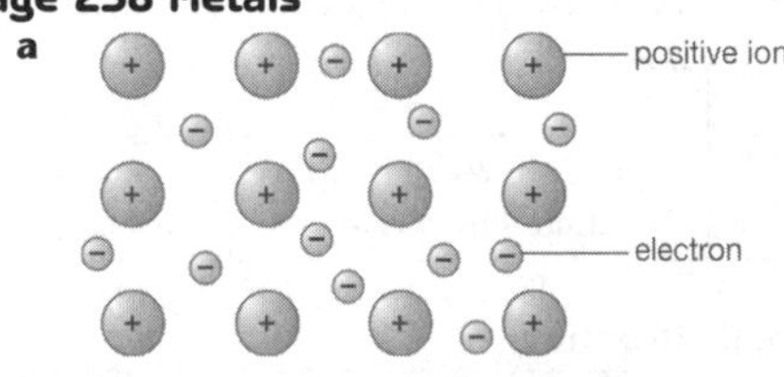

(2)

  b Electrons can move *(1)*
2 a Atoms in the metal slide over each other *(1)*
  b i Conducts electricity *(1)*; conducts heat *(1)*
    ii Delocalised electrons are free to move *(1)*

Page 238 Alkali metals

3 a Have only one or two electrons in outer shell *(1)*
  b Atom A *(1)*; has one electron in outer shell *(1)*
  c Atom B *(1)*; numbers of protons are the same as number of electrons in an atom *(1)*

Page 239 Halogens

1 a

(1)

  b –1 *(1)*
  c Sodium *(1)*
2 a i

Solution of compound	Does it react with fluorine gas?	Colour after reaction
Sodium chloride	Yes	Very pale green
Sodium bromide	Yes	**Orange/brown**
Sodium iodide	Yes	**Brown**

(2)

    ii Sodium fluoride; chlorine *(1)*
  b

Solution of compound	Does it react with bromine?
Sodium fluoride	**No**
Sodium chloride	**No**
Sodium iodide	**Yes**

All 3 correct = (2) 2 or 1 correct = (1)

Page 239 Nanoparticles

3 a Particles containing a few hundred atoms *(1)*; very small *(1)*
  b They are in the form of hollow tubes *(1)*; with a very high surface area *(1)*
  c *(Any 3:)* Companies invest in new research that might make profits in the future; nanoparticles have many uses that can be sold to make money; examples of uses of nanoparticles (example of uses of nanoparticles:) biosensors; harder wearing or stain resistant materials; information processors; catalysts

Page 240 Smart materials

1 Spectacles: A; car dials: D; helmet: B; mugs C
All correct = (4) 2 or 3 correct = (2) 1 correct = (1)

Page 240 Compounds

2 a Elements contain only one type of atom *(1)*; compounds contain more than one type of atom chemically joined together *(1)*
  b *(Any 3:)* In a mixture the elements are not chemically joined together; in a compound the elements are chemically joined together; mixtures have different properties to the compounds of the same elements; for example: hydrogen and oxygen are gases, water is a liquid
  c The formula of water is H2O *(1)*; which shows that hydrogen and oxygen always react in a 2:1 ratio *(1)*
3 a N_2 + **3** $H_2 \rightarrow$ **2** NH_3
  b Mg + **2** $HNO_3 \rightarrow Mg(NO_3)2$ + **H_2** *(4)*

Page 241 Percentage composition

1 a Compound A is CO *(1)*; compound B is CO2 *(1)*
  b Compound B contains a higher percentage than A *(1)*

Page 241 Moles

2 a 32 g *(1)*; 64 g *(1)*
  b Mass of oxygen in SO_2 = 32
  Percentage mass of sulfur in SO_2 = (64 ÷ 32) x 100 *(1)*
  = 50 % *(1)*
  c 640 tonnes *(2)*

Page 242 Yield of product

1 a *(Any 2:)* Makes more product; less reactants wasted; saves energy / fuel
  b *(Any 2:)* Most of the atoms in the reactants are used up; to form products; few atoms are left unreacted
  c Theoretical yield is the maximum calculated yield *(1)*; actual yield is the amount obtained in practice *(1)*
2 a Mass of 1 mole magnesium = 24 g; mass of 1 mole magnesium oxide = 40 g *(1)*
  Mass of magnesium oxide made = $\frac{40}{24}$ x 2.4 or $\frac{40}{10}$ *(1)*
  = 4.0 g *(1)*
  b percentage yield = $\frac{\text{actual yield}}{\text{theoretical yield}}$
  percentage yield = $\frac{3}{4}$ x 100 *(1)*
  = 75% *(1)*

Page 242 Reversible reactions

3 a $C_2H_4 + H_2O \rightleftharpoons C_2H_5OH$
  Correct layout and formulae = (1) reversible sign correct = (1)
  b i H_2O / CH_5OH *(1)*
    ii C_2H_4 *(1)*

Page 243 Equilibrium 1

1 a A *(1)*
  b *(Any 2:)* reaction is reversible; do not get 100% yield; recycling gases means more will react / better yield
  c

Change to reaction condition	Effect on yield (increases/ decreases/stays the same)
A higher temperature	**Decreases**
A higher pressure	**Increases**
Using less catalyst	**Stays the same**

(3)

Page 243 Haber process

2 a

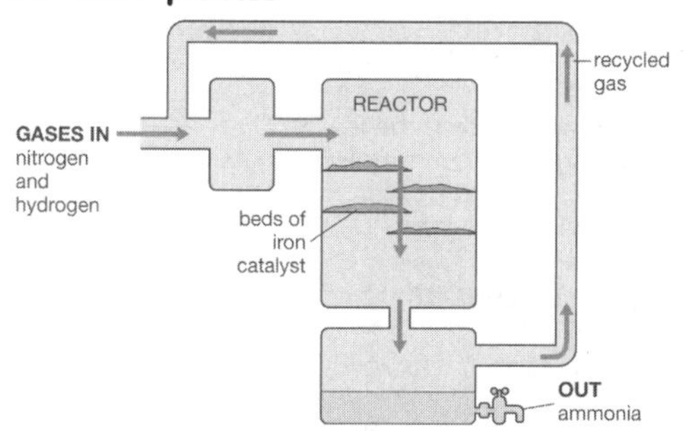

(3)

b i Increases both rate *(1)*; and yield *(1)*

ii High pressures are expensive / difficult to maintain / equipment needs to be specially designed / danger of leaks / not as safe to work with *(1)*

c *(Any 2:)* Reaction is reversible; so do not get a high yield; recycles unreacted hydrogen and nitrogen; increases yield of ammonia

C2b Discover electrolysis!

Page 245 Rates of reactions

1 a

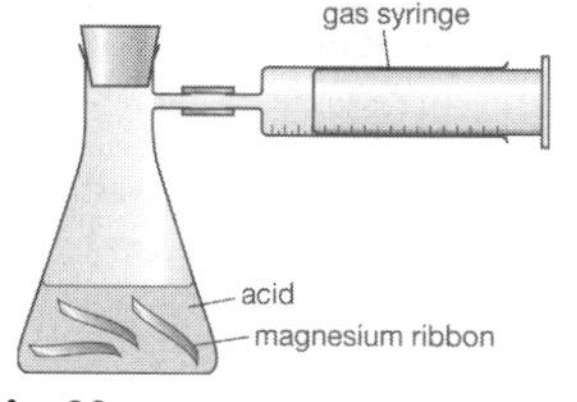

(3)

b i $\frac{90}{50} = 1.8$ *(2)*

ii Acid B *(1)*; fastest reaction *(1)*

c 1 temperature *(1)*; 2 mass of magnesium *(1)*; 3 volume of acid *(1)*

Page 245 Following the rate of reaction

2 a Time *(1)*; volume of gas *(1)*

b No more gas will be made *(1)*

c Mass *(1)*

Page 246 Collision theory

1 a i Acid particles are closer together (not just more acid particles) higher temperature *(1)*

ii The particles move faster *(1)*

b *(Any 2:)* Increased surface area of zinc; higher frequency of collisions / more successful collisions (not just 'more collisions); higher temperature / higher concentration / smaller lumps

2 a At a higher temperature, the concentration decreases *(1)*; particles move away from each other and the gas expands *(1)*

b At a higher pressure, the concentration increases *(1)*; gas particles are pushed together *(1)*

Page 246 Heating things up

3 a Flask loses mass / gets lighter *(1)*; because reaction produces carbon dioxide *(1)*; which is a gas / leaves the flask / is lost *(1)*

b *(Any 3:)* Collisions are more frequent *(1)*; more particles have enough energy to react *(1)*; double the number of successful collisions occur per second *(1)*

Page 247 Grind it up, speed it up

1 a Large surface area *(1)*; most concentrated acid *(1)*; highest temperature *(1)*

b i Any value less than 4.5 cm^3/s *(1)*; acid is less concentrated *(1)*; so reaction is slower *(1)*

ii Any value higher than 4.5 cm^3/s (with units) *(1)*; higher temperature *(1)*; increases reaction rate *(1)*

Page 247 Concentrate now

2 a The acid *(2)*

b Acid is being used up *(1)*; concentration of acid falls *(1)*; reactions are slower at lower concentrations *(1)*

c Experiment 2 is less concentrated acid *(1)*; so is at a slower rate *(1)*

d i 5 cm^3 *(1)*

ii Because this would exactly half the concentration of the acid *(1)*; this will cause the rate to also halve *(1)*

Page 248 Catalysts

1 a $2H_2O + O_2$ *Formulae correct = (1) balancing correct = (1)*

b

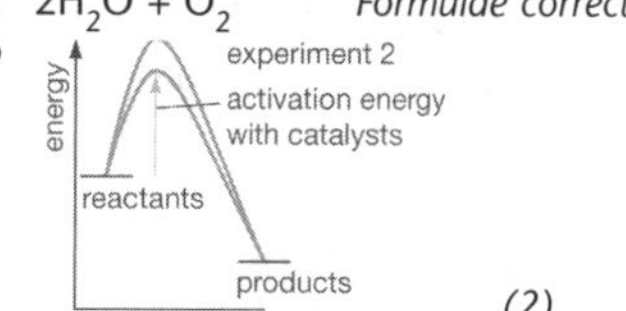

(2)

c 0.2g / the same mass *(1)*; because a catalyst is not used up in the reaction *(1)*

2 a It is not used up *(1)*

b *(Any 2:)* They have a very long lifetime / are not used up; they make reactions faster; link to cost: more products made quickly / do not need to buy more catalyst very often / works out cheaper over time

Page 248 Energy changes

3 a

Reaction	Temperature change	Exothermic or endothermic?
Dissolving ammonium nitrate in water	Decreases	**Endothermic**
Adding zinc powder to copper	**Increases**	Exothermic
Adding magnesium ribbon to an acid	Increases	**Exothermic**

(2)

b Oxidation *(1)*; neutralisation *(1)*

Page 249 Equilibrium 2

1 a Carbon monoxide *(1)*

b The forward reaction is endothermic *(1)*; therefore a low temperature will increase the forward reaction *(1)*

c Higher temperatures increase the rate of reaction *(1)*

Page 249 Industrial processes

2 a Nitrogen + hydrogen $\rightleftharpoons$ ammonia

Correct names = (1) reversible sign correct = (1)

b i Iron acts as a catalyst *(2)*

ii Increases surface area; speeds up rate of reaction *(2)*

c i Nitrogen *(1)*; hydrogen *(1)*

ii Reaction is reversible / reaches equilibrium *(1)*

3 a Less than 15% *(1)*

b i A low temperature gives too low a reaction rate *(1)*; a high temperature gives too low a yield *(1)*

ii A higher pressure gives a higher yield *(1)*; and a faster rate of reaction *(1)*

iii Optimum conditions are a compromise *(1)*; between yield and rate *(1)*

Page 250 Free ions

1 a Contains charged ions *(1)*; in a regular arrangement *(1)*

b Ions cannot move in solid *(1)*; ions move freely in solution *(1)*

c i Chlorine *(1)*

ii *(Any 2:)* Positive electrode / cathode; because it forms from chloride ions; which are negatively charged

Page 250 Electrolysis equations

2 a i Hydrogen *(1)*

ii Sodium more reactive than hydrogen / sodium very reactive *(1)*

b To melt the sodium chloride *(1)*; sodium chloride has a high melting point *(1)*

c i Na^+ + e^- *(1)* $\rightarrow$ Na *(1)*

ii Reaction involves gain of electrons *(1)*

Page 251 Uses for electrolysis

1 a

hydrogen gas used for fuel/making ammonia/ hydrogenating vegetable oils

chlorine gas used for making bleach/ solvents PVC/treating water

IN sodium chloride solution

OUT solution of sodium hydroxide used for making soaps, detergents, paper/purify bauxite

+ −

(6)

2 a Copper *(1)*; positive *(1)*; gaining *(1)*

b

Positive electrode
Cu(s) g Cu^{2+}(aq) + $2e^-$

Negative electrode
Cu^{2+}(aq) + $2e^-$ g Cu(s)

Page 251 Acids and metals

3 a Add both metals to samples of the acid *(1)*; zinc reacts / fizzes because it is more reactive than hydrogen *(1)*; copper does not react because it is less reactive than hydrogen *(1)*

Page 252 Making salts from bases

1 a Zinc chloride *(1)*
b i Copper is too unreactive to react with an acid *(1)*
ii Copper carbonate *(1)*

Page 252 Acids and alkalis

2 a From **blue** to **red** *(1)*
b Because it only has two colours / needs to be different colours at different pHs *(1)*
c i Use the indicator to find out how much acid he needs to add *(1)*; repeat without indicator *(1)*; using same amounts of acid and alkali *(1)*
ii By evaporating the solution *(1)*
3 a All contain nitrogen / all contain ammonium ions *(1)*
b Ammonium nitrate *(1)*
c Sulfuric acid *(1)*; H_2SO_4 *(1)*
d Compound C *(1)*

Page 253 Neutralisation

1 a

Solution	Type of positive ion	Type of negative ion
NaOH	Na^+	OH^-
H_2SO_4	H^+	SO_4^{2-}
Na_2SO_4	Na^+	SO_4^{2-}
HBr	H^+	**Br^-**

(2)

b i NaOH / Sodium hydroxide *(1)*
ii Copper carbonate *(1)*
2 a HCl(aq) → H^+(aq) + Cl^-(aq)
Ions correct = (1) state symbols correct = (1)
b i H^+(aq) + OH^-(aq) g H_2O(l) *(1)*
ii All acids contain H^+ *(1)*; all alkalis contain OH^- *(1)*; neutralisation reaction is the same every time *(1)*
iii Lithium chloride *(1)*

Page 253 Percipitation

3 a Lead carbonate; lead hydroxide *(1)*
b Calcium phosphate *(1)*
c They are precipitates / solids *(1)*; which can be easily filtered off *(1)*

P2a Discover forces!

Page 255 See how it moves!

1 a i Travelling forward *(1)*; at a steady speed *(1)*
ii Stopped *(1)*
b Section C
c Curved line getting steeper *(2)*
2 a Average speed = $\frac{\text{total distance}}{\text{time}}$ *(1)*
$= \frac{90}{2}$ *(1)*
= 45 km/h *(1)*
b The coach will slow down at junctions / traffic lights *(1)*; and speed up on clear roads / or his speed is constantly changing *(1)*

Page 255 Speed isn't everything

3 a Speed *(1)*; in a certain direction *(1)*
b Speed is constant *(1)*; but the direction changes so velocity changes *(1)*
c Tension in the string stops the bung flying away *(1)*
4 a 0 m/s *(1)*
b Acceleration = $\frac{\text{change in speed}}{\text{time taken}}$ *(1)*
$= \frac{150}{0.5}$ *(1)*
= 300 m/s^2 *(1)*

Page 256 Velocity-time graphs

1 a i Steady speed *(1)*
ii Accelerating *(1)*
b Section D *(1)*
c The line would reach the x-axis *(1)*
2 a Draw tangent to curve at appropriate time *(1)*; gradient of the tangent is the speed at that time *(1)*
b Read value off the y-axis *(1)*

Page 256 Let's force it!

3 a i 400 N *(1)*
ii Accelerating *(1)*; forwards *(1)*
b Balanced *(1)*
c Steady speed *(1)*; forwards *(1)*
d Draw an arrow to repesent the size and direction of each force *(1)*; draw each arrow nose to tail and add another line to make complete a triangle *(1)*; this line represents the size + direction of the resultant force *(1)*

Page 257 Force and acceleration

1 a It increases
b Force = mass x acceleration *(1)*
= 20 x 4 *(1)*
= 80 N *(1)*
c Use the force meter to apply a known force *(1)*; and use the accelerometer to measure the resulting acceleration *(1)*;
mass = $\frac{\text{force}}{\text{acceleration}}$ *(1)*
2 a Yes; it does prove this *(1)*; as force doubles, acceleration doubles *(1)*
b 4N reading *(1)*; doesn't follow the pattern *(1)*

Page 257 Balanced forces

3 a They are the same
b The tractor's force is bigger
c The force from the tractor is making the car accelerate *(1)*; as well as lifting the car from the ditch *(1)*

Page 258 Terminal velocity

1 a It increases *(1)*
b Top speed reached when forces balance *(1)*
c *(Any 3:)* Forward force gets greater as he accelerates *(1)*; Air resistance increases *(1)* because his speed increases *(1)* at his top speed, forces are balanced *(1)*
2 a Air resistance *(1)*
b Air resistance acts in the opposite direction to motion *(1)* and makes him decelerate *(1)*
c Larger surface area = slower speed *(1)*; some parachutes may be modified to reduce turbulence *(1)*

Page 258 Stop!

3 a Braking distance *(1)*; thinking distance *(1)*
b Slower reaction time / less experience / concentrating on car control not road conditions *(1)*
c The stopping distance is longer due to slower reaction time *(2)*
d The tread on tyres ensures that water on the road squirts out sideways *(1)*; so the rubber keeps in good contact with a wet road *(1)*; and maximises frictional forces *(1)*

Page 259 Moving through fluids

1 a As it falls it accelerates *(1)*; and the drag increases with speed until drag equals weight *(1)*
b It gets smaller *(1)* because the viscosity of the fluid is greater *(1)* (more drag at lower speeds)
c A liquid exerts a drag force: the falling marble displaces a greater mass of the honey compared with water *(1)*

Page 259 Energy to move

2 a Kinetic energy *(1)*; heat energy *(1)*
b Electrical energy *(1)*; heat energy *(1)*
3 a Spreads to surroundings / moving parts of car *(1)*; which become hotter *(1)*
b More work done *(1)*; against friction *(1)*
c Kinetic energy can transferred to the flywheel making it spin as the vehicle brakes *(1)*; when the vehicle needs to restart, the energy in a flywheel can immediately be used to generate electricity *(1)*

Page 260 Working hard

1 a 12 J

b Weight (allow gravity) *(1)*

c Work done = force x distance *(1)*
= 25 x 2 *(1)*
= 50 J *(1)*

d Energy transferred equals distance travelled along the floor (overcoming friction) *(1)*; but there is now also a vertical component lifting the box against gravity *(1)*

Page 260 How much energy?

2 Andrea as has a large elastic harness attached to her when she does a bungee jump from the top of a bridge. As she jumps, the **gravitational potential** energy she has on the bridge changes into **kinetic** energy as she falls. At the bottom of the jump, the rope is fully extended and it has gained **elastic potential** energy. Eventually she stops moving because all the energy has spread as **thermal/heat** energy to the surroundings. *(4)*

3 a Energy something has because it is moving *(1)*

b Gravitational potential energy *(1)*; and a small amount of heat *(1)*

c kinetic energy = $\frac{1}{2}mv^2$ *(1)*
= $\frac{1}{2}$ x 0.3 x 100 *(1)*
= 15 J *(1)*

Page 261 Momentum

1 a i Speed / mass *(1)*

ii Velocity / force *(1)*

b Momentum = mass x velocity *(1)*
= 0.5 x 8 *(1)*
= 4 kg m/s *(1)*

2 a Momentum = mass x velocity *(1)*
= 2 x 3 *(1)*
= 6 kg m/s *(1)*

b 6 kg m/s *(1)*

c Velocity = $\frac{\text{momentum}}{\text{mass}}$ *(1)*
= $\frac{6}{4}$ *(1)*
= 1.5 m/s *(1)*

Page 261 Off with a bang!

3 a Balloon moves one way *(1)*; air moves in the opposite direction *(1)*

b Total momentum equals zero *(1)*

c Momentum of balloon equals momentum of air so it stays at zero *(1)*

d i Swimmer pushes water backwards *(1)*; so she must move forwards *(1)*; to conserve momentum *(1)*

ii Water is pushed in one direction *(1)*; and the swimmer spins in the opposite direction *(1)*
to conserve momemtum *(1)*

Page 262 Keep it safe

1 a The force changes the velocity of the object *(1)*

b i Zero *(1)*

ii Same change in momentum *(1)*; but less force is felt *(1)*; because the change is spred over a longer time

c The crumple zones slow down the rate of change of momentum *(2)*

2 a Zero

b i Sarah's phone felt a smaller force due to longer time of impact / the carpet absorbs some KE *(1)*

ii Force = $\frac{\text{change in momentum}}{\text{time}}$ *(1)*
= $\frac{1}{0.02}$ *(1)*
= 50 N *(1)*

Page 262 Static electricity

3 a Electrons *(1)*

b Electrons are rubbed off the cloth *(1)* and onto the balloon *(1)* so there are more negative charges than positive charges *(1)*

c The two balloons move apart *(1)* because like charges repel *(1)*

Page 263 Charge

1 a Charges cannot move throughout insulators *(1)*; in conductors, charge can flow *(1)*

b i Electrons rubbed off the wool *(1)* cannot flow away / through the balloon *(1)*; so they build up in one place *(1)*

ii *(Any 3:)* The plaster of a wall is also an insulator; the charge on the balloon induces positive charge to move towards its surface; and attract the negatively charged balloon; and the negative charge in the wall is repelled away from the surface; this is called electrostatic induction

2 a Electric charge *(1)*

b Leave the electroscope uncharged; bring a comb near it *(1)*; if the comb is charged, the gold leaf will move away from the support *(1)*; if it is not charged – the gold leaf is unaffected *(1)*

Page 263 Van de Graaff generator

3 a Insulators / plastic *(1)*

b Belt moves / rubs against roller *(1)*; transferring electrons *(1)*

c To stop the charge flowing away through the bench / prevent discharge *(1)*

Page 264 Sparks will fly!

1 a The cloud's negative static charge induces a positive charge on the earthed lightning conductor *(1)*; this charge streams up towards the cloud *(1)*; discharging it and reducing the chance of a lightning strike *(1)*

b You see a spark *(1)*

c Earthing prevents build-up of charge *(1)*; in an oil refinery, volatile gases could ignite if there is a spark from a discharge *(1)*

2 a Opposite charges attract / the toner and the drum are oppositely charged *(1)*

b Negative charge *(1)*

3 a Electrons are rubbed off the paint particles *(1)*; as they rub on the sides of the spray gun *(1)*

b Paint droplets gain positive charge in the spray gun *(1)*; the car is earthed, so a negative charge is induced which attracts the paint droplets *(1)*

4 High voltage wires in the chimney are negatively charged *(1)*; air particles are ionised between the wires and positively charged air particles stay near the wires *(1)*; negatively charged particles are attracted to the earthed metal plates lining the chimney and pick up the ash and dust particles on their way *(1)*

P2b Discover nuclear fusion!

Page 266 Circuit diagrams

1 a Shine a light on it / cover it up *(1)*

b 2 *(1)*

c Fuse *(1)*

2 a Variable resistor *(1)*

b A slider is used to reduce the variable resistor's resistance *(1)*; this increases the current *(1)*;
which increases the brightness *(1)*

c Can't control bulbs separately *(1)*; bulbs are dimmer *(1)*

d The energy the cell delivers *(1)* = the energy the components use *(1)*

Page 266 Resistance 1

3 a Flow of electrons

b Resistance of conductor is low *(1)*; resistance of insulator is high *(1)*

c Metals contain electrons that are free to move around but in plastics, the electrons cannot flow *(1)*

Page 267 Resistance 2

1 a Resistance increases if electrons *(1)*; have more collisions *(1)*; with nuclei *(1)*

b Different material have a different atomic structure / arrangement so there is a different chance of collisions with electrons

Page 267 Ohm's Law

2 a As voltage increases current increases *(1)*; voltages and current are directly proportional *(1)*

b 1.5 A *(1)*

c Resistance = $\frac{\text{voltage}}{\text{current}}$ *(1)*
= $\frac{6}{1.5}$ *(1)*
= 4 ohms *(1)*

d Type of metal *(1)*; temperature of wire *(1)*

e Temperature of wire increases / is not constant *(1)* so the resistance of the wire increases with voltage *(1)*

f 0 V *(1)*

Page 268 More components

1 a Heat it up / cool it down *(1)*
  b More electrons are released *(1)*; so the resistance falls *(1)*
  c A digital thermometer *(1)*; includes a thermistor *(1)*; as the temperature changes, the resistance changes *(1)*
  d It decreases
  e Automatic light control systems use an LDR *(1)*; at night low light levels reduce the current and turns the lighting on in the morning the reverse happens *(1)*

Page 268 Components in series

2 a They are the same *(1)*
  b 3 V *(1)*
  c 1 1V; 2 1V; 3 1V *(1)*
  d It doubles *(1)*

Page 269 Components in parallel

1 a 1.5 A *(1)*
  b 3 V *(1)*
  c 3 V *(1)*
  d Can control bulbs separately/ each bulb is brighter than in a series circuit *(1)*

Page 269 The three-pin plug

2 a Plastic is an insulator (allow for safety) *(1)*
  b If the fuse blows the current will stop *(1)*; it isolates the live wire *(1)*
  c The cable consists of three metal wires *(1)*; each enclosed in plastic *(1)*; and held together in one plastic / insulating layer *(1)*
  d It is double insulated / it has a plastic outer casing *(1)*; which means live parts cannot be touched *(1)*

Page 270 Domestic electricity

1 a Y *(1)*
  b The current changes direction *(1)*; 50 times per second *(1)*
  c (Cathode ray) Oscilloscope (or CRO) *(1)*
  d Maximum of +230 volts *(1)*; decreases to 0V (Earth) and down to –230 volts *(1)*; then increases to 0V and back up to +230 V *(1)*

Page 270 Safety at home

2 a When the current was too big *(1)*; the fuse blew / melted / broke and broke the circuit *(1)*
  b Different equipment uses different sizes of current *(1)*
  c Radio double insulated / covered in plastic / no metal parts accessible (or vice versa for heater) *(1)*
  d If the metal casing becomes live earth *(1)*; wire provides safe / low resistance route for charge to flow away *(1)* which blows the fuse *(1)*
  e Responds more quickly *(1)* if there is a fault / responds to much lower current differences *(1)*

Page 271 Which fuse?

1 a i 3A *(1)*
    ii It will blow / the fuse wire will melt *(1)*; if the current is higher than *(1)*; 13 A *(1)*
  b i Current = $\frac{\text{power}}{230\text{ V}}$ *(1)*
    $= \frac{2000}{230}$ *(1)*
    = 8.7 A *(1)*
    ii The fuse would melt / blow *(1)*; because the current was too large *(1)*; so the equipment would stop working *(1)*
    iii Charge = $\frac{\text{energy transformed}}{\text{p.d.}}$ *(1)*
    $= \frac{2000}{230}$ *(1)*
    = 8.69 Coulombs *(1)*

Page 271 Radioactivity

2 a i Neutron *(1)*
    ii Neutron *(1)*; proton *(1)*
  b Charged particle / atom which has lost / gained an electron *(1)*
  c i 9 *(1)*
    ii 5 *(1)*
    iii 2 *(1)*
  d i The number of neutrons in the nucleus *(1)*
    ii The number of protons in the nucleus *(1)*

Page 272 Alpha, beta and gamma rays 2

1 a i The atom will decay / will not stay the same *(1)*
    ii An isotope that is radioactive *(1)*
  b Something that is ionising can knock electrons off nearby atoms *(1)*; leaving them as charged ions *(1)*
  c Gamma rays have no mass or charges *(1)*; and so they are not very ionising *(1)*
  d X = 95 *(1)*; Y = 237 *(1)*

Page 272 Background radiation 2

2 a Ionising radiation that *(1)*; surrounds us all the time *(1)*
  b Cosmic rays *(1)*
  c Medical e.g. X-rays / fall-out from nuclear weapon testing/discharge from nuclear power / air travel *(1)*
  d Natural sources *(1)*
3 a Granite is more radioactive than other rocks *(1)*; granite is only found in certain places *(1)*
  b Yes *(1)*; *(Any 2:)* levels of radioactivity are low / have been naturally present for many years and can be reduced with vents / do not pose a great health risk

Page 273 Inside the atom

1 a Positively charged *(1)*; central nucleus *(1)*; surrounded by negatively charged electrons *(1)*
  b i Some positive charged alpha particles are deflected during Rutherford's experiment *(1)*
    ii Many alpha particles passed straight through *(1)*
    iii Electrons are relatively easily removed from atoms, leaving charged ions *(1)*
  c Like-charged protons repel each other (electric repulsion) *(1)*; neutrons do not feel this force *(1)*; the strong nuclear force attracts neutrons and protons keeping them together *(1)*

Page 273 Nuclear fission

2 a When one large nucleus splits into two or more smaller parts *(1)* releasing energy *(1)*
  b Nuclear power stations / nuclear bombs *(1)*
  c i 3 *(1)*
    ii At each stage *(1)* a bigger number of nuclei get involved / each neutron produces more than one more neutron *(1)*
  d The nucleus must be large and unstable for fission to be possible *(2)*
  e There are two neutrons provided, but only one needed to trigger the reaction *(1)*

Page 274 Nuclear power station

1 a Nuclear fission *(1)*
  b Graphite rods moderate (or slow down) neutrons *(1)*; so that fuel nuclei will absorb them / control rods absorb the neutrons *(1)*
  c Kinetic energy → electrical energy *(1)*
  d i Heat is given out by nuclear fission which changes water into steam *(3)*
    ii Steam spins the turbine which spins the generator *(2)*
  e Some uranium is enriched and reused but *(1)*; plutonium has to be disposed of safely e.g. underground *(1)*

Page 274 Nuclear fusion

2 a Two small nuclei join up forming a larger nucleus *(1)* releasing heat *(1)*
  b In a star
  c Helium *(1)*; energy *(1)*
  d The heat and pressure is needed to overcome the strong forces *(1)*; inside the atom which repel the positively charged nuclei *(1)*
  e Scientists have so far reached the huge temperatures needed only for a fraction of a second *(1)*; the reaction is hard to control *(1)*